Model Cellulosic Surfaces

ACS SYMPOSIUM SERIES **1019**

Model Cellulosic Surfaces

Maren Roman, Editor
Virginia Polytechnic Institute and State University

Sponsored by the
ACS Division of Cellulose and Renewable Materials

American Chemical Society, Washington DC

Library of Congress Cataloging-in-Publication Data

Model cellulosic surfaces / Maren Roman, editor ; sponsored by the ACS Division of Cellulose and Renewable Materials.
       p. cm. -- (ACS symposium series ; 1019)
   Includes bibliographical references and index.
   ISBN 978-0-8412-6965-1 (alk. paper)
   1. Cellulose esters--Congresses. 2. Cellulose--Surface--Congresses. I. Roman, Maren. II. American Chemical Society. Cellulose and Renewable Materials Division.
   QD323.M63 2009
   660.6'3--dc22
                                                    2009031660

The paper used in this publication meets the minimum requirements of American National Standard for Information Sciences—Permanence of Paper for Printed Library Materials, ANSI Z39.48−1984.

Foreword

The ACS Symposium Series was first published in 1974 to provide a mechanism for publishing symposia quickly in book form. The purpose of the series is to publish timely, comprehensive books developed from the ACS sponsored symposia based on current scientific research. Occasionally, books are developed from symposia sponsored by other organizations when the topic is of keen interest to the chemistry audience.

Before agreeing to publish a book, the proposed table of contents is reviewed for appropriate and comprehensive coverage and for interest to the audience. Some papers may be excluded to better focus the book; others may be added to provide comprehensiveness. When appropriate, overview or introductory chapters are added. Drafts of chapters are peer-reviewed prior to final acceptance or rejection, and manuscripts are prepared in camera-ready format.

As a rule, only original research papers and original review papers are included in the volumes. Verbatim reproductions of previous published papers are not accepted.

ACS Books Department

Contents

Cellulosic Surfaces

Indexes

Preface

This book is based on a symposium entitled *Model Cellulosic Surfaces*, which was held by the Cellulose and Renewable Materials Division of the American Chemical Society (ACS) on March 25–26, 2007 in Chicago, IL during the 233rd ACS National Meeting. The symposium comprised 22 oral presentations on the preparation, properties, and applications of model cellulosic surfaces.

Cellulose, the main structural component of plant cell walls, has been used by humans as flax, hemp, and cotton fibers since prehistoric times. Today, cellulose, in the form of cotton and bleached wood pulp, forms the basis for a large number of products routinely used in our daily lives, including paper; cotton, rayon, and cellulose acetate textiles; cigarette filters; cellophane film for food and decorative packaging; solid pharmaceutical formulations containing cellulose-based excipients and coatings; cotton wound dressings; regenerated cellulose household and medical sponges; and regenerated cellulose and cellulose acetate membranes, to name a few. Many applications of cellulose and chemical cellulose derivatives, such as esters and ethers, involve contact of the material with a liquid phase that contains dissolved or dispersed matter. Examples of such situations include the processes of papermaking and recycling, the dyeing and laundering of cellulose-based textiles, the use of cellulosic membranes and cotton gauze in hemodialysis and wound healing, respectively, and the enzymatic conversion of lignocellulosic biomass to ethanol. For most of these applications, detailed knowledge of the interactions of the cellulosic material with the dissolved or dispersed matter and the factors that govern these interactions is desirable, yet often difficult to acquire, in large part because of the complexity of the material's surface with respect to composition and morphology. Smooth model surfaces for cellulosic materials offer the possibility to study the interactions in a simplified environment. Literature reports involving model cellulosic surfaces date back to the 1930s. However, recent technological advances in thin film preparation and characterization methods have spurred rapid growth of research in this area, making a symposium on this topic timely.

The book contains twelve chapters, grouped by topic into Introduction (Chapter 1), Cellulose Surfaces (Chapter 2–8), and Cellulosic Surfaces (Chapters 9–12). The first four chapters and Chapter 12 are original reviews on different aspects of model cellulosic surfaces. The remaining chapters are original reports of research data. Chapter 1 is a comprehensive review of the early and recent literature on model cellulosic surfaces, including both model surfaces of cellulose and cellulose derivatives. Chapter 2 provides a detailed overview of the methods of preparation of model surfaces of cellulose, with an emphasis on the challenges involved. Chapter 3 reviews the literature on model cellulose I surfaces, a special case of model cellulose surfaces with a native cellulose morphology. Chapter 4 is a review of the literature on polyelectrolyte multilayer films containing cellulose derivatives or cellulose nanocrystals, and films that use cellulose fibers as substrates. Chapter 5 presents two methods for achieving preferred orientation of cellulose whiskers in thin cellulose whisker films. Chapter 6 describes the application of multiple incident media ellipsometry for measuring simultaneously the refractive index and film thickness of thin cellulose films. Chapter 7 reports the use of inkjet technology for the deposition of cellulose nanocrystals onto flat substrates. Chapter 8 describes the properties of hydroxypropyl xylan and its adsorption behavior on model cellulose surfaces and hydroxyl- and methyl-terminated self assembled monolayers. Chapter 9 discusses several chemical modification strategies for polysaccharides that induce self-assembly behavior of the polysaccharide on gold and other surfaces. Chapter 10 presents a study on the properties of model cellulose ester surfaces. Chapter 11 demonstrates the potential of oxidized cellulose as a substrate for the environmental remediation of heavy metals in groundwater. And Chapter 12 reviews the surface properties of cellulose, cellulose ethers, and cellulose esters, with a focus on the surface free energy, the Lewis acid–base properties, and the Hamaker constant.

The book is targeted at industrial scientists and engineers in the pulp and paper, textile, food, pharmaceutical, and bioethanol industries, among others, as well as academic scientists and engineers, and graduate students who are engaged in research involving cellulose and cellulosic interfaces.

Acknowledgments

I would like to take this opportunity to thank Steve Eichhorn for initiating and co-chairing this symposium, and for his consultation on the title for the symposium and this book. Furthermore, I would like to thank all symposium presenters and their co-authors, as well as all attendees, for making the symposium a success. Finally, I would like to acknowledge sponsorship of the symposium by the ACS Cellulose and Renewable Materials Division and by Q-Sense.

Regarding the book, I am deeply indebted to all authors and co-authors, and anyone who has assisted them, for their efforts in providing and revising the chapters as well as their patience during the editing and peer review process. It has been an invaluable experience and opportunity of growth and maturation. I am very grateful to my reviewers, Stephanie Beck, Joe Bozell, Nicole Robitaille Brown, Derek Budgell, Kevin Edgar, Al French, Gil Garnier, Scott Renneckar, and Bill Winter, for providing excellent and timely peer reviews. Furthermore, I would like to thank the current and former staff of the Books Division of ACS Publications, Jessica Rucker, Bob Hauserman, and Dara Moore, of Journal Production & Manufacturing

Operations, Joe Yurvati, Esther Ober, and Cynthia Porath, and of Copyright, Permissions & Licensing, C. Arleen Courtney, for their guidance in the editing and publication process. Special thanks also go to Margaret Brown, Production Editor in the Books Division, and Jay Cherniak, Elsevier, for editorial assistance with one of the chapters.

Maren Roman
Department of Wood Science and Forest Products
Virginia Polytechnic Institute and State University
Blacksburg, VA 24061

Introduction

Chapter 1

Model Cellulosic Surfaces: History and Recent Advances

Maren Roman

Macromolecules and Interfaces Institute and Department of Wood Science and Forest Products, Virginia Polytechnic Institute and State University, Blacksburg, VA 24061

Many applications of native cellulose, regenerated cellulose, and chemical cellulose derivatives, such as esters and ethers, involve contact of the material with a liquid phase that contains dissolved or dispersed matter. For most of these applications, detailed knowledge of the interactions of the cellulosic material with the dissolved or dispersed matter and the factors that govern these interactions is desirable, yet often difficult to acquire, in large part because of the complexity of the material's surface with respect to composition and morphology. Smooth model surfaces for cellulosic materials offer the possibility to study the interactions in a simplified environment and have been used for a number of years. This literature review aims to provide an overview of the initial and most recent developments in the preparation and use of model cellulosic surfaces.

Introduction

Cellulose is the main structural component of plant cell walls and, as such, the most abundant polymer on earth. It has been estimated that nature produces ~180 billion tons of cellulose annually (*1*). Humans have used cellulose in the form of flax, hemp, and cotton fibers since prehistoric times. Today, cotton and bleached wood pulp, a fibrous mass of more or less pure cellulose produced from tree trunks, form the basis for a large number of products routinely used in our daily lives, including cotton, rayon, and cellulose acetate (CA) textiles, a

4

variety of paper products, CA cigarette filters, cellophane wrapping materials (clear films), solid pharmaceutical formulations (drug tablets), cotton wound dressings, household and medical sponges, and CA- and regenerated-cellulose membranes, to name a few. Many applications of native cellulose, regenerated cellulose, and chemical cellulose derivatives, such as esters and ethers, involve contact of the material with a liquid phase that contains dissolved or dispersed matter. Examples of such situations include the processes of papermaking and recycling, the dyeing and laundering of cellulose-based textiles, the use of cellulosic membranes and cotton gauze in hemodialysis and wound healing, respectively, and the enzymatic conversion of lignocellulosic biomass to ethanol. For most of these applications, detailed knowledge of the interactions of the cellulosic material with the dissolved or dispersed matter and the factors that govern these interactions is desirable, yet often difficult to acquire, in large part because of the complexity of the material's surface with respect to composition and morphology. Smooth model surfaces for cellulosic materials offer the possibility to study the interactions in a simplified environment and have been used for a number of years. This literature review aims to provide an overview of the initial and most recent developments in the preparation and use of model cellulosic surfaces. For the purpose of this review, model cellulosic surfaces are defined as more or less thin, free standing or supported films of cellulose or a cellulose derivative prepared or used for surface property or interaction studies.

History of Cellulose and Cellulose Derivatives

The Molecular Structure of Cellulose

Cellulose was discovered in 1838 by Anselme Payen, a Parisian professor of industrial and agricultural chemistry at the *École centrale des arts et manufactures*, who observed that the composition of non-lignified plant matter (on average 44% carbon, 6% hydrogen, and 50% oxygen) differed significantly from that of lignified plant matter, specifically oak and beech wood (54% carbon, 6% hydrogen, and 40% oxygen) (*2*). To further analyze this difference, he studied the partial digestion of ground oak and beech wood by concentrated nitric acid (*3*). The chemical composition of the residue (43.85% carbon, 5.86% hydrogen, and 50.28% oxygen), led him to conclude that lignified plant matter was composed of two different materials, one of which had the molecular formula of a carbohydrate and the same composition as starch and the other one had a lower oxygen content. The starch-resembling substance, which seemed to consist solely of glucose residues and be present universally in both lignified and non-lignified plant cells, he called cellulose (*4*).

Many years later, in 1921, another critical piece of the puzzle of the molecular structure of cellulose was put in place by Karrer and Widmer, who, based on results from acetolysis studies on cotton, cellobiose, and octacetyl-cellobiose, proposed that cellulose was composed to at least 50% of cellobiose (*5, 6*).

A chain-like molecular structure comprised of covalently linked anhydroglucose units (AGUs) was first postulated by Sponsler and Dore in 1926 based on x-ray crystallography data of ramie cellulose (*7, 8*). However, the cellulose model of Sponsler and Dore predicted the AGUs to be linked (1→1) and (4→4). The correct type of linkage, (1→4), of AGUs in cellulose was proposed two years later independently by Freudenberg and Braun (*9*), and Haworth (*10*). Haworth's structure for cellulose took into account the findings of Sponsler and Dore of a six-atom ring form for the AGU. Haworth also supported the notion of a β-linkage between the AGUs, as opposed to the α-linkage found in starch (*11*). It is now known that cellulose is a linear polysaccharide of (1→4)-linked β-D-glucopyranose units as shown in Figure 1. The cellobiose unit is commonly accepted as the repeat unit, based on the dimension in chain direction of the crystal lattice unit cell.

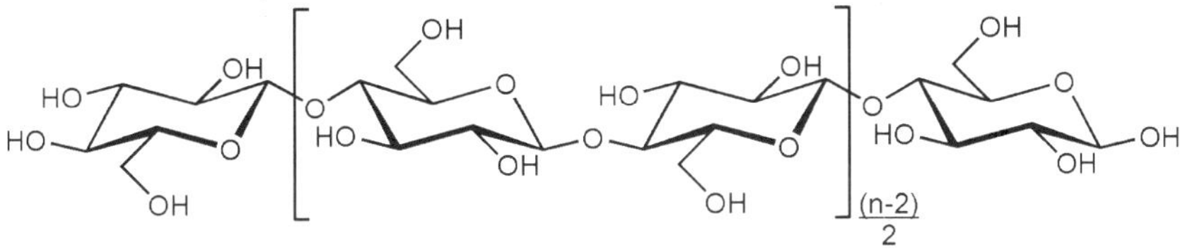

Figure 1. Molecular structure of cellulose

In the late 1920s, the number of AGUs that constituted one cellulose molecule, i.e. the degree of polymerization, DP, or *n* in Figure 1, was still a much debated topic. It was commonly believed that cellulose consisted of low-molecular-weight chains, which aggregated and formed micelles, similar to those found in soap solutions. The radical idea of macromolecules that were comprised of several hundreds of covalently linked smaller units was still widely rejected and it was not until the early 1930s that the macromolecular nature of cellulose could be proven beyond doubt. The micellar theory for cellulose was refuted by Staudinger and Schweitzer through viscosity measurements on dilute solutions of cellulose in Schweitzer's reagent, which showed that the viscosity of the solutions was independent of temperature (*12*). In the same report, the authors described molecular weight measurements on a number of different cellulose materials. A calibration curve for the specific viscosity–molecular weight relationship had been established with partially depolymerized cellulose samples, whose molecular weight had been determined by end-group analysis. The highest molecular weight, corresponding to a DP of 1200, was observed for gently purified cotton. However, the authors noted that the DP of native cellulose was probably higher because Schweitzer's reagent, which was used as the solvent in the viscosity measurements, was known to slowly degrade cellulose in solution. Nevertheless, the enormous size of cellulose molecules had successfully been demonstrated at that point. It is now known that the DP of native cellulose can be as high as 27,000 in certain algae (*13*). A more detailed history of cellulose research can be found in ref 14.

History of Cellulose Esters

The three hydroxyl groups of the AGU interact with those of neighboring AGUs in the same or an adjacent cellulose molecule via hydrogen bonding. The extensive intra- and intermolecular hydrogen bonding and related molecular rigidity renders native cellulose highly crystalline, insoluble in common solvents, and unmeltable, i.e., its melting temperature lies above its decomposition temperature. The unprocessable nature of cellulose motivated scientists early on to study ways to chemically alter the material. In 1833, long before the chemical structure of cellulose was known, Henri Braconnot, a French chemist and pharmacist, discovered that saw dust, linen, and cotton wool reacted with nitric acid to form a highly inflammable compound which he called xyloïdine (*15*). The reaction was refined in 1846 by Christian Friedrich Schönbein, a German chemist, who found that a mixture of nitric and sulfuric acid was more effective in nitrating cotton fibers than nitric acid alone (*16*). The discovery of readily soluble nitric esters of cellulose shortly after the industrial revolution and the mechanization of spinning and weaving in the cotton fiber industry, was the stepping stone for several new industries, including the guncotton industry (explosives and propellants based on highly nitrated cellulose, practically insoluble in ether-alcohol), the collodion industry (photographic film and topical pharmaceutical formulations prepared from solutions of less highly nitrated cellulose in an ether/ethyl alcohol solvent system), the celluloid industry (moldable plastics based on camphor-containing cellulose nitrates (CNs)), and the pyroxylin industry (lacquers, waterproofing solutions, imitation leather, and artificial silk filaments prepared from CN solutions in amyl acetate or commercial wood alcohol).

The value and variety of commercial products that could be developed from CN impelled scientists to investigate the synthesis of other cellulose esters. The possibility of obtaining information about the molecular structure of cellulose, of which only the chemical formula ($C_6H_{10}O_5$) was known at that time, was another important reason for studying other chemical derivatives. The cellulose ester that received the most attention, after CN, was CA (Figure 2). The development of CA was spurred by governmental regulations issued after a series of fires in early movie theaters, stipulating the use of non-flammable cinematographic film materials. Like CN, CA *per se* was unmoldable but could be made plastic by the addition of camphor and other organic chemicals. Thus, at the beginning of the 20th century, it was possible to produce CA-based materials that combined incombustibility, plasticity, and solubility in inexpensive organic solvents. Yet, the higher cost of the materials used to produce CA, as compared to those used in the production of CN, restricted the use of CA to applications where the superior properties of CA justified the higher production costs. Other organic cellulose esters, such as cellulose propionate, cellulose butyrate, and mixed esters, were investigated as well. A very detailed account of the early history of cellulose esters can be found in a two-volume compendium on the nitrocellulose industry by Edward C. Warden, published in 1911, and a ten-volume book on the technology of cellulose esters by the same author, published in 1921, both available online (*17, 18*).

Though in many of its early applications as a moldable plastic, CA has been replaced with less expensive petroleum-based thermoplastics, it is still widely used in textiles, cigarette filters, certain injection-molded objects, display packaging and extruded plastic films, dialysis membranes, sheeting, lacquers, protective coatings, and protective films in liquid crystal displays. Other organic cellulose esters, in particular the mixed esters cellulose acetate propionate (CAP, Figure 2) and cellulose acetate butyrate (CAB, Figure 2) have since their first appearance in the patent literature in 1899 (*19*) found widespread application in sheeting, molding plastics, film products, lacquer coatings, and melt dip coatings. The recent advances in cellulose ester performance and application have been reviewed by Edgar et al. (*20*).

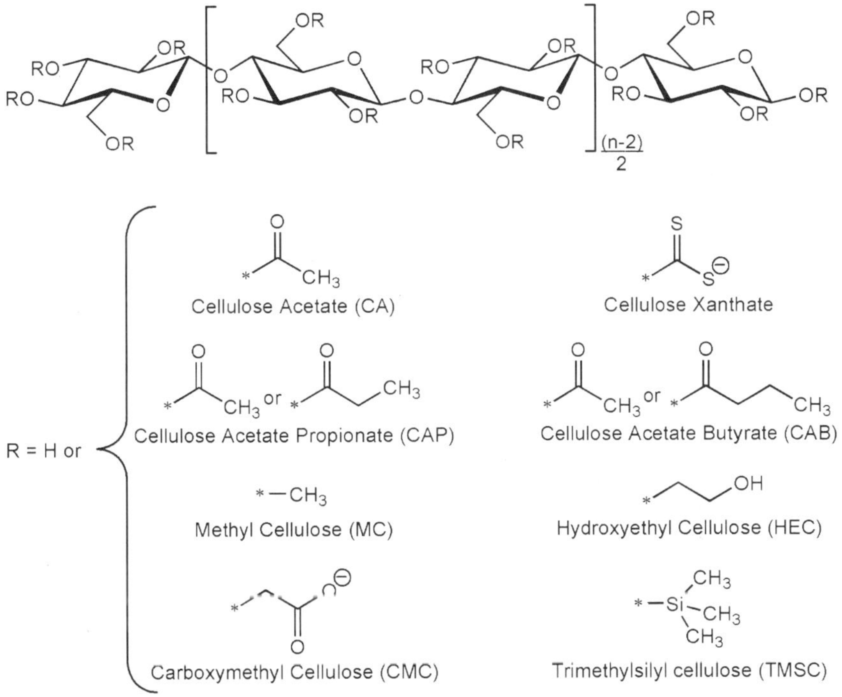

Figure 2. Molecular structures of selected cellulose derivatives, mentioned in the text (The asterisk indicates where the substituent connects.)

History of Regenerated Cellulose

Some applications of CN, in particular those of CN silk in textiles, involved a denitration step, converting the CN back to cellulose, to reduce the flammability of the final product. At the end of the 19th and beginning of the 20th century, several competing technologies existed for the production of regenerated cellulose fibers, some of which are still in use today (*21*). The so-called *cuprammonium silk* was produced by dissolving cellulose in an ammonia-containing solution of $Cu(OH)_2$, $CuCO_3$, or $CuCl_2$, and spinning filaments from the cuprammonium solution of cellulose into a water bath, containing sulfuric acid for the neutralization of ammonia and regeneration of cellulose. *Viscose silk* was produced by treating alkali cellulose, formed upon swelling of cellulose in strong NaOH solutions (17–19%), with CS_2, resulting in the formation of a soluble cellulose xanthate (Figure 2). A solution of the xanthate in dilute NaOH was extruded into an aqueous coagulation bath containing $(NH_4)_2SO_4$ and then run through dilute H_2SO_4 for the regeneration of cellulose. Today, regenerated cellulose prepared by essentially the same process finds application as clear, readily biodegradable films in food and decorative packaging (cellophane), as semipermeable membranes in dialysis, including hemodialysis, and as versatile fibers in woven textiles (rayon).

Like the cellulose esters, regenerated cellulose has been replaced in some of its earlier applications by less expensive, or better performing petroleum-based polymers. Increasingly stringent environmental legislation requiring the recovery of CS_2 and the malodorous breakdown product H_2S in the viscose process imposed additional limitations on the competitiveness of regenerated cellulose. More environmentally benign than the viscose process, the lyocell process, developed in the 1980s, offers an improved cost–performance ratio and thus promises to be more competitive (*22*). In the lyocell process, shredded wood pulp is mixed with a 76–78% aqueous solution of *N*-methyl morpholine *N*-oxide (NMMO) and subsequently heated under vacuum to reduce the water content, resulting in complete dissolution of the pulp. The viscous solution is then extruded through an air gap into a spin bath containing dilute NMMO solution and washed with hot demineralized water in a series of wash baths. The relatively high cost of the solvent is partially offset by a greater than 99% recovery rate, which together with recycling of the process water minimizes the environmental impact of the lyocell process.

Given the dwindling petroleum resources coupled with rising worldwide demands, and the related emphasis on the sustainability of our economy, cellulose-based materials may experience a come-back into the markets of petroleum-based polymers on the basis of the renewable nature of their starting materials (cotton and wood pulp) and the more or less retained biodegradability.

History of Cellulose Ethers

Ethers of cellulose were first synthesized by Wilhelm Suida in 1905 in search of a method to enhance the retention of dyestuffs by cellulose-based textile fibers (*23*). Suida's research was followed by the studies of Dreyfus, Lilienfeld, and Denham and Woodhouse in the 1910s (*18*), which led to the commercialization of cellulose ethers in the 1920s. The first cellulose ether to gain commercial importance was carboxymethyl cellulose (CMC, Figure 2). Methyl cellulose (MC, Figure 2) and hydroxyethyl cellulose (HEC, Figure 2) became economically relevant ten years later. These three cellulose ethers, CMC, MC, and HEC, still dominate the market today together with a series of mixed ethers, such as hydroxyalkyl methyl celluloses. Most cellulose ethers are water soluble, which forms the basis for the majority of their applications. They are primarily used to control the rheology of water-based formulations. Other functions include retention of water, stabilization of colloidal suspensions, enhancement of film formation, lubrication, and gelation (*24*). Cellulose ethers are safe for oral consumption and topical application and, as a result, are widely used in food products, cosmetics, personal care products, and pharmaceutical formulations. In addition, they are used in oil recovery, adhesives, coatings, printing, ceramics, textiles, building materials, paper, and agriculture.

Early Model Surfaces

Studies involving model cellulosic surfaces, as defined here, date back to the 1930s. As a reference point, Irving Langmuir's work on the chemical forces in solids, liquids, and surface films, for which he received the Nobel Prize in Chemistry in 1932, was performed in the period 1916–1943. Using Langmuir's method of ref 25 for the determination of contact angles of liquids in air against solid surfaces, Sheppard and Newsome, in 1935, measured the work of adhesion of water to model surfaces of a homologous series of cellulose triesters, from the acetate to the myristate (*26*). The authors prepared thin films of the cellulose triesters on glass plates by solvent casting from chloroform solutions. The contact angles in air of large drops of water of "at least 1.5 cm in diameter" deposited onto the cellulose triester films were calculated from the height of the drop, measured with a spherometer. The authors found that the work of adhesion decreased steadily with the length of the hydrocarbon chain of the substituent, ultimately approaching that of paraffin.

A similar study by Bartell and Ray in 1952 compared the contact angles, measured by three different methods, of water and different organic liquids in air against various commercial and non-commercial cellulose esters and EC (*27*). In the *vertical rod method*, a drawn fiber of the polymer, rod or plate coated with the polymer, or a free-standing film of the polymer were partially, vertically immersed in the liquid. The contact angle was measured from a magnified silhouette of the liquid–solid interface, projected onto a screen or photographic plate. In the *tilting plate method*, a coated plate or rod was partially immersed in the liquid and tilted until the liquid–air interface on one side of the rod or plate remained horizontal up to the liquid–air–solid contact line. The contact angle in these experiments was given by the angle of tilt of the rod or plate. In the *controlled-drop-volume method*, a refined form of the sessile drop method, the contact angle of a drop of liquid deposited with a fine-tipped pipette onto a flat polymer film, free standing or supported, was measured by either erecting a tangent upon the enlarged image of the drop or by calculations involving drop shape factors.

As opposed to the earlier study by Sheppard and Newsome, the study by Bartell and Ray took into account the hysteresis effect, i.e. the difference between the contact angle of an advancing drop of liquid and that of a receding one. The authors confirmed the findings of Sheppard and Newsome of a systematic increase in hydrophobic character with the length of the hydrocarbon chain of the substituents. Furthermore, they observed an increase in hydrophobic character in CA with increasing degree of acetylation, i.e. average number of acetyl groups per AGU, and little influence of the DP. The vertical rod method and the sessile drop method were found to be superior to the tilting plate method, because they allowed enclosure of the system to prevent evaporation.

A subsequent report by the same authors (*28*), using the same three analytical methods and cellulose derivatives, focused on the interfacial contact angles in solid–water–organic liquid systems and proposed a theoretical approach for calculating the interfacial contact angle in solid–liquid–liquid systems from the advancing and receding contact angles of each of the liquids in air.

Early Studies on Cellophane Surfaces

Because of the inherent insolubility of cellulose, model cellulose surfaces could not be prepared in the same, easy way as model surfaces of soluble cellulose derivatives. Early studies involving model cellulose surfaces relied on commercially available films of regenerated cellulose (cellophane). The studies by McLaren et al. (*29–35*), dating back to 1947, on adhesion of high polymers to cellulose will only be discussed briefly as they lie at the edge of the scope of this review. In these studies, two strips of cellophane were coated with a polymer, including different petroleum-based polymers and cellulose derivatives, and fused together for a certain amount of time at elevated temperature and under pressure. The shear strength at breakage, determined with a tensile tester, was used as a measure for the strength of adhesion and the results for polymers with different compositions, molecular weights, and tack temperatures, were discussed in terms of the heat of adsorption and the energy of cohesion.

The 1950s brought about significant advances in surface chemistry through the work of William A. Zisman and coworkers on the wetting of high- and low-energy surfaces (*36*), for which Zisman received the Kendall Award of the American Chemical Society in 1963. He showed that for many polymer surfaces, linear relationships hold between the cosines of the contact angles, θ, and the surface tensions of the respective liquids, allowing reliable predictions as to the wettability of a polymer by a liquid of known surface tension. Zisman further introduced the concept of *critical surface tension of wetting (of the solid)*, which is the surface tension of the liquid at which wetting of the solid by the liquid can be expected. The critical surface tension of wetting is determined from a Zisman plot, i.e. a plot of $\cos\theta$ versus the surface tension of the test liquids, by extrapolating the regression line to $\cos\theta = 1$ ($\theta = 0$).

Zisman's approach to characterizing wettability was first applied to model cellulose surfaces by Ray et al. in 1958 (*37*). The authors used the sessile-drop method in combination with free standing cellophane films and the vertical-rod method for raw and α-cellulose cotton as well as rayon fibers. The study confirmed the linear relationship, observed for petroleum-based polymers by Zisman et al., for regenerated cellulose and the liquids water, tetrabromoethane, and methylene iodide. The critical surface tension of wetting of cellophane surfaces was found to be 44 mN/m.

V. R. Gray, in 1962, reported a study on the wettability of wood in which he also measured the advancing and receding contact angles of extracted cellophane films (*38*). For those films, he obtained a critical surface tension of wetting of 45 mN/m, in close agreement with the results of Ray et al. (*37*). Gray proposed the concepts of *surface tension for optimum wetting* and *surface tension for optimum adhesion*, for which the values for cellophane were 64.8 and 64.9 mN/m, respectively.

The commercial availability of transmission electron microscopes and the development of surface replica methods in the 1940s set the stage for surface morphology studies. Jayme and Balser, in 1964, obtained the first micro-scale images of self- and machine-cast cellophane surfaces and found that machine-cast cellophane films exhibited striations in the machine direction whereas self-cast films showed no marked orientation (*39–41*).

Early Studies on Self-Cast Cellulose Films

The first study using self-cast model cellulose surfaces was an extensive study by Karl Borgin, entitled "The Properties and the Nature of the Surface of Cellulose" and reported in a series of papers between 1959 and 1962 (*42–45*). Borgin evaluated three preparation methods, namely regeneration of cellulose films cast from cellulose solutions in a cuprammonium solvent system, regeneration of cellulose films cast from cellulose xanthate solutions, and saponification of CA films by alkali. He observed no difference in the appearance of the films prepared by different methods and from different cellulose starting materials, and chose the cuprammonium method for his study as the easiest and most convenient one. He used both the sessile-drop or controlled-drop-volume method in combination with films cast onto glass slides and the vertical-rod method with films cast onto glass rods and observed no difference between the results from the two methods.

Borgin found that the contact angle of water in air against a cellulose surface depended on the water content of the substrate, the relative humidity at which the film was equilibrated and the measurement performed, the temperature, and the contact time. Based on contact angle measurements at different relative humidities and substrate water contents, he proposed a hydration model for cellulose in which cellulose with a water content below 12% contained only tightly bound water that did not alter the surface properties. Any water in excess of 12% was present as free water, the first layer of which had the greatest effect on the surface properties and contact angle. This theory of Borgin was supported by the results of a separate study by P. H. Hermans, who determined by x-ray diffraction that regenerated cellulose fibers contained 12.1% chemically bound water (*46*).

Contact angle measurements in air with different organic liquids and in cellulose–water–hydrocarbon systems showed that the interfacial affinity between cellulose and water was much higher than between cellulose and hydrocarbons and that water was able to displace a hydrocarbon from the surface of cellulose but not the opposite (*44*). Surface-active agents were shown to increase the surface wetting power of water with respect to cellulose but to reduce its capillary wetting power.

Borgin's studies included contact angle measurements of water in air against several cellulose ethers and esters (*45*). He pointed out that cellulose

12

behaved oppositely to what was expected with respect to solubility in that the solubility of cellulose seemed to increase with decreasing hydrophilicity brought about by hydroxyl substitution. He speculated correctly that the methyl and ethyl groups did not cause water solubility directly but that they prevented the mutual attraction between the cellulose hydroxyl groups, thus making them more accessible to the water molecules.

Advances in surface chemistry in the early 1960s, based on the work of Girifalco and Good (*47, 48*) and Fowkes (*49, 50*), allowed a more detailed interpretation of solid–liquid contact angles. Girifalco and Good, in 1957, hypothesized that the free energy of adhesion at the interface between a solid and a liquid (or between two immiscible liquids) is equal to the geometric mean of the free energies of cohesion of the individual phases multiplied by a parameter Φ (*47*). For two-phase systems of the "regular" ($\Phi = 1$) and "pseudo-regular" ($\Phi \approx 1$) type, this 'geometric mean' combining rule fairly accurately predicted the interfacial tension from the surface tensions of the individual phases. A few years later, in 1964, Fowkes proposed that the surface tension of a liquid (or the surface energy of a solid) can be considered the sum of contributions resulting from different intermolecular forces, such as London dispersion forces, metallic bonds, and hydrogen bonds (*49*). For the special case that the interacting forces are entirely dispersion forces, e.g. two-phase systems involving hydrocarbons, he developed an equation that related the interfacial tension to the surface tensions of the individual phases. Like Girifalco and Good, Fowkes used the geometric mean for the interfacial attraction term. Fowkes's theory allowed the determination of the London dispersion force contribution to the surface free energy of a solid and, by subtracting the surface free energy due to London dispersion forces from the total surface free energy, the determination of the *excess energy* due to polar interactions.

Fowkes's approach to interpreting contact angle data was first applied to model cellulose surfaces by Luner and Sandell in 1968 (*51, 52*). To evaluate the effect of film preparation, the authors used four different methods for preparing regenerated cellulose films, namely a two-bath viscose casting method, a one-bath viscose casting method, a dry viscose casting method, and saponification of solvent-cast CA films. In addition, the effect of stretching of the films, to introduce a preferred orientation, was investigated. Contact angles of various liquids against these surfaces were determined by the sessile drop method. The critical surface tension of wetting of the model cellulose surfaces, determined by the Zisman-plot method, ranged from 35.6 to 49.0 Nm/m and depended on the cellulose starting material and the method of film preparation. The values for the London dispersion force contribution to the surface free energy of the cellulose films ranged from 39 to 51 mN/m and the excess energy due to polar interactions was 57, 38, and 30% of the total surface energy, for the liquids water, glycerol, and formamide, respectively.

The 1970s

Surface chemistry in the 1970s was influenced by the theories of Owens and Wendt (*53*), Kaelble (*54*), Wu (*55*), and Neumann et al. (*56*). The theories of Owens and Wendt, and Kaelble, generally combined into what is known as the *geometric mean approximation*, treat interactions as due only to dispersion forces and polar interactions and predict both through geometric mean equations. Wu's theory, referred to as the *harmonic mean approximation*, proposed the use of the harmonic mean instead of the geometric mean for deducing the polar and dispersive force contributions to the surface free energy. A detailed review of these and the above mentioned theories of the wetting of low-energy surfaces can be found in ref 48. An extensive compilation of literature values for the surface parameters of cellulose and cellulose derivatives, defined by the above mentioned and more recent theories, is given in Chapter 12 of this volume, "Surface Properties of Cellulose and Cellulose Derivatives: A Review".

The theory of Owens and Wendt was first applied to cellophane surfaces by Philip F. Brown in his Ph.D. research at the Institute of Paper Chemistry in Appleton, Wisconsin (*57*, *58*). The research objective was to determine the effect of treatment in a corona discharge on the surface free energy and its polar and non-polar components of cellulose and relate the changes in surface free energy to the bond strength between cellulose surfaces. The morphology of the untreated and treated cellophane surfaces was studied by transmission electron microscopy of surface replicas. Furthermore, the surface roughness and conformability, and the effect of humidity on these properties, were measured with a modified version of the Chapman smoothness tester. The contact angles of water, methylene iodide, and tetrabromoethane against untreated and treated cellophane surfaces were measured by the sessile drop method. Brown found that treatment of cellulose in a corona discharge, causing oxidation of the cellulose surface, resulted in an increase in the surface free energy, due to an increase in polarity accompanied by a small decrease in the magnitude of dispersion forces. The increase in polarity was considered the reason for the observed increase in water-induced bond strength between two treated cellophane films.

A few years later at the same institution, students James L. Ferris and Ronald E. Swanson used self-cast cellulose films in their Ph.D. research related to paper sizing (*59*, *60*). The dissertation of Ferris studied the effect of stearic acid, adsorbed from the vapor phase, on the wettability of cellulose surfaces. Swanson's dissertation was a continuation of this work and focused on fractional surface coverage and the effect of fatty-acid chain length and branching. Both students used the two-bath viscose method of Luner and Sandell (*52*) to prepare the model cellulose surfaces and the sessile drop method for contact angle measurements. The morphology of the surfaces was determined by transmission electron microscopy of surface replicas. The contact angle data were analyzed according to the Owens–Wendt theory and both students observed a decrease in the surface free energy of cellulose with increasing amount of adsorbed fatty acid, primarily due to a decrease in the polar contribution.

With the goal to characterize wood fibers from a fundamental surface chemical point of view, Lee and Luner, in 1972, determined the surface properties of lignin, cellulose, and CA surfaces from contact angle measurements in air using the theories of Zisman, Fowkes, and Owens and Wendt (*61*). In addition, they used a theory by Tamai et al. (*62*), which extended Fowkes's theory to solid–liquid–liquid systems.

The 1980s and Early 1990s

Matsunada and Ikada, in 1981, evaluated the two-liquid (solid–liquid–liquid) approach for the determination of London dispersion force contributions to the surface free energy and of non-dispersive interactions of hydrophilic polymers, including cellulose (extracted cellophane films) (*63*). The authors considered the two-liquid approach superior to the water-in-air approach for hydrophilic surfaces because the contact angles were stable with respect to time and equilibria were reached instantaneously. Furthermore, the measurements were not affected by air humidity because the films were completely immersed in a hydrophobic liquid. Their results confirmed that the geometric mean approximation accurately described the dispersion force interactions but was inaccurate with respect to interactions due to other forces. Non-dispersive interactions were more accurately described by a then recent theory of Fowkes (*64*), which considered interactions to be dominated by dispersion forces and acid–base interactions.

In the same year, Katz and Gray used inverse gas chromatography to measure the London dispersion force contribution to the surface free energy of cellophane (*65–67*). Vapor phase adsorption was considered superior to contact angle measurements for the determination of surface free energy components because it avoided the complications related to film swelling and contact angle hysteresis, arising from surface roughness, heterogeneity, and porosity. The authors derived an equation that related the work of adhesion according to Fowkes (*49, 68*) for a hydrocarbon and a second phase with the adsorption free energy for alkanes per mole of methylene groups. Using the incremental free energy of adsorption, determined experimentally using the n-alkanes heptane to decane, the authors estimated the London dispersion force component of the cellophane surface free energy to be ~46 mN/m and ~42 mN/m for the geometric mean and arithmetic mean approximation, respectively. The London dispersion force component was found to decrease slightly with increasing relative humidity.

In a study in 1983, Edward Sacher investigated the question whether "standard" surface free energy values might exist (*69*). He used Kaelble's graphical method to interpret contact angle data from a variety of test liquids on different polymer surfaces, including a commercial cellophane sample. His values for the dispersive and polar components of the surface free energy of cellulose of 20.2 mN/m and 33.7 mN/m are not reliable, however, because he apparently did not extract any plasticizers, e.g. glycerol, which might have been present in the cellulose films, prior to the measurements. Furthermore, he reported "substrate cockling" for the contact angle measurements with water.

In order to gain a better understanding of detergency phenomena, Saito and Yabe studied the surface free energies of CAs of different degree of substitution, i.e. different average number of substituents per AGU (*70*). The authors hypothesized that if the details of the three interfacial systems involved in the removal of soil from substrates (soil/substrate, soil/liquid, and substrate/liquid) were known, the work of adhesion at the soil/substrate/liquid interface could be estimated. CA films of four different degrees of acetylation between 38.4 and 43.8% were cast from solution onto glass slides. Extracted commercial cellophane films were used for a degree of acetylation of 0%. The data from sessile drop contact angle measurements with various liquids were analyzed according to the geometric and harmonic mean approximations. The authors found that the dispersion force component increased and the polar force component decreased with increasing degree of acetylation and that the trend was more pronounced for the geometric mean approximation than the harmonic mean approximation.

Toussaint and Luner, in 1988, evaluated various theories, including the Zisman critical surface tension of wetting, the geometric mean approximation, the harmonic mean approximation, and a theory derived by David and Misra (*71*), based on a two-liquid approach using water and n-octane, which happen to have equal dispersion force contributions to the surface tension, for the determination of surface properties of cellulose, CA, and cellulose surfaces modified with alkyl ketene dimer (AKD), prepared from a mixture of stearic acid (55%) and palmitic acid (45%) (*72*). Model CA surfaces were prepared by casting films from CA solutions in acetone onto glass slides. Model cellulose surfaces were obtained by regeneration from CA films with sodium methoxide in methanol. Cellulose surfaces were treated with AKD solutions in toluene at different concentrations, followed by oven heating, during which AKD reacted with the cellulose surface. The critical surface tension of wetting was found to decrease with increasing amount of AKD. Similarly, the total surface free energy was found to decrease with increasing amount of AKD, due to a decrease in the polar component. The values for the dispersive and polar components of the surface free energy obtained by the geometric and harmonic mean approximations, as well as those obtained in air and immersed in hydrocarbons were found to be in good agreement.

In a subsequent study (*73*), the authors measured the contact angles of hydrocarbons against the same model surfaces immersed in water or ethylene glycol to compare the surface properties of the dry and fully hydrated state. The dispersive components of the surface free energy were found to be equal for different AKD contents as well as hydration states and only when the proportion of the dispersive component relative to the total surface free energy was considered was an increase with AKD content for the dry state apparent. For the hydrated state, dispersion forces accounted for ~50% of the work of adhesion, regardless of the AKD content. The extent of non-dispersive interactions was found to decrease with AKD content and the effect was more pronounced in the dry state than in the hydrated state, which was attributed to conformational differences in the AKD chains when immersed in octane or water. The authors further interpreted their results according to a then recent theory by van Oss, Good, and Chaudhury (*74*), now commonly referred to as *Lifshitz–van der*

16

Waals/acid–base (LW/AB) approach, which allows the determination of the electron-acceptor and electron-donor contributions to the surface free energy. The obtained values showed that the percentage of the surface free energy due to acid–base interactions decreased with increasing AKD content and that it was much lower than the percentage of the work of adhesion due to dispersive interactions.

To enhance our understanding of the relationship between the physico-chemical properties and blood compatibility of cellulose-based dialysis membranes, Kamusewitz et al. conducted contact angle and contact-angle hysteresis measurements on pristine and surface-modified regenerated-cellulose dialysis membranes (*75*). The solid–vapor interfacial tension values, calculated using the geometric mean approximation, indicated that the surface of the water-swollen, pristine cellulose membrane was morphologically smooth and covered by a closed water film. The surface of the surface-modified membrane, at which some of the cellulose hydroxyl groups had been masked by copolymer grafts, exhibited areas with significantly lower interfacial tension.

The Era of Modern Thin Film Techniques

Two independent developments set the stage for a new era of model cellulosic surfaces. One was in the 1980s the successful application of the Langmuir–Blodgett (LB) technique (*76*) to polymeric systems, i.e. the successful transfer of polymer monolayers at the air–water interface onto solid substrates (*77, 78*). The second one was the development of spin coating technology for high-throughput coating of silicon wafers with thin photoresist films in the electronics industry. Furthermore, technological advances in surface and thin film characterization methods significantly facilitated or enabled quantitative measurements of properties including thickness, refractive index, surface roughness, surface forces, and adsorption. Techniques such as ellipsometry, reflectometry, atomic force microscopy (AFM), colloidal probe microscopy, and quartz crystal microgravimetry, are now routinely used in studies involving model cellulosic surfaces. A journal article providing a tabular overview of the preparation and characterization methods applied to model cellulosic surfaces is in preparation by the author of this review.

Overview of Film Preparation Techniques

The techniques that have been used for the preparation of cellulosic thin films can be categorized into dip coating, solvent casting, LB deposition, and spin coating.

Dip coating involves dipping a substrate into a solution of the coating material and allowing the solvent to evaporate. The thickness of the film is primarily controlled by the concentration and viscosity of the solution. A few studies have used a process of dipping the substrate into a solution of the coating material, allowing the coating material to adsorb onto the substrate, driven by intermolecular forces, and rinsing off the excess material and solvent in a subsequent step before drying. This adsorption-based procedure will be considered a dip coating method in this review.

By some definitions, solvent casting is the same as dip coating. However, here solvent casting will be defined as a method in which a solution of the coating substance is spread or deposited onto a flat solid substrate or measured into a mold, followed by controlled evaporation of the solvent. The thickness of the film is controlled by the concentration of the solution and the thickness of the liquid film before evaporation of the solvent.

The LB technique involves depositing the coating material into a monolayer on the surface of an aqueous phase, compressing the monolayer to a desired surface pressure, and then transferring the monolayer onto a solid substrate by moving a vertically oriented substrate through the air–liquid interface. By repeating the transfer step, thicker films can be created in monolayer increments. A variation of the LB technique is the Langmuir–Schaefer technique in which the substrate is lowered with a horizontal orientation onto the air–liquid interface (*79*). In this review, these two methods will not be distinguished and will both be denoted as LB technique.

In spin coating, a small, fixed volume of a solution of the coating material is deposited into the center of a flat substrate, which is then accelerated to a rotational speed of a few thousand revolutions per minute. The rotation of the substrate causes the solution to spread out more or less evenly, after which the solvent evaporates leaving a thin, relatively smooth film. The thickness of the film is controlled by a complex combination of factors including the concentration of the solution, the spinning velocity of the substrate, which affects the evaporation rate of the solvent, and the viscosity of the solution, which for polymers depends on the molar mass and its distribution (*80*).

The early studies, described in the preceding section, were based primarily on the dip coating method. Since the early 1990s, LB deposition and spin coating have become the most widely used preparation techniques for cellulosic thin films. The advantages and disadvantages of each method are listed in Table 1. A more detailed review of the preparation methods for model cellulose surfaces is given by Kontturi and Österberg in Chapter 2 of this volume, "Cellulose Model Films: Challenges in Preparation".

Table 1. Comparison of preparation techniques

Technique	Advantages	Disadvantages
Dip coating	• Does not require special equipment, though automated equipment facilitates the process • Linear relationship of film thickness and number of applied layers • May allow high throughput (if evaporation based)	• Requires a solid substrate to support the film • May be time consuming (particularly if adsorption based)
Solvent casting	• Does not require special equipment • Produces free standing films	• Produces relatively thick films with high surface roughness • Film thickness more difficult to control • May be time consuming (depending on the evaporation rate of the solvent)
LB deposition	• Produces ultrathin, smooth films • Linear relationship of film thickness and number of monolayers • May produce linear alignment in the film	• Requires special equipment • Requires a liquid or solid substrate to support the film • Time consuming
Spin coating	• Produces relatively thin and smooth films • Allows high throughput • May produce radial alignment in the film	• Requires special equipment • Requires a solid substrate to support the film • Film thickness more difficult to control • Film smoothness inversely related to film thickness

Model Cellulosic Surfaces in the Era of Modern Thin Film Techniques

This review covers both model surfaces for cellulose and model surfaces for cellulose derivatives. Since cellulose derivatives are usually soluble in common solvents, the preparation of cellulose-derivative model surfaces is straightforward.

With respect to model surfaces for cellulose, two types of model surfaces can be distinguished. The first type, which here will be called *regenerated-cellulose model surfaces*, are prepared either from a cellulose solution in a suitable solvent system or by converting a thin film of a cellulose derivative through a chemical reaction into a cellulose film. In either case, the native cellulose morphology is lost in the process and the thin cellulose film will have a non-native morphology. Model surfaces prepared from commercial regenerated-cellulose products, such as cellophane film or dialysis membranes, will be considered under the heading of regenerated-cellulose model surfaces. The second type, which here will be called *native-cellulose model surfaces*, are prepared from aqueous colloidal suspensions of nanoscale cellulose fibrils or fragments thereof. The morphology of these films more closely resembles that of native cellulose.

Cellulose-Derivative Model Surfaces

The new era of model cellulosic surfaces began with the application of the LB technique to cellulose derivatives. Literature reports involving LB films of cellulose derivatives are numerous and their objectives can be broadly divided into five thematic areas:

(I)   elucidation of the molecular arrangement in the films (e.g. *81–97*);

(II)   determination of the bulk properties of the films (e.g. *98–101*);

(III)   incorporation of chromophores into the films (e.g. *102–114*);

(IV)   (bio)technological applications of the films (e.g. *115–137*); and

(V)   the fabrication of supramolecular constructs (e.g. *138, 139*).

Despite the large number of studies involving cellulose-derivative LB films, only very few studies reported the use of these films as model surfaces as defined here. Instead, most cellulose-derivative model surfaces of this new era have been prepared by spin coating. However, a few studies have used dip coating or have employed commercially available films. The recent studies involving cellulose-derivative model surfaces are briefly described below.

Diamantoglou et al. studied the effect of the substituent type and degree of substitution on the blood compatibility of solvent-cast cellulosic surfaces from various cellulose esters, ethers, and carbamates (*140–145*). The blood compatibility was evaluated in terms of complement activation (C5a), number of platelets in the supernatant, and thrombin, measured as a complex with antithrombin III (TAT).

Rankl et al., in an effort to further our understanding of the relationship between the properties of a surface and its interaction with biological molecules, measured the contact angles of a number of different probe liquids on LB-films of cellulose derivatives that are commonly used in biological assays, such as trimethylsilyl cellulose (TMSC, Figure 2), aminopropyl TMSC, and cinnamate TMSC, and calculated the surface free energy parameters of these cellulose ethers by the geometric mean approximation and the LW/AB approach (*146*).

Oh and Luner determined the surface free energy parameters of EC films prepared by dip coating, using contact angle measurements in combination with the LW/AB approach (*147*). In a subsequent study (*148*), using the same methods, Luner and Oh determined the surface free energy parameters of spin-coated films of hydroxypropylmethyl cellulose (HPMC), MC, hydroxypropyl cellulose, and HEC and compared the obtained values with those from EC and cellulose, determined in previous studies (*73, 147*).

Lua et al. studied the surface structure of spin-coated films of EC, HPMC, and EC/HPMC blends by optical microscopy, AFM, Raman mapping, and Raman spectroscopy (*149*).

Micic et al. used commercial CA replicating sheets as a model surface for cellulose in a study of the self-assembly of a lignin model compound on cellulose surfaces by environmental scanning electron microscopy (*150*).

Using a quartz crystal microbalance, Gotoh studied the detergency of a model aqueous detergent solution with respect to arachidic acid LB-layers on spin-coated CA films (*151*), and, in a subsequent study (*152*), the adsorption of polyethylene and nylon-12 microspheres onto these films in order to gain a better understanding of the redeposition of particulate soils in the detergent process.

Castro et al. studied the adsorption of concanavalin A, a mannose and glucose specific lectin, onto spin-coated CMC films (*153*). The analytical techniques that were used in this study, and which are common to the studies described below from the same laboratory, were ellipsometry, contact angle measurements, and AFM. Kosaka et al. prepared model surfaces from a series of cellulose esters using adsorption-based dip coating as well as spin coating and studied the effect of annealing on film morphology as well as the adsorption of bovine serum albumin and lipase onto the films (*154*); immobilization of lipase on the films (*155*); surface properties and dewetting of the films (*156*); and the Hamaker constants of the cellulose esters (Chapter 10 of this volume). Also from the same laboratory, Amim et al. studied the preparation of model surfaces of CA, CAP, CAB, and CMCAB by spin coating from acetone and ethyl acetate solutions (*157*). In a subsequent study, Amim et al. studied the effect of the solvent on the surface parameters, according to the geometric and harmonic mean approximations, and dewetting of annealed CAP and CAB films, prepared by adsorption-based dip coating (*158*). In addition to ellipsometry, contact angle measurements, and AFM, in this latest study sum frequency generation spectroscopy was used to obtain information on the interfacial molecular orientation.

A number of groups studied the surface forces of model surfaces made from HEC derivatives, such as ethyl HEC (EHEC) (*159–165*), non-ionic and cationic HEC derivatives (*166, 167*), hydrophobically modified EHEC (HM-EHEC) (*168*), and a cationic HM-HEC derivative (*169*), as well as the effect of the anionic surfactant sodium dodecyl sulfate on these forces (*162, 166, 168, 169*) by surface force apparatus. The model surfaces were generally prepared by adsorption of the cellulose derivative from solution onto mica substrates.

Pincet et al. used the surface force apparatus to study the forces between a model CA surface and a ribonuclease A- or human serum albumin-layer adsorbed onto a mica substrate in an effort to elucidate the phenomenon of flux decline in membrane filtrations of protein solutions (*170*). The model CA surfaces were prepared by dipping a mica substrate into a CA solution in acetone.

Regenerated-Cellulose Model Surfaces

The literature on model cellulose surfaces, both native and regenerated, has recently been reviewed in great detail by Kontturi et al. (*171*), and will here therefore only be summarized.

Spin coating was first applied to regenerated-cellulose model surfaces by Neuman et al., who spin coated solutions of cellulose in trifluoroacetic acid onto mica substrates (*172*). Concurrently, Schaub et al. investigated the use of the LB technique to deposit mono- and multilayers of TMSC onto various substrates. By subsequently cleaving off the trimethylsilyl groups in humid HCl vapor, the cellulose molecular structure could be regenerated, yielding very smooth cellulose films. The thickness of these films was directly proportional to the number of monolayers (*173, 174*). The elegance of the method by Schaub et al. has motivated numerous studies involving model cellulose surfaces prepared in this manner. In fact, to this day, all studies involving model cellulose surfaces prepared by the LB technique, which are about 50% of all studies involving regenerated-cellulose model surfaces, follow the protocol by Schaub et al., with minor modifications in regard to solvent and concentration of the spreading solution, film pressure, or type of substrate.

Since the report by Schaub et al., several groups have studied in detail the properties of model cellulose surfaces prepared by the LB technique (*174, 175*). Furthermore, LB model cellulose surfaces have been used to:

(I)   determine the Hamaker constant of cellulose (*176*);

(II)   measure the surface and friction forces of cellulose (*177–183*);

(III)   understand moisture-related swelling (*175, 177–180, 184*);

(IV)   measure the work of adhesion with respect to poly(dimethylsiloxane) (PDMS) caps (*185*);

(V)   measure the surface adsorption of dye molecules (*174*), polyelectrolytes (*175, 186, 187*), carboxymethylated cellulose nanofibrils (*188*), xylan (*189*), water-soluble xylan derivatives (*190*), a model carbohydrate–lignin copolymer complex (*191*), cationic starch (*192*), and cationic surfactants (*193*); and recently

(VI)   study the build-up of polyelectrolyte multilayers (*194*)

The spin coating method by Neuman et al. has received less attention than the LB method by Schaub et al. However, in 2000, Geffroy et al. reported a combined approach that used TMSC solutions and spin coating to prepare model cellulose surfaces on silicon wafers (*195*). This method was subsequently optimized by Kontturi et al. (*196, 197*) and has since become a popular method.

In a recent study, Rossetti et al. elucidated and compared the nanoscale structures of TMSC-based model cellulose surfaces prepared by the LB technique and by spin coating (*198*). The authors used non-contact AFM (NC-AFM) and grazing incidence small-angle x-ray scattering (GI-SAXS) to measure the characteristic in-plane length scales, as well as x-ray specular reflectometry to obtain information in the direction normal to the surface. The two preparation methods were found to yield identical root-mean-square surface roughness values (0.7–0.8 nm), which were independent of the scan area, indicating a smooth and uniform topography. Both methods produced similar characteristic in-plane length scales of 32 nm and 50 nm for NC-AFM and GI-SAXS, respectively, which were attributed to the formation of bundles of cellulose molecules during cellulose regeneration. X-ray reflectometry revealed for both methods the absence of periodic structures in the z-direction but a slightly lower density for the spin coated model surface. Swelling experiments showed that significant swelling of the model surfaces occurred only at relative humidities above 97% and that, especially in the thicker films, swelling was accompanied by an out-of-plane rearrangement of the cellulose bundles. In summary, both methods seemed to produce equivalent model surfaces in terms of surface roughness and characteristic in-plane scale, but spin coating appeared to yield a slightly less compact structure in the z-direction.

As an alternative to the TMSC-based spin coating method, methods that involve direct dissolution of cellulose in either NMMO diluted with dimethyl sulfoxide (DMSO) (*199–201*) or dimethyl acetamide (DMAc) combined with lithium chloride (*202, 203*) have been developed and are becoming more popular.

Spin-coated cellulose films are often anchored to the substrate by physisorption to a thin polymer coating on the substrate. In 2005, Freudenberg et al. introduced a method for covalent attachment of the cellulose film to the substrate (*204*). The method uses maleic anhydride copolymers that are covalently attached to aminosilane-modified glass or silicon oxide surfaces. Covalent bonding of the spin-coated cellulose film to the copolymer layer is achieved by reacting the cellulose hydroxyl groups with the maleic anhydride groups of the anchoring layer.

The applications of spin-coated regenerated-cellulose model surfaces are similar to those prepared by the LB technique. Namely, spin-coated regenerated-cellulose model surfaces have been used to:

(I)   determine the Hamaker constant of cellulose (*205*);

(II)   measure the surface forces of cellulose (*172, 206–211*);

(III)   understand moisture-related swelling (*184, 199, 202, 212, 213*);

(IV)   measure the work of adhesion with respect to PDMS caps (*214*) and ink varnish-coated PDMS caps (*215*)

(V)   measure the surface adsorption of polyelectrolytes (*195, 211, 216*), surfactants (*217, 218*), water-soluble xylan derivatives (Chapter 8 of this volume), and cationic nanosized latex particles (*219*); and

(VI)   study the surface adsorption of cellulolytic enzymes and enzymatic degradation of cellulose (*202, 203, 220–224*).

Furthermore, Klash et al. have recently used a spin-coated model cellulose surface derived from a cellulose solution in DMAc/LiCl as a reference in a study on the use of chemical force microscopy for mapping of the chemical components on the surfaces of untreated and pulped wood fibers (*225*).

In another recent study, Notley et al. used model cellulose surfaces, prepared by spin coating of a cellulose solution in NMMO/DMSO, to study, by colloidal probe microscopy, the adhesion between cellulose surfaces with adsorbed layers of phenylboronic acid-derivatized polyvinylamine (*226*).

Despite these advances in the preparation of model cellulose surfaces, a few recent studies have still employed conventional solvent casting or have used commercially available regenerated cellulose materials, specifically cellulose dialysis membranes, as model cellulose surfaces: James E. Bradbury used cellulose dialysis membranes in his Ph.D. research on vapor phase adsorption of thermal AKD decomposition products onto cellulose, aiming to explain AKD sizing reversion (*227*). Garnier et al. used model cellulose surfaces, prepared by transesterification of dip coated cellulose acetate films with sodium methoxide, to study the wetting of cellulose by molten AKD wax (*228*). Using surfaces prepared in the same manner, Dickson et al. studied the hydrophobization of cellulose by alkenyl succinic anhydride (*229*). Yang et al. used cellulose dialysis membranes as a model paper surface in a study of the effects of lignin and extractives on the effectiveness of fluorochemicals to reduce wetting by oils and hydrophobic liquids (*230*). Many more studies from the same laboratory involve the use of cellulose membranes as model paper surfaces (*231–240*). However, since these studies focus on wet delamination strengths they lie outside the scope of this review. Zauscher and Klingenberg used cellulose dialysis membranes to study normal and friction forces between cellulose surfaces (*241–243*). Nigmatullin et al. prepared model cellulose surfaces by solvent casting a CA solution onto a porous polypropylene substrate followed by alkaline saponification and used these surfaces to study the interactions between two cellulose surfaces in the presence of cellulose binding domains (*244*). Linder and Gatenholm used commercial cellophane films and cellulose dialysis membranes in a study of the effect of cellulose substrate morphology on xylan deposition during autoclaving (*245*). And Sudeep Vaswani used cellulose dialysis membranes in his Ph.D. research on the surface modification of paper by plasma enhanced chemical vapor deposition of fluorocarbon precursors (*246*).

Some studies, on the other hand, have focused on the surface properties of regenerated-cellulose and cellulose-derivative membranes for their application in hemodialysis (*247–254*). These studies are part of a larger effort to correlate the physico-chemical surface properties of biomedical polymers with their blood compatibility and plasma protein adsorption behavior. Specifically, streaming potential measurements (*247–250*) and contact angle measurements (*248, 251*) have been used to determine the interfacial charge and surface energy of cellulosic membranes, respectively, and in situ attenuated total internal reflectance Fourier transform infrared spectroscopy has been used to monitor the adsorption of proteins (*252, 253*) and surfactants (*254*) onto cellulosic membranes. A survey of the physico-chemical techniques applied to the characterization of biomedical polymers is given in ref 255.

A study that deserves a more detailed discussion is that published by Yamane et al. in 2006 (*256*). These authors prepared regenerated-cellulose model surfaces with different degrees of crystallinity and crystal lattice orientations to study the effect of these two parameters on the wettability of cellulose by water. The authors found a direct correlation between the orientation of the crystal lattice in the films and the wettability of the films. The contact angles were lowest when the crystal lattice orientation exposed the equatorial hydroxyl groups. A crystal lattice orientation that exposed the axial hydrogen atoms was expected to show very high contact angles, which was confirmed by a low surface energy value in computer simulations. Furthermore, high degrees of crystallinity were correlated with low contact angles. The authors concluded that the wettability of cellulose was governed primarily by its crystalline properties.

A separate body of literature, which partially overlaps with the one reviewed here, is based on the use of colloidal regenerated-cellulose spheres as model cellulose surfaces (*177, 181–183, 185, 189, 208–211, 226, 241–244, 257–275*). The cellulose spheres are used as probes in surface or friction force measurements by colloidal probe microscopy. In a colloidal probe force measurement, a cellulose sphere is attached with an adhesive to the end of an AFM cantilever and then either lowered toward and subsequently retracted or moved across an opposing planar or curved surface for normal surface force or lateral friction force determination, respectively. Only those colloidal probe microscopy studies that involve planar model cellulose surfaces have been included in this review.

Native-Cellulose Model Surfaces

The literature on native-cellulose model surfaces has been reviewed in detail by Cranston and Gray in Chapter 3 of this volume, "Model Cellulose I Surfaces: A Review". Therefore, only a few selected studies will be outlined here.

One of the first studies involving a native-cellulose model surface was that reported by Linder et al. in 2003, investigating the deposition of xylan from alkaline solution onto cellulose substrates at elevated temperatures (*276*). The authors used purified bacterial cellulose pellicles as model pulp fiber substrates and studied the mechanism by which deposits of alkali-extracted birch xylan form on these model surfaces under "pulping-like" conditions. The proposed mechanism involved the aggregation of xylan molecules during autoclaving, as a result of a reduction in glucuronic acid content, followed by adsorption of the xylan aggregates onto the cellulose surface.

Around the same time, Edgar and Gray reported the preparation of native-cellulose model surfaces from aqueous suspensions of cellulose nanocrystals (277). Cellulose nanocrystals are short, rod-like fragments of native cellulose microfibrils, generally obtained by acid hydrolysis of a purified cellulose starting material (278, 279). Edgar and Gray tested several film preparation methods and found that the bottom sides of free-standing cellulose nanocrystal films, prepared by solvent casting in polystyrene Petri dishes, presented very smooth surfaces. For applications in aqueous media, the films had to undergo a mild thermal treatment to prevent disintegration of the film upon submersion. The authors also studied model surfaces prepared by spin coating of cellulose nanocrystal suspensions onto glass and mica substrates and observed that the resulting surfaces were slightly rougher than those prepared by the solvent-casting method and that these surfaces showed radial, shear-induced alignment of the nanocrystals.

The preparation of native-cellulose model surfaces by spin coating of cellulose nanocrystal suspensions was further studied by Lefebvre and Gray (280). In this study, the authors optimized the preparation conditions for spin coating of thin films of cellulose nanocrystals onto silicon wafers and used these films to investigate the surface forces before and after deposition of a polyelectrolyte layer or multilayer by AFM deflection–distance measurements.

Stiernstedt et al. (183) used similarly prepared films to study the effect of cellulose morphology on the surface and friction forces of cellulose films. The normal surface forces and lateral friction forces of four different types of model cellulose surfaces were measured by colloidal probe microscopy, using a regenerated cellulose sphere as the colloidal probe. The friction coefficient was highest when the opposing surface was that of another colloidal cellulose sphere and lowest for the native-cellulose model surface, in direct correlation with the surface roughness values. The friction forces were significantly reduced in the presence of xyloglucan. Regarding the normal surface forces, all four surfaces exhibited long-range double layer interactions, indicative of a permanent surface charge. The measured surface potentials, ranges of steric repulsion, and adhesion values did not differ significantly for the four cellulose surfaces.

Notley et al. (208) took a similar approach, also involving colloidal probe microscopy, to study the effect of cellulose morphology on the normal forces of cellulose surfaces. The authors compared model cellulose surfaces with three different morphologies, namely amorphous, semicrystalline with a (native) cellulose I crystal structure, and semicrystalline with a (non-native) cellulose II crystal structure. The authors observed three very distinct force profiles: the profile of the amorphous cellulose surface was dominated by steric repulsion forces, a result of significant film swelling, whereas that of the cellulose I surface was dominated by long-range repulsive double-layer interactions. The profile of the cellulose II surface was most sensitive to the surrounding pH and was dominated by attractive van der Waals forces at low pH values and by repulsive double-layer forces at higher pH values. The long-range repulsive double-layer forces of the cellulose I surface were attributed to negatively charged sulfate groups on the surface of the cellulose nanocrystals, introduced during the preparation by sulfuric acid hydrolysis. This report therefore indicates a serious drawback of native-cellulose model surfaces prepared from sulfuric-acid hydrolyzed cellulose nanocrystals: the surfaces display highly acidic groups not present in native cellulose. A possible solution to this problem could be the use of hydrochloric-acid hydrolyzed cellulose nanocrystals. However, model surfaces prepared from such nanocrystals are less smooth than those prepared from sulfuric-acid hydrolyzed cellulose nanocrystals (Figure 3), due to a less uniform particle size distribution.

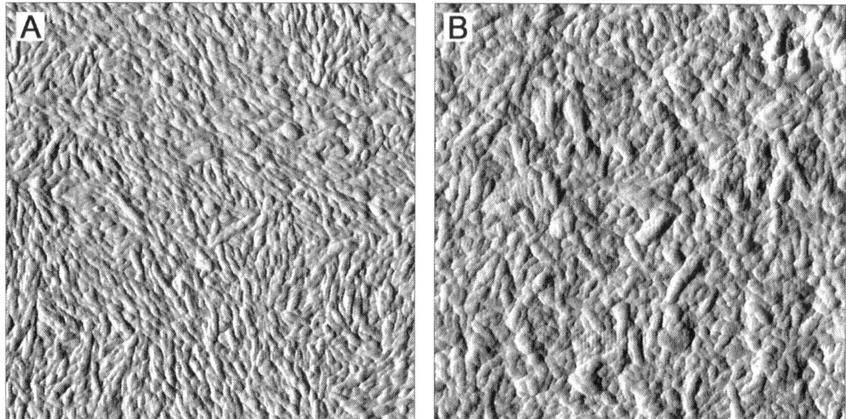

Figure 3. AFM amplitude images (2 μm × 2 μm) of model cellulose surfaces from cellulose nanocrystals prepared with (A) sulfuric acid (64 wt %, 10 mL/g pulp, 45 °C, 45 min) and (B) hydrochloric acid (4 N, 30 mL/g pulp, 80 °C, 225 min). The root-mean-square roughness is 2.1 nm for A and 6.2 nm for B.

In a subsequent study (*214*), the authors compared the adhesion properties of these three model surfaces at two different relative humidities with respect to PDMS caps. The adhesion properties were measured with a microadhesion measurement apparatus and analyzed according to the Johnson-Kendall-Roberts theory. The authors found the work of adhesion from loading to be similar for all three surfaces. Contact angle measurements with water and methylene iodide and interpretation of the data according to the LW/AB approach revealed that the adhesive interactions were dominated by dispersion forces. The magnitude of the adhesion hysteresis was significantly smaller for the cellulose I surface than for the other two surfaces, which was attributed to the reduced ability of this surface to rearrange its chemical groups. The magnitude of the adhesion hysteresis was found to increase with increasing relative humidity in accordance with an increase in molecular mobility.

A different approach to the preparation of native-cellulose model surfaces was taken by Habibi et al. in 2007 (ref 281 and Chapter 5 of this volume). The authors studied the use of the LB technique to transfer monolayers of cellulose nanocrystals onto silicon wafer substrates. The formation of cellulose nanocrystal monolayers at the air–water interface of an LB trough was achieved with the help of a cationic amphiphile, which was believed to draw the nanocrystals from the subphase to the surface through ionic interactions. The amphiphile was later removed from the transferred LB film by chloroform treatment. Because this preparation method relies on the presence of negatively charged groups on the surface of the nanocrystals, these model surfaces will have the same drawback as those prepared by spin-coating of sulfuric-acid hydrolyzed cellulose nanocrystals. Though the authors claimed to have removed the sulfate groups from the film surface through treatment with 1% NaOH, they did not provide evidence for the successful removal. Such treatment might merely convert the acid form of the sulfate to the sodium form but not cleave the sulfate ester bonds. Nevertheless, LB films of cellulose nanocrystals may have some merit based on the high degree of alignment of the rod-like nanoparticles in these films. In a subsequent study, Habibi et al. demonstrated that highly oriented cellulose nanocrystal films could also be prepared by solvent casting in an electric field (ref 282 and Chapter 5).

Ahola et al. investigated the preparation of native-cellulose model surfaces from suspensions of cellulose nanofibrils as opposed to nanocrystals (*283*). The nanofibrils were prepared by mechanical disintegration of bleached sulfite pulp with a high-pressure fluidizer and two different pretreatments, producing a low-charge type and a partially-carboxymethylated type. Model surfaces from these two types of cellulose nanofibrils were obtained by spin coating aqueous suspensions onto silica substrates. The resulting films were very thin (4–6 nm), relatively smooth, and the nanofibrils appeared to be randomly oriented in the AFM pictures presented. The model films were used to study film swelling and surface interactions in water in the absence and presence of electrolytes.

Recent Advances

The recent advances in the area of model cellulosic surfaces can be thematically divided into:

(I) advances in analytical techniques applied to model cellulosic surfaces,

(II) advances in the preparation of smooth model cellulosic surfaces, and

(III) advances towards more complex model surfaces

Advances in Analytical Techniques Applied to Model Cellulosic Surfaces

The most common methods used to characterize model cellulosic surfaces are AFM, for measuring surface roughness and film thickness, ellipsometry, for measuring film thickness, and x-ray photoelectron spectroscopy, for determining the chemical composition. The surface forces of model cellulosic surfaces are frequently studied by surface force apparatus, colloidal probe microscopy, or force–distance measurements with an atomic force microscope. Adsorption studies usually employ a quartz crystal microbalance with or without dissipation monitoring. Several new developments in the area of analytical techniques applied to model cellulosic surfaces are outlined below.

Ellipsometry is an optical technique that measures the changes in the state of polarization of light upon reflection from a film-covered surface. Karabiyik et al., in Chapter 6 of this volume, have used a variation of ellipsometry, namely multiple incident media (MIM) ellipsometry, to measure simultaneously the thickness and refractive index of model cellulose surfaces. MIM ellipsometry involves measuring the ellipticity at Brewster's angle in two different inert, ambient media. By assuming that the thickness of the film is the same in both media, the dielectric constant of the film and subsequently the thickness and refractive index can be calculated from the ellipticity data in the two media. The advantage of this method is that it does not require prior knowledge or modeling of the film's refractive index. The authors compared the values obtained for thin LB- and spin-coated films of TMSC and regenerated cellulose, as well as spin-coated films of cellulose nanocrystals. The refractive indices for the three types of films were 1.46 ± 0.01, 1.51 ± 0.01, and 1.51 ± 0.01, respectively. The refractive index and thickness values were in quantitative agreement with values obtained by spectroscopic ellipsometry and multiple angle of incidence ellipsometry.

Thickness measurements of supported thin films by reflectometry are based on interference effects of radiation reflected at the air–film and film–substrate interfaces. Cranston and Gray developed a method, based on angle-dependent optical reflectometry, to quantify the thickness uniformity and birefringence of cellulose nanocrystal-containing polyelectrolyte multilayer films (*284*). Although the films that were used in this study are not model cellulose surfaces as defined here, the study is included in this discussion based on its close relation and relevance to the field. The method developed by the authors involves measuring the angle dependence of the film's reflectivity at four different set-ups: the reflectivity as a function of incident angle is measured twice in the center and twice at the edge of the spin-coated film; at each location, the reflectivity as a function of incident angle is measured in two orthogonal directions, i.e. the planes of the incident angles are orthogonal in the two measurements. The local film thickness and refractive index are obtained by fitting the reflectivity curves from the four set-ups to a two-layer model. The local film birefringence, arising from shear-induced alignment of the rod-like cellulose nanocrystals during spin coating, is calculated using the refractive indices from the two orthogonal measurements. The thickness values obtained by the new method agreed well with those from AFM scratch-height analysis, ellipsometry, and wavelength-dependent reflectometry. The difference between center and edge thickness increased with the number of bilayers, but the center-to-edge thickness ratio stayed roughly the same. The local birefringence was zero in the center of the film and as high as 0.065 at the edge, indicating a very high degree of alignment of the crystals. Besides allowing the determination of local birefringence, the new method was considered an easier and more convenient technique for quantifying thickness uniformity than AFM scratch-height analysis.

Kaya et al., in Chapter 8 of this volume, have reported the use of surface plasmon resonance (SPR) spectroscopy to measure adsorption of hydroxypropyl xylan onto model cellulose surfaces. The application of SPR spectroscopy to model cellulose surfaces is not new. In 1996, Buchholz et al. demonstrated the utility of SPR spectroscopy in quantifying adsorption of dyes and polyelectrolytes on regenerated-cellulose model surfaces (*174*). However, because this first report did not provide any details on the analytical technique and because SPR spectroscopy has not yet found broad application in the area of model cellulosic surfaces, the study of Kaya et al. and two preceding studies from the same laboratory (*190, 191*), which had been omitted from the previous literature review (*171*), will be included in this discussion.

In SPR spectroscopy, light in the near infrared region is reflected off of a metal surface, generally gold. Changes in the resonance angle, which is the incident angle at which the reflected light intensity is a minimum, correlate with refractive index changes in the vicinity of the metal surface. Adsorption of molecules onto the metal surface or onto a thin film covering the metal surface, resulting in local changes in the refractive index, can thus be monitored through changes in the resonance angle. To study adsorption onto model cellulose surfaces by SPR spectroscopy, gold SPR sensor slides were coated with thin films of regenerated cellulose using the LB-technique. Esker et al. studied the adsorption of anionic, cationic, and fluorine-containing, water-soluble xylan derivatives onto these surfaces and found that adsorption of xylan was promoted by cationic and fluorine-containing substituents (*190*). Gradwell et al. investigated the adsorption of a pullulan abietate, mimicking a lignin–carbohydrate complex, onto these surfaces and observed significant, spontaneous, and apparently irreversible adsorption (*191*). Surface modification of cellulose fibers by irreversible adsorption was suggested as a biomimetic approach to strengthening the filler–matrix interface in wood–plastic composites. Kaya et al. compared the adsorption of hydroxypropyl xylan onto these surfaces with that observed on hydroxyl-terminated and methyl-terminated self-assembled monolayers (SAMs) (Chapter 8). The authors observed submonolayer coverage on the regenerated cellulose and hydroxyl-terminated SAM but monolayer coverage on the methyl-terminated SAM. The preference for the methyl-terminated SAM was interpreted as an indication that hydrogen-bonding interactions did not play a significant role in hydroxypropyl xylan adsorption onto these solid substrates.

Schaffner et al. used evanescent wave video microscopy (EWVM) to study the adsorption of anionic, carboxyl-functionalized polystyrene latex spheres in the presence of a cationic surfactant onto regenerated-cellulose model surfaces (*285*). The technique is similar to SPR spectroscopy in that it is sensitive to changes near a solid surface due to a radiation-induced field that decays rapidly away from the interface. In EWVM, a laser beam is directed through a prism onto a glass slide. As long as the angle of incidence of that laser beam is larger than the angle of total internal reflection, an evanescent wave occurs on the opposite side of the glass, the intensity of which decays exponentially with distance. Because of the limited range of the evanescent wave, only objects in the vicinity of the glass surface are illuminated. The decay profile of the evanescent wave can be controlled by the angle of incidence of the laser beam. Schaffner et al. deposited a regenerated-cellulose layer on the evanescent-wave side of the glass slide, which functioned as the top wall of a parallel-plate flow cell with laminar shear flow, perpendicular to the cellulose surface. Adsorption of latex spheres from the flowing stream onto the cellulose surface was indicated by fluorescence emission by the particles, which was recorded through the optical path of an inverted optical microscopy with a CCD camera. The method was sensitive enough to detect absorption and desorption of individual particles (~ 230 nm in diameter) and to allow particle counting in the observation area. The rate of adsorption and final surface coverage were found to depend on the ratio of latex particle and surfactant concentrations.

Advances in the Preparation of Smooth Model Cellulosic Surfaces

The discovery of several new solvent systems for cellulose, namely precooled aqueous solutions of NaOH/thiourea (*286*), NaOH/urea (*287, 288*), and LiOH/urea (*289*), motivated the development of a new preparation method for model cellulose surfaces. Yan et al. dissolved microcrystalline cellulose and cellulose isolated from wheat straw in aqueous solutions of 6 wt % NaOH and 5 wt % thiourea and spin coated the solutions onto polyacrylamide-coated mica substrates (*290*). The authors found that the wheat-straw cellulose gave smoother model surfaces than the microcrystalline cellulose. A concentration of 0.3% wheat-straw cellulose in the spin-coating solution resulted in cellulose films with a height of ~5 nm and a root-mean-square surface roughness of ~2 nm.

An entirely new type of model cellulose surfaces was recently developed by Yokota et al. (*291*). The authors exploited the chemical difference of the reducing end of the cellulose chain relative to the remaining hydroxyl groups and reacted the aldehyde group with thiosemicarbazide. As intended, the resulting cellulose thiosemicarbazone spontaneously chemisorbed from NMMO solution onto gold-coated silicon wafers. Electron diffraction patterns of the cellulosic SAMs, after removal of any physisorbed molecules through several washing steps, showed reflections of a native-cellulose crystal structure. This new approach has the potential to provide very smooth model surfaces with a native cellulose morphology and without the drawbacks of a negative surface charge.

Advances Towards More Complex Model Surfaces

Precise control of the nanostructure, with respect to composition and morphology, of potentially multi-component thin films would enable us to fabricate more complex model surfaces that more closely resemble the original, for example the surface of a lignocellulosic fiber.

One of the earliest studies, reporting the preparation of microstructured cellulose films, is that by Wiegand et al., published in 1997 (*292*). The authors used the LB technique to generated thin films of thexyldimethylsilyl-cellulose cinnamate on hydrophobized cover slips. The films were covered with electron microscopy grids and irradiated for 1–5 min with UV light of > 200 nm, leading to cross-linking of the film through photocycloaddition reaction between the cinnamoyl groups. Next, the non-exposed (non-cross-linked) domains were removed with chloroform, leaving a negative pattern on the cover slip. Finally, the silyl groups were removed by repeated exposure to concentrated HCl vapor, each time followed by immersion of the film in chloroform. A positive pattern was obtained by cross-linking the entire film and then irradiating the cross-linked film through the mask with UV light in the 200–300 nm region, resulting in photooxidative decomposition in the UV-exposed domains and the formation of low-molecular-weight fragments. Removal of the decomposed material with chloroform, followed by the removal of the silyl groups with HCl vapor resulted in a positive pattern of cross-linked cellulose cinnamate. The use of an additional photocrosslinkable polymer, based on polyglutamate and containing hydrophobic side chains, gave the authors additional control over the microstructure of the resulting film. A similar approach to microstructured cellulose films was taken by Rehfeldt and Tanaka in 2003 (*184*). These authors exposed thin cellulose films, regenerated from LB-deposited or spin-coated TMSC films, on silicon wafers through a Cu electron microscopy grids to deep UV light (λ < 220 nm). Both groups focused on the use of the created cellulosic micropatterns in biotechnological applications (*292, 293*).

In 2004, two groups reported independently the preparation of cellulosic films with breath figure patterns (*294, 295*). The term "breath figure" denotes the pattern that is formed when water vapor condenses on a cold surface (*296*). The breath figure method for the creation of porous films relies on the condensation of water vapor on the surface of an evaporating polymer solution. Provided that the solvent is water-immiscible and that the polymer has amphiphilic properties, the condensed water droplets arrange into a hexagonal lattice in the liquid polymer film and, upon evaporation of the solvent, result in a porous film with honeycomb structure. Park and Kim studied the formation of breath figure patterns in CAB films, prepared by solvent casting or spin coating in a humid environment, and by spin coating in a dry environment from polymer solutions containing water-miscible solvents and a small amount of water (*294*). The authors found that spin casting from a two-component solvent system gave similar patterns as the other two, more tedious methods. Kasai and Kondo studied the preparation of honeycomb-patterned cellulose films from an emulsion of water in a chloroform solution of CA (*295*). Films were solvent cast onto glass slides and dried slowly at 100% relative humidity. The formed honeycomb-patterned CA films were subsequently deacetylated by immersing the films into ammonium hydroxide solution. The pore size of the films could be controlled by varying the time between forming the emulsion and casting the film.

Kontturi et al., in a study published in 2005, exploited the phase separation of binary polymer blends to prepare homogenous cellulose films with uniform, protruding features. The authors spin coated mixtures of TMSC and polystyrene (PS), co-dissolved in toluene, onto silicon wafers. The obtained TMSC/PS blend films were uniform in height. However, conversion of the TMSC to cellulose, through exposure to HCl vapor, caused concave features corresponding to the cellulose domains. Subsequent removal of the PS with toluene turned the concave features into protruding ones by decreasing the average height of the film. Further investigation of the films revealed that the areas from which the PS was removed still had a thin cellulose layer, indicating the coverage of the silicon wafer by cellulose was continuous. This polymer blend-based method by Kontturi et al. is a promising approach as it offers a certain degree of control over both the morphology and the composition of multicomponent films.

Another study, published in the same year by the same laboratory, introduced a method for the preparation of "open" regenerated-cellulose films (*297*). "Open films" are discontinuous films that provide incomplete coverage of the substrate. The open cellulose films, which could be described as a collection of individual, elongated cellulose patches protruding from the flat silicon substrate, were prepared by spin coating a dilute (20 mg/L) solution of TMSC in toluene onto untreated silicon wafers and subsequent regeneration of cellulose by vapor phase acid hydrolysis. In addition to the preparation of the films, the authors studied the behavior of these films upon wetting and subsequent drying. In a subsequent study, Kontturi et al. introduced the preparation of open native-cellulose films from dilute aqueous suspensions of cellulose nanocrystals (*298*). Three different substrates were tested, namely silicon wafers with a native oxide layer, silicon wafers with a surface layer of titanium dioxide, and regenerated-cellulose-coated silicon wafers. The most uniform open films, i.e. distributions of cellulose nanocrystals, were obtained on the titania substrate, due to attractive electrostatic interactions between the positively charged substrate and negatively charged particles. The deposition of open cellulose films onto closed films of different morphology or chemical nature represents another promising tool for the preparation of advanced model surfaces with controlled microstructure.

A very different approach to microstructured cellulose films has been reported by Roman and Navarro in Chapter 7 of this volume, "Deposition of Cellulose Nanocrystals by Inkjet Printing". The authors explored the use of inkjet technology for the deposition of cellulose nanocrystals onto flat substrates. A conventional drop-on-demand inkjet printer with piezoelectric print heads was modified to accommodate microscope slides as printing substrates. The printing solutions, aqueous cellulose nanocrystal suspensions of different concentration and a 0.05% chitosan solution in 0.1 N hydrochloric acid, were filled into refillable, spongeless cartridges. The printing of thin, continuous films and continuous micropatterns, such as square grid patterns, was found to be complicated by two processes: (1) dewetting of the glass substrate and (2) the coffee-drop-effect, resulting in transport of the nanoparticles to the edges of the droplets. Several solutions were proposed, including modification of the substrate's surface chemistry to promote wetting and electrostatic interactions with the negatively charged cellulose nanocrystals. Co-deposition of cellulose nanocrystals and chitosan from separate cartridges resulted in the formation of a uniform layer, in contrast to deposition of the individual components, which resulted in contact-line deposits for the cellulose nanocrystals and a network of ridges for the chitosan solution. Despite the initial difficulties related to poor substrate wetting and undesirable mass transport in the drying deposits, inkjet technology holds great promise for micropatterning and co-deposition of cellulose nanocrystals onto flat substrates based on the numerous ways by which the physico-chemical properties of the substrate's surface can be adjusted.

Whereas the above methods focus mainly on advanced morphologies, advances with respect to the composition of model cellulosic surfaces have also been made. The basis for these advances is a method that dates back to 1966 (*299*) and that allows the fabrication of well-defined, composite, multi-layered films on solid substrates by consecutive adsorption of oppositely charged species. Since this first report of the *layer-by-layer (LbL) assembly method*, which involved oppositely charged colloidal particles, the method has been adapted to the use of other materials, including synthetic polymers, proteins, enzymes, lipids, organic and inorganic nanoparticles, and stacks of dye molecules, to name a few. Furthermore, the method has been extended to non-ionic interactions, such as hydrogen bonding, charge transfer interactions, stereo complex formation, metal–ligand coordination, π–π interactions, and even specific biological substrate–ligand interactions, as well as to deposition methods other than solution-based adsorption, such as spin coating, spraying, and photolithograph (*300*). A review of the recent literature on LbL assembly is given in ref 300.

The potential of the LbL assembly method to provide advanced, composite model surfaces for the study of cellulose-based systems has recently been recognized and several groups have studied the preparation of multilayered films involving either cellulose nanocrystals or water-soluble, charged cellulose derivatives, such as CMC. The literature on cellulose-containing multilayered films has been reviewed by Cranston and Gray in Chapter 4 of this volume, "Polyelectrolyte Multilayer Films Containing Cellulose: A Review". Whereas most of the studies to date have relied on ionically driven adsorption from solution, LbL-assembled composite model surfaces for natural systems, such as lignocellulosic fibers, will have to be based primarily on non-ionic interactions, such as hydrogen bonding and van der Waals forces. The following section summarizes two related, recent articles on cellulose nanocrystal-containing multilayered films, reporting, for the first time, the use of neutron reflectometry for the characterization of cellulose-containing LbL-assembled films and the preparation of non ionic films consisting of alternating layers of cellulose and xyloglucan, representing advanced model surfaces for primary plant cell walls.

The first of these two studies involved thin films consisting of alternating layers of cellulose nanocrystals and poly(allylamine hydrochloride) and focused on the use of neutron reflectometry and AFM to elucidate the fine structure of the films (*301*). The authors found that the cellulose nanocrystal layers consisted of two layers of which the bottom layer had a higher packing density (50%) than the top layer (25%). Furthermore, the thickness of the films was found to increase linearly with the number of bilayers and the film surface was found to become increasingly rough. In the second study, the authors used the LbL assembly method to deposit alternating layers of cellulose nanocrystals and xyloglucan onto negatively charged silicon wafers that were covered with very thin polyelectrolyte-based multilayered primers (*302*). The authors characterized the non-ionic cellulose–xyloglucan films using the same two methods as in the previous study and found that the cellulose nanocrystal layers in this multilayer system were monolayers with a packing density of 40–45% and that the xyloglucan layers were very thin (a few nanometers). As in the previous case, the film thickness was found to increase linearly with the number of bilayers but the surface roughness was found to stay constant and to be fairly low (3.5 nm). The structural differences between the two multilayered films were attributed to the absence of an entropic gain related to the release of counterions for the non-ionic system.

Perspectives

Model cellulosic surfaces have come a long way from the first solvent-cast cellulose ester surfaces for the determination of the work of adhesion with respect to water to the molecularly smooth, micropatterned, or multilayered surfaces of today used for in-situ monitoring of molecular adsorption and enzymatic degradation events. Model surfaces of cellulose esters and ethers are readily prepared by dip coating, solvent casting, or spin coating solutions of the cellulose derivative in common solvents. For model surfaces of cellulose, several different preparation methods are available, each having advantages and disadvantages. Future advances in the area of model cellulosic surfaces are likely to include the development of new characterization techniques and the refinement of existing ones, and the development of new preparation methods that provide better control of the morphology and composition of the model surface.

Acknowledgement

I thank my student Feng Jiang for providing the AFM images shown in Figure 3.

References

1. Engelhardt, J. Sources, industrial derivatives and commercial application of cellulose. *Carbohydr. Eur.* **1995,** *12,* 5–14.
2. Payen, A. Mémoire sur la composition du tissu propre des plantes e du ligneux. *C. R. Hebd. Seances Acad. Sci.* **1838,** *7,* 1052–1056.
3. Payen, A. Sur un moyen d'isoler le tissue élémentaire des bois. *C. R. Hebd. Seances Acad. Sci.* **1838,** *7,* 1125.
4. Payen, A. Rapport sur un Mémoire de M. Payen, relativ à la composition de la matière ligneuse. *C. R. Hebd. Seances Acad. Sci.* **1839,** *8,* 51–53.
5. Karrer, P.; Widmer, F. Polysaccharide III. Beitrag zur Kenntnis der Cellulose. *Helv. Chim. Acta* **1921,** *4,* 174–184.
6. Karrer, P. *Einführung in die Chemie der Polymeren Kohlenhydrate. Ein Grundriß der Chemie der Stärke, des Glykogens, der Zellulose u. a. Polysaccharide.* Akademische Verlagsgesellschaft: Leipzig, 1925.
7. Sponsler, O. L.; Dore, W. H. The structure of ramie cellulose as derived from X-ray data. *Colloid Symp. Monogr.* **1926,** *4,* 174–265.

8. Sponsler, O. L. The Cellulose Space Lattice of Plant Fibres. *Nature* **1927,** *120,* 767.

9. Freudenberg, K.; Braun, E. Methylcellulose. *Liebigs Ann. Chem.* **1928,** *460,* 288–304.

10. Haworth, W. N. The structure of carbohydrates. *Helv. Chim. Acta* **1928,** *11,* 534–548.

11. Freudenberg, K. *Tannin, Cellulose, Lignin.* Springer Verlag: Berlin, 1933.

12. Staudinger, H.; Schweitzer, O. Über hochpolymere Verbindungen, 48. Mitteil.: Über die Molekülgröße der Cellulose. *Ber. Dtsch. Chem. Ges.* **1930,** *63,* 3132–3154.

13. Conner, A. H. Size Exclusion Chromatography of Cellulose and Cellulose Derivatives. In *Handbook of Size Exclusion Chromatography;* Wu, C.-S., Ed.; Chromatographic Science Series 69; Marcel Dekker: New York, 1995; Chapter 13, pp 331–352. http://www.fpl.fs.fed.us/documnts/pdf1995/conne95a.pdf (accessed March 9, 2009).

14. Zugenmaier, P. History of Cellulose Research. In *Crystalline Cellulose and Derivatives: Characterization and Structures;* Springer Series in Wood Science; Springer-Verlag: Berlin, 2008; pp 7–51.

15. Braconnot, H. De la transformation de plusieurs substances végétales en un principe nouveau. *Ann. Chim. Phys.* **1833,** *52,* 290–294.

16. Schönbein, C. F. On the Discovery of Gun-Cotton. *Philos. Mag.* **1847,** *31,* 7–12.

17. Worden, E. C. *Nitrocellulose Industry. A compendium of the history, chemistry, manufacture, commercial application and analysis of nitrates, acetates, and xanthates of cellulose as applied to the peaceful arts. With a chapter on gun cotton, smokeless powder and explosive cellulose nitrates.* D. Van Nostrand Company: New York, 1911. http://books.google.com/books?id=n6aEAAAAIAAJ, http://books.google.com/books?id=bKqEAAAAIAAJ (accessed March 9, 2009).

18. Worden, E. C. *Technology of Cellulose Esters, A theoretical and practical treatise on the origin, history, chemistry, manufacture, technical application and analysis of the products of acylation and alkylation of normal and modified cellulose, including nitrocellulose, celluloid, pyroxylin, collodion, celloidin, guncotton, acetylcellulose and viscose, as applied to technology, pharmacy, microscopy, medicine, photography, and the warlike and peaceful arts.* E. & F. N. Spon, Ltd. : Haymarket, S. W. 1, 1921. http://books.google.com/books?id=7h9LAAAAMAAJ, http://books.google.com/books?id=NhdKAAAAMAAJ, http://books.google.com/books?id=fiRLAAAAMAAJ (accessed March 9, 2009).

19. Weber, C. O.; Cross, C. F. Process of Making Cellulose Esters. U.S.P. 632,605, September 5, 1899.

20. Edgar, K. J.; Buchanan, C. M.; Debenham, J. S.; Rundquist, P. A.; Seiler, B. D.; Shelton, M. C.; Tindall, D. Advances in cellulose ester performance and application. *Prog. Polym. Sci.* **2001,** *26,* 1605–1688.

21. Woodings, C. A brief history of regenerated cellulosic fibres. In *Regenerated Cellulose Fibres;* Woodings, C., Ed.; CRC Press: Boca Raton, FL, 2001; pp 1–21.

22. White MBE, P. Lyocell: The production process and market development. In *Regenerated Cellulose Fibres;* Woodings, C., Ed.; CRC Press: Boca Raton, FL, 2001; pp 62–87.

23. Suida, W. Über den Einfluß der Aktiven Atomgruppen in den Textilfasern auf das Zustandekommen von Färbungen. *Monatsh. Chem.* **1905,** *26,* 413–427.

24. Thielking, H.; Schmidt, M. Cellulose Ethers. In *Ullmann's Encyclopedia of Industrial Chemistry;* 7th ed.; Wiley-VCH: Weinheim, Germany, 2006.

25. Langmuir, I. The Mechanism of the Surface Phenomena of Flotation. *Trans. Faraday Soc.* **1920,** *15,* 62–74.

26. Sheppard, S. E.; Newsome, P. T. Some properties of cellulose esters of homologous fatty acids. *J. Phys. Chem.* **1935,** *39,* 143–152.

27. Bartell, F. E.; Ray, B. R. Wetting Characteristics of Cellulose Derivatives. I. Contact Angles Formed by Water and by Organic Liquids. *J. Am. Ceram. Soc.* **1952,** *74,* 778–783.

28. Ray, B. R.; Bartell, F. E. Wetting Characteristics of Cellulose Derivatives. II. Interrelations of Contact Angles. *J. Am. Ceram. Soc.* **1953,** *57,* 49–56.

29. McLaren, A. D.; Hofrichter, C. H., Jr. Theory of adhesion of high polymers to cellulose. *Pap. Trade J.* **1947,** *125,* 96–100.

30. Hofrichter, C. H., Jr.; McLaren, A. D. Temperature dependence of the adhesion of high polymers to cellulose. *Ind. Eng. Chem.* **1948,** *40,* 329–331.

31. McLaren, A. D. Adhesion of high polymers to cellulose. Influence of structure, polarity, and tack temperature. *J. Polym. Sci.* **1948,** *3,* 652–662.

32. McLaren, A. D. Influence of physical properties on adhesion of high polymers to cellulose. *Pap. Trade J.* **1948,** *126,* 139–140.

33. Seiler, C. J.; McLaren, A. D. Theories of adhesion as applied to the adhesive properties of high polymers. *ASTM Bull.* **1948,** *155,* 50–52.

34. McLaren, A. D.; Seiler, C. J. Adhesion. III. Adhesion of polymers to cellulose and aluminum. *J. Polym. Sci.* **1949,** *4,* 63–74 & 408.

35. McLaren, A. D. Adhesion of polymers to cellulose. *Mod. Plast.* **1954,** *31,* 114–116 & 181.

36. Zisman, W. A. Relation of the Equilibrium Contact Angle to Liquid and Solid Constitution. In *Contact Angle, Wettability and Adhesion;* Fowkes, F. M., Ed.; Advances in Chemistry Series 43; American Chemical Society: Washington, DC, 1964; pp 1–51.

37. Ray, B. R.; Anderson, J. R.; Scholz, J. J. Wetting of Polymer Surfaces. I. Contact Angles of Liquids on Starch, Amylose, Amylopectin, Cellulose and Polyvinyl Alcohol. *J. Phys. Chem.* **1958,** *62,* 1220–1227.

38. Gray, V. R. The Wettability of Wood. *For. Prod. J.* **1962,** *12,* 452–461.

39. Jayme, G.; Balser, K. Clarification of the structure of viscose cellophane. I. Electron microscopic studies of its surface. *Papier* **1964,** *18,* 746–758.

40. Jayme, G.; Balser, K. Electron optical studies on the differences in the surface texture of native and regenerated cellulose. *Papier* **1965,** *19,* 741–749.

41. Jayme, G.; Balser, K. Comparison of cuprammonium and viscose cellulose films by electron microscopy. *Papier* **1967,** *21,* 678–88.
42. Borgin, K. The Properties and the Nature of the Surface of Cellulose, Part One: Theoretical and Experimental. *Norsk Skogind.* **1959,** *13,* 81–92.
43. Borgin, K. The Properties and the Nature of the Surface of Cellulose, Part Two: Cellulose in Contact with Water. Experimental Results and their Interpretation. *Norsk Skogind.* **1959,** *13,* 429–442.
44. Borgin, K. The Properties and the Nature of the Surface of Cellulose, Part Three: Cellulose in Contact with Organic Liquids and Aqueous Solutions of Reduced Surface Tension. *Norsk Skogind.* **1960,** *14,* 485–495.
45. Borgin, K. The Properties and the Nature of the Surface of Cellulose, Part Four: The Effect of Changing or Modifying the Surface of Cellulose. *Norsk Skogind.* **1691,** *15,* 384–391.
46. Hermans, P. H. *Physics and Chemistry of Cellulose Fibres.* Elsevier: New York, 1949.
47. Girifalco, L. A.; Good, R. J. A Theory for the Estimation of Surface and Interfacial Energies. I. Derivation and Application to Interfacial Tension. *J. Phys. Chem.* **1957,** *61,* 904–909.
48. Good, R. J. Contact-Angle, Wetting, and Adhesion – A Critical-Review. *J. Adhes. Sci. Technol.* **1992,** *6,* 1269–1302.
49. Fowkes, F. M. Attractive Forces at Interphases. *Ind. Eng. Chem.* **1964,** *56,* 40–52.
50. Fowkes, F. M. Dispersion Force Contributions to Surface and Interfacial Tensions, Contact Angles, and Heats of Immersion. In *Contact Angle, Wettability and Adhesion;* Fowkes, F. M., Ed.; Advances in Chemistry Series 43; American Chemical Society: Washington, DC, 1964.
51. Sandell, M. A. Wetting of Wood Polysaccharides. M.S. Thesis, State University College of Forestry at Syracuse University, Syracuse, NY, 1968.
52. Luner, P.; Sandell, M. Wetting of cellulose and wood hemicelluloses. *J. Polym. Sci., Polym. Symp.* **1969,** *No. 28,* 115–142.
53. Owens, D. K.; Wendt, R. C. Estimation of the surface free energy of polymers. *J. Appl. Polym. Sci.* **1969,** *13,* 1741–1747.
54. Kaelble, D. H. Dispersion-Polar Surface Tension Properties of Organic Solids. *J. Adhes.* **1970,** *2,* 66–81.
55. Wu, S. Calculation of interfacial tension in polymer systems. *J. Polym. Sci., Part C: Polym. Symp.* **1971,** *34,* 19–30.
56. Neumann, A. W.; Good, R. J.; Hope, C. J.; Sejpal, M. An equation-of-state approach to determine surface tensions of low-energy solids from contact angles. *J. Colloid Interface Sci.* **1974,** *49,* 291–304.
57. Brown, P. F. The Role of Surface Chemistry in the Bonding of a Cellulose Substrate Treated in a Corona Discharge. Ph.D. Thesis, Lawrence University, Appleton, WI, January 1971. http://etd.gatech.edu/theses/available/ipstetd-81/brown_pf.pdf (accessed March 8, 2009).
58. Brown, P. F.; Swanson, J. W. Wetting properties of cellulose treated in a corona discharge. *Tappi J.* **1971,** *54,* 2012–2018.
59. Ferris, J. L. The Wettability of Cellulose Film as Affected by Vapor-Phase Adsorption of Amphipathic Molecules. Ph.D. Thesis, Lawrence University,

Appleton, WI, June 1974. http://etd.gatech.edu/theses/available/ipstetd-140/ferris_jl.pdf (accessed March 8, 2009).

60. Swanson, R. E. The Influence of Molecular Structure of Vapor Phase Chemisorbed Fatty Acids Present in Fractional Monolayer Concentrations on the Wettability of Cellulose Film. Ph.D. Thesis, Lawrence University, Appleton, WI, January 1976. http://etd.gatech.edu/theses/available/ipstetd-292/swanson_re.pdf (accessed March 8, 2009).

61. Lee, S. B.; Luner, P. The Wetting and Interfacial Properties of Lignin. *Tappi J.* **1972**, *55*, 116–121.

62. Tamai, Y.; Makuuchi, K.; Suzuki, M. Experimental Analysis of Interfacial Forces at the Plane Surface of Solids. *J. Phys. Chem.* **1967**, *71*, 4176–4179.

63. Matsunaga, T.; Ikada, Y. Dispersive Component of Surface Free-Energy of Hydrophilic Polymers. *J. Colloid Interface Sci.* **1981**, *84*, 8–13.

64. Fowkes, F. M.; Mostafa, M. A. Acid-Base Interactions in Polymer Adsorption. *Ind. Eng. Chem. Prod. Res. Dev.* **1978**, *17*, 3–7.

65. Katz, S.; Gray, D. G. The Adsorption of Hydrocarbons on Cellophane .1. Zero Coverage Limit. *J. Colloid Interface Sci.* **1981**, *82*, 318–325.

66. Katz, S.; Gray, D. G. The Adsorption of Hydrocarbons on Cellophane .2. Finite Coverage Region. *J. Colloid Interface Sci.* **1981**, *82*, 326–338.

67. Katz, S.; Gray, D. G. The Adsorption of Hydrocarbons on Cellophane .3. Effect of Relative-Humidity. *J. Colloid Interface Sci.* **1981**, *82*, 339–351.

68. Fowkes, F. M. Predicting attractive forces at Interfaces – Analogy to solubility parameter. In *Chemistry and Physics of Interfaces – II;* Gushee, D. E., Ed.; American Chemical Society: Washington, DC, 1971; pp 153–167.

69. Sacher, E. The Possibility of Standard Surface-Tension Values for Polymers. *J. Colloid Interface Sci.* **1983**, *92*, 275–276.

70. Saito, M.; Yabe, A. Dispersion and Polar Force Components of Surface-Tension of Some Polymer-Films. *Text. Res. J.* **1983**, *53*, 54–59.

71. David, D. J.; Misra, A. Surface Energetics Characterization and Relationship to Adhesion Using a Novel Contact-Angle Measuring Technique. *J. Colloid Interface Sci.* **1985**, *108*, 371–376.

72. Toussaint, A. F.; Luner, P. The wetting properties of hydrophobically modified cellulose surfaces. In *Cellulose and Wood – Chemistry and Technology;* Schuerch, C., Ed.; John Wiley & Sons: New York, 1988; pp 1515–1530.

73. Toussaint, A. F.; Luner, P. The Wetting Properties of Grafted Cellulose Films. *J. Adhes. Sci. Technol.* **1993**, *7*, 635–648.

74. Van Oss, C. J.; Good, R. J.; Chaudhury, M. K. Additive and Nonadditive Surface-Tension Components and the Interpretation of Contact Angles. *Langmuir* **1988**, *4*, 884–891.

75. Kamusewitz, H.; Malsch, G.; Paul, D. Contact-Angle Measurements on Surface Modified Cellulose Membranes. *Acta Polym.* **1991**, *42*, 454–457.

76. Blodgett, K. Films Built by Depositing Successive Monomolecular Layers on a Solid Surface. *J. Am. Chem. Soc.* **1935**, *57*, 1007–1022.

77. Breton, M. Formation and Possible Applications of Polymeric Langmuir-Blodgett Films – A Review. *J. Macromol. Sci.-Rev. Macromol. Chem. Phys.* **1981**, *C21*, 61–87.

78. Tredgold, R. H.; Winter, C. S. Langmuir-Blodgett Monolayers of Preformed Polymers. *J. Phys. D-Appl. Phys.* **1982,** *15,* L55–L58.
79. Langmuir, I.; Schaefer, V. J. Built-up films of proteins and their properties. *Science* **1937,** *85,* 76–80.
80. Schubert, D. W.; Dunkel, T. Spin coating from a molecular point of view: its concentration regimes, influence of molar mass and distribution. *Mater. Res. Innov.* **2003,** *7,* 314–321.
81. Kawaguchi, T.; Nakahara, H.; Fukuda, K. Monolayers and Multilayers of Amphiphilic Cellulose Esters and Some Novel Comblike Polymers. *J. Colloid Interface Sci.* **1985,** *104,* 290–293.
82. Kawaguchi, T.; Nakahara, H.; Fukuda, K. Monomolecular and Multimolecular Films of Cellulose Esters with Various Alkyl Chains. *Thin Solid Films* **1985,** *133,* 29–38.
83. Itoh, T.; Tsujii, Y.; Fukuda, T.; Miyamoto, T.; Ito, S.; Asada, T.; Yamamoto, M. Fluorescence Spectroscopy for a Cellulose Trioctadecanoate Monolayer at the Air-Water-Interface. *Langmuir* **1991,** *7,* 2803–2807.
84. Itoh, T.; Tsujii, Y.; Suzuki, H.; Fukuda, T.; Miyamoto, T. Monolayer and Multilayer Langmuir-Blodgett-Films of Cellulose Tri-N-Alkyl Esters Studied by Transmission Electron-Microscopy. *Polym. J.* **1992,** *24,* 641–652.
85. Yanusova, L.; Stiopina, N.; Feigin, L.; Baklagina, Y.; Khripunov, A.; Lavrentiev, V. X-Ray Characterization of Cellulose LB Films. *Physica B* **1994,** *198,* 138–139.
86. Yanusova, L.; Stiopina, N.; Feigin, L.; Baklagina, Y.; Khripunov, A. Langmuir-Blodgett-Films of Substituted Cellulose Acetomyristate – Fabrication and X-Ray-Diffraction Study. *Mater. Sci. Eng. C-Biomimetic Mater. Sens. Syst.* **1995,** *2,* 225–227.
87. Khripunov, A. K.; Baklagina, Y. G.; Stepina, N. D.; Yanusova, L. G.; Feigin, L. A.; Denisov, V. M.; Volkov, A. Y.; Lavrent'ev, V. K. Model of packing of cellulose acetomyristinate in Langmuir-Blodgett films. *Crystallogr. Rep.* **2000,** *45,* 318–322.
88. Feigin, L.; Klechkovskaya, V.; Stepina, N.; Tolstikhina, A.; Khripunov, A.; Baklagina, Y.; Volkov, A.; Antolini, R. On the supramolecular organization of Langmuir-Blodgett cellulose acetovalerate films. *Colloid Surf. A-Physicochem. Eng. Asp.* **2002,** *198,* 13–19.
89. Stepina, N. D.; Klechkovskaya, V. V.; Yanusova, L. G.; Feigin, L. A.; Tolstikhina, A. L.; Sklizkova, V. P.; Khripunov, A. K.; Baklagina, Y. G.; Kudryavtsev, V. V. Formation of Langmuir-Blodgett films in solutions of comblike polymers. *Crystallogr. Rep.* **2005,** *50,* 614–624.
90. Schaub, M.; Fakirov, C.; Schmidt, A.; Lieser, G.; Wenz, G.; Wegner, G.; Albouy, P. A.; Wu, H.; Foster, M. D.; Majrkzak, C.; Satija, S. Ultrathin Layers and Supramolecular Architecture of Isopentylcellulose. *Macromolecules* **1995,** *28,* 1221–1228.
91. Kasai, W.; Kuga, S.; Magoshi, J.; Kondo, T. Compression behavior of Langmuir-Blodgett monolayers of regioselectively substituted cellulose ethers with long alkyl side chains. *Langmuir* **2005,** *21,* 2323–2329.
92. Kasai, W.; Tsutsumi, K.; Morita, M.; Kondo, T. Orientation of the alkyl side chains and glucopyranose rings in Langmuir-Blodgett films of a

regioselectively substituted cellulose ether. *Colloid Polym. Sci.* **2008**, *286*, 707–712.

93. Ifuku, S.; Kamitakahara, H.; Takano, T.; Tsujii, Y.; Nakatsubo, F. Preparation and characterization of 6-*O*-(4-stearyloxytrityl)cellulose acetate Langmuir-Blodgett films. *Cellulose* **2005**, *12*, 361–369.

94. Ifuku, S.; Nakai, S.; Kamitakahara, H.; Takano, T.; Tsujii, Y.; Nakatsubo, F. Preparation and characterization of monolayer and multilayer Langmuir-Blodgett films of a series of 6-*O*-alkylcelluloses. *Biomacromolecules* **2005**, *6*, 2067–2073.

95. Maaroufi, A.; Mao, L. J.; Ritcey, A. M. Structural Characterization of Langmuir-Blodgett-Films of a Cellulose Derivative by X-Ray Photoelectron-Spectroscopy. *Macromol. Symp.* **1994**, *87*, 25–33.

96. Zhang, Y. B.; Tun, Z.; Ritcey, A. M. X-ray and neutron reflectometry investigation of Langmuir-Blodgett films of cellulose ethers. *Langmuir* **2004**, *20*, 6187–6194.

97. Roth, S. V.; Artus, G. R. J.; Rankl, M.; Seeger, S.; Burghammer, M.; Riekel, C.; Muller-Buschbaum, P. Lateral structural variations in thin cellulose layers investigated by microbeam grazing incidence small-angle X-ray scattering. *Physica B* **2005**, *357*, 190–192.

98. Kimura, S.; Kusano, H.; Kitagawa, M.; Kobayashi, H. TEM characterization of cellulose Langmuir-Blodgett films. *Appl. Surf. Sci.* **1999**, *142*, 579–584.

99. Kimura, S.; Kusano, H.; Kitagawa, M.; Kobayashi, H. Layer-by-layer characterization of cellulose Langmuir-Blodgett monolayer films. *Appl. Surf. Sci.* **1999**, *142*, 585–590.

100. Kimura, S.; Kitagawa, M.; Kusano, H.; Kobayashi, H. Electrical and dielectric properties of cellulose LB films. *Polym. Adv. Technol.* **2000**, *11*, 723–726.

101. Kimura, S. I.; Kitagawa, M.; Kusano, H.; Kobayashi, H. In and out plane X-ray diffraction of cellulose LB films. In *Novel Methods to Study Interfacial Layers;* Möbius, D., Miller, R., Eds.; Studies in Interface Science 11; Elsevier: Amsterdam, The Netherlands, 2001; pp 255–264.

102. Tsujii, Y.; Itoh, T.; Fukuda, T.; Miyamoto, T.; Ito, S.; Yamamoto, M. Multilayer Films of Chromophoric Cellulose Octadecanoates Studied by Fluorescence Spectroscopy. *Langmuir* **1992**, *8*, 936–941.

103. Ifuku, S.; Tsujii, Y.; Kamitakahara, H.; Takano, T.; Nakatsubo, F. Preparation and characterization of redox cellulose Langmuir-Blodgett films containing a ferrocene derivative. *J. Polym. Sci. Pol. Chem.* **2005**, *43*, 5023–5031.

104. Sakakibara, K.; Ifuku, S.; Tsujii, Y.; Kamitakahara, H.; Takano, T.; Nakatsubo, F. Langmuir-Blodgett films of a novel cellulose derivative with dihydrophytyl group: The ability to anchor beta-carotene molecules. *Biomacromolecules* **2006**, *7*, 1960–1967.

105. Sakakibara, K.; Kamitakahara, H.; Takano, T.; Nakatsubo, F. Redox-active cellulose Langmuir-Blodgett films containing beta-carotene as a molecular wire. *Biomacromolecules* **2007**, *8*, 1657–1664.

106. Sakakibara, K.; Ogawa, Y.; Nakatsubo, F. First cellulose Langmuir-Blodgett films towards photocurrent generation systems. *Macromol. Rapid Commun.* **2007,** *28,* 1270–1275.

107. Sakakibara, K.; Nakatsubo, F. Fabrication of anodic photocurrent generation systems by use of 6-*O*-dihydrophytylcellulose as a matrix or a scaffold of porphyrins. *Cellulose* **2008,** *15,* 825–835.

108. Sakakibara, K.; Nakatsubo, F. Effect of fullerene on photocurrent performance of 6-*O*-porphyrin-2,3-di-*O*-stearoylcellulose Langmuir-Blodgett films. *Macromol. Chem. Phys.* **2008,** *209,* 1274–1281.

109. Mao, L.; Ritcey, A. M. Langmuir-Blodgett-Films from Cellulose Derivatives Containing Carbazole. *Thin Solid Films* **1994,** *242,* 263–266.

110. Mao, L. J.; Ritcey, A. M. Excimer formation and excitation energy transfer in Langmuir-Blodgett films of cellulose derivatives containing carbazole. *Thin Solid Films* **1996,** *284,* 618–621.

111. Mao, L. J.; Ritcey, A. M. Preparation of cellulose derivatives containing carbazole chromophore. *J. Appl. Polym. Sci.* **1999,** *74,* 2764–2772.

112. Mao, L. J.; Ritcey, A. M. Langmuir-Blodgett films of cellulose ethers containing carbazole. *Macromol. Chem. Phys.* **2000,** *201,* 1718–1725.

113. Panambur, G.; Robert, C.; Zhang, Y. B.; Bazuin, C. G.; Ritcey, A. M. Langmuir monolayers of polyion complexes of soluble quaternary ammonium-functionalized azobenzene chromophores. *Langmuir* **2003,** *19,* 8859–8866.

114. Panambur, G.; Zhang, Y. B.; Yesayan, A.; Galstian, T.; Bazuin, C. G.; Ritcey, A. M. Preparation and characterization of polyion-complexed Langmuir-Blodgett films containing an NLO chromophore. *Langmuir* **2004,** *20,* 3606–3615.

115. Seufert, M.; Fakirov, C.; Wegner, G. Ultrathin Membranes of Molecularly Reinforced Liquids on Porous Substrates. *Adv. Mater.* **1995,** *7,* 52–55.

116. Sigl, H.; Brink, G.; Seufert, M.; Schulz, M.; Wegner, G.; Sackmann, E. Assembly of polymer/lipid composite films on solids based on hairy rod LB-films. *Eur. Biophys. J. Biophys. Lett.* **1997,** *25,* 249–259.

117. Furch, M.; Ueberfeld, J.; Hartmann, A.; Bock, D.; Seeger, S. Ultrathin oligonucleotide layers for fluorescence based DNA-sensors. In *Biomedical Systems and Technologies;* Croitoru, N. I., Frenz, M., King, T. A., Pratesi, R., Scheggi, A. M. V., Seeger, S., Wolfbeis, O. S., Katzir, A., Eds.; Proceedings of the Society of Photo-Optical Instrumentation Engineers (SPIE) 2928; SPIE Optical Engineering Press: Bellingham, WA, 1996; pp 220–226.

118. Hartmann, A.; Bock, D.; Jaworek, T.; Kaul, S.; Schulze, M.; Tebbe, H.; Wegner, G.; Seeger, S. Cellulose-antibody films for highly specific evanescent wave immunosensors. In *Biomedical Optoelectronics in Clinical Chemistry and Biotechnology;* Andersson Eagels, S., Corti, M., Kroo, N., Kertesz, I., King, T. A., Pratesi, R., Seeger, S., Weber, H. P., Katzir, A., Eds.; Proceedings of the Society of Photo-Optical Instrumentation Engineers (SPIE) 2629; SPIE Optical Engineering Press: Bellingham, WA, 1996; pp 108–118.

119. Löscher, F.; Hartmann, A.; Ueberfeld, J.; Ruckstuhl, T.; Bock, D.; Jaworek, T.; Wegner, G.; Seeger, S. Cellulose-protein films for highly specific

evanescent wave immunosensors. In *Biomedical Systems and Technologies;* Croitoru, N. I., Frenz, M., King, T. A., Pratesi, R., Scheggi, A. M. V., Seeger, S., Wolfbeis, O. S., Katzir, A., Eds.; Proceedings of the Society of Photo-Optical Instrumentation Engineers (SPIE) 2928; SPIE Optical Engineering Press: Bellingham, WA, 1996; pp 209–219.

120. Löscher, F.; Ruckstuhl, T.; Jaworek, T.; Wegner, G.; Seeger, S. Immobilization of biomolecules on Langmuir-Blodgett films of regenerative cellulose derivatives. *Langmuir* **1998,** *14,* 2786–2789.

121. Löscher, F.; Ruckstuhl, T.; Seeger, S. Ultrathin cellulose-based layers for detection of single antigen molecules. *Adv. Mater.* **1998,** *10,* 1005–1009.

122. Zahn, M.; Kurzbuch, D.; Pampaloni, F.; Seeger, S. Optical tweezers as tools for studying molecular interactions at surfaces. In *Optical Diagnostics and Living Cells II;* Farkas, D. L., Leif, R. C., J., T. B., Eds.; Proceedings of the Society of Photo-Optical Instrumentation Engineers (SPIE) 3604; SPIE Optical Engineering Press: Bellingham, WA, 1999; pp 90–99.

123. Laib, S.; Krieg, A.; Rankl, M.; Seeger, S. Supercritical angle fluorescence biosensor for the detection of molecular interactions on cellulose-modified glass surfaces. *Appl. Surf. Sci.* **2006,** *252,* 7788–7793.

124. Jung, S.; Angerer, B.; Löscher, F.; Niehren, S.; Winkle, J.; Seeger, S. Biotin-functionalized cellulose-based monolayers as sensitive interfaces for the detection of single molecules. *ChemBioChem* **2006,** *7,* 900–903.

125. Guiomar, A. J.; Evans, S. D.; Guthrie, J. T. Immobilization of glucose oxidase onto a Langmuir-Blodgett ultrathin film of a cellulose derivative deposited on a self-assembled monolayer. *Supramol. Sci.* **1997,** *4,* 279–291.

126. Kusano, H.; Kimura, S.; Kitagawa, M.; Kobayashi, H. Application of cellulose Langmuir-Blodgett films as humidity sensors, and characteristics of the sorption of water molecules into polymer monolayers. *Thin Solid Films* **1997,** *295,* 53–59.

127. Kusano, H.; Kitagawa, M. Novel Humidity and Gas Detector Using Langmuir-Blodgett Cellulose-Thin-Film Coated Quartz Crystal Oscillator. *IEICE Trans. Electron.* **2008,** *E91C,* 1876–1880.

128. Hillebrandt, H.; Wiegand, G.; Tanaka, M.; Sackmann, E. High electric resistance polymer/lipid composite films on indium-tin-oxide electrodes. *Langmuir* **1999,** *15,* 8451–8459.

129. Tanaka, M.; Kaufmann, S.; Nissen, J.; Hochrein, M. Orientation selective immobilization of human erythrocyte membranes on ultrathin cellulose films. *Phys. Chem. Chem. Phys.* **2001,** *3,* 4091–4095.

130. Hillebrandt, H.; Tanaka, M.; Sackmann, E. A novel membrane charge sensor: Sensitive detection of surface charge at polymer/lipid composite films on indium tin oxide electrodes. *J. Phys. Chem. B* **2002,** *106,* 477–486.

131. Plagge, A.; Stratmann, M.; Kowalik, T.; Adler, H. J. Thin layers of cellulose derivatives – Selected analytical aspects. *Macromol. Symp.* **1999,** *145,* 103–111.

132. Plagge, A.; Stratmann, M.; Kowalik, T.; Adler, H. J. Selected aspects of surface engineering with cellulose polymers. *Advanced Engineering Materials* **2000,** *2,* 376–378.

133. Kowalik, T.; Adler, H. J. P.; Plagge, A.; Stratmann, M. Ultrathin layers of phosphorylated cellulose derivatives on aluminium surfaces. *Macromol. Chem. Phys.* **2000,** *201,* 2064–2069.

134. Jaehne, E.; Kowalik, T.; Adler, H. J. P.; Plagge, A.; Stratmann, M. Ultrathin layers of phosphorylated cellulose derivatives on metal surfaces. *Macromol. Symp.* **2002,** *177,* 97–109.

135. Plagge, A.; Adler, H. J.; Jaehne, E.; Paliwoda, G.; Rohwerder, M.; Stratmann, M.; Eichhorn, K. J. Single and double polymer layer arrangements of acid groups containing cellulose and basic groups containing polyethyleneimine on steel. *Macromol. Mater. Eng.* **2007,** *292,* 1245–1255.

136. Plagge, A.; Adler, H. J.; Rohwerder, M.; Stratmann, M. Deposition of functional cellulose films on titanium alloy surfaces. *Thin Solid Films* **2008,** *516,* 2130–2137.

137. Cohen-Atiya, M.; Vadgama, P.; Mandler, D. Preparation, characterization and applications of ultrathin cellulose acetate Langmuir-Blodgett films. *Soft Matter* **2007,** *3,* 1053–1063.

138. Wegner, G. Nanocomposites of hairy-rod macromolecules: Concepts, constructs, and materials. *Macromol. Chem. Phys.* **2003,** *204,* 347–357.

139. Woo, C. K.; Schiewe, B.; Wegner, G. Multilayered assembly of cellulose derivatives as primer for surface modification by polymerization. *Macromol. Chem. Phys.* **2006,** *207,* 148–159.

140. Diamantoglou, M.; Lemke, H. D.; Vienken, J. Cellulose-Ester as Membrane Materials for Hemodialysis. *Int. J. Artif. Organs* **1994,** *17,* 385–391.

141. Diamantoglou, M.; Nywlt, M.; Vienken, J. Cellulose derivatives as membrane materials for haemodialysis structure properties relations. *Papier* **1995,** *49,* 757–764.

142. Vienken, J.; Diamantoglou, M.; Hahn, C.; Kamusewitz, H.; Paul, D. Considerations on Developmental Aspects of Biocompatible Dialysis Membranes. *Artif. Organs* **1995,** *19,* 398–406.

143. Diamantoglou, M.; Meilssner, C.; Sollinger, S.; Hahn, C.; Kayser, H.; Baurmeister, U. Introduction and optimal adjustment of synthetic substituents as key concept in the development of new hemocompatible dialyzer membranes of modified cellulose. *J. Am. Soc. Nephrol.* **1997,** *8,* A0739–A0739.

144. Diamantoglou, M.; Platz, J.; Vienken, J. Cellulose carbamates and derivatives as hemocompatible membrane materials for hemodialysis. *Artif. Organs* **1999,** *23,* 15–22.

145. Vienken, J.; Diamantoglou, M.; Henne, W.; Nederlof, B. Artificial dialysis membranes: From concept to large scale production. *Am. J. Nephrol.* **1999,** *19,* 355–362.

146. Rankl, M.; Laib, S.; Seeger, S. Surface tension properties of surface-coatings for application in biodiagnostics determined by contact angle measurements. *Colloid Surf. B-Biointerfaces* **2003,** *30,* 177–186.

147. Oh, E.; Luner, P. E. Surface free energy of ethylcellulose films and the influence of plasticizers. *Int. J. Pharm.* **1999,** *188,* 203–219.

148. Luner, P. E.; Oh, E. Characterization of the surface free energy of cellulose ether films. *Colloid Surf. A-Physicochem. Eng. Asp.* **2001,** *181,* 31–48.

149. Lua, Y. Y.; Cao, X. P.; Rohrs, B. R.; Aldrich, D. S. Surface characterizations of spin-coated films of ethylcellulose and hydroxypropyl methylcellulose blends. *Langmuir* **2007**, *23*, 4286–4292.

150. Micic, M.; Radotic, K.; Jeremic, M.; Leblanc, R. M. Study of self-assembly of the lignin model compound on cellulose model substrate. *Macromol. Biosci.* **2003**, *3*, 100–106.

151. Gotoh, K. The role of liquid penetration in detergency of long-chain fatty acid. *J. Surfactants Deterg.* **2005**, *8*, 305–310.

152. Gotoh, K.; Nakata, Y.; Tagawa, M. Evaluation of particle deposition in aqueous solutions by the quartz crystal microbalance method. *Colloid Surf. A-Physicochem. Eng. Asp.* **2006**, *272*, 117–123.

153. Castro, L. B. R.; Petri, D. F. S. Assemblies of concanavalin A onto carboxymethylcellulose. *J. Nanosci. Nanotechnol.* **2005**, *5*, 2063–2069.

154. Kosaka, P. M.; Kawano, Y.; Salvadori, M. C.; Petri, D. F. S. Characterization of ultrathin films of cellulose esters. *Cellulose* **2005**, *12*, 351–359.

155. Kosaka, P. M.; Kawano, Y.; El Seoud, O. A.; Petri, D. F. S. Catalytic activity of lipase immobilized onto ultrathin films of cellulose esters. *Langmuir* **2007**, *23*, 12167–12173.

156. Kosaka, P. M.; Kawano, Y.; Petri, D. F. S. Dewetting and surface properties of ultrathin films of cellulose esters. *J. Colloid Interface Sci.* **2007**, *316*, 671–677.

157. Amim, J.; Kosaka, P. M.; Petri, D. F. S. Characteristics of thin cellulose ester films spin-coated from acetone and ethyl acetate solutions. *Cellulose* **2008**, *15*, 527–535.

158. Amim, J.; Kosaka, P. M.; Petri, D. F. S.; Maia, F. C. B.; Miranda, P. B. Stability and interface properties of thin cellulose ester films adsorbed from acetone and ethyl acetate solutions. *J. Colloid Interface Sci.* **2009**, *332*, 477–483.

159. Malmsten, M.; Claesson, P. M.; Pezron, E.; Pezron, I. Temperature-Dependent Forces between Hydrophobic Surfaces Coated with Ethyl(Hydroxyethyl)Cellulose. *Langmuir* **1990**, *6*, 1572–1578.

160. Pezron, I.; Pezron, E.; Claesson, P. M.; Malmsten, M. Temperature-Dependent Forces between Hydrophilic Mica Surfaces Coated with Ethyl(Hydroxyethyl)Cellulose. *Langmuir* **1991**, *7*, 2248–2252.

161. Malmsten, M.; Claesson, P. M. Temperature-Dependent Adsorption and Surface Forces in Aqueous Ethyl(Hydroxyethyl)Cellulose Solutions. *Langmuir* **1991**, *7*, 988–994.

162. Claesson, P. M.; Malmsten, M.; Lindman, B. Forces between Hydrophobic Surfaces Coated with Ethyl(Hydroxyethyl)Cellulose in the Presence of an Ionic Surfactant. *Langmuir* **1991**, *7*, 1441–1446.

163. Claesson, P. Poly(Ethylene Oxide) Surface-Coatings – Relations between Intermolecular Forces, Layer Structure and Protein Repellency. *Colloid Surf. A-Physicochem. Eng. Asp.* **1993**, *77*, 109–118.

164. Blomberg, E.; Claesson, P. M. Interactions between poly(ethylene oxide) coated surfaces and between such surfaces and proteins. *J. Dispersion Sci. Technol.* **1998**, *19*, 1107–1126.

165. Freyssingeas, E.; Thuresson, K.; Nylander, T.; Joabsson, F.; Lindman, B. A surface force, light scattering, and osmotic pressure study of semidilute aqueous solutions of ethyl(hydroxyethyl)cellulose-long-range attractive force between two polymer-coated surfaces. *Langmuir* **1998,** *14,* 5877–5889.
166. Ananthapadmanabhan, K. P.; Mao, G. Z.; Goddard, E. D.; Tirrell, M. Surface Forces Measurements on a Cationic Polymer in the Presence of an Anionic Surfactant. *Colloids Surf.* **1991,** *61,* 167–174.
167. Argillier, J. F.; Ramachandran, R.; Harris, W. C.; Tirrell, M. Polymer Surfactant Interactions Studied with the Surface Force Apparatus. *J. Colloid Interface Sci.* **1991,** *146,* 242–250.
168. Joabsson, F.; Thuresson, K.; Blomberg, E. Interfacial interaction between sodium dodecyl sulfate and hydrophobically modified ethyl(hydroxyethyl)cellulose. A surface force study. *Langmuir* **2001,** *17,* 1506–1510.
169. Shubin, V. Adsorption of Cationic Polymer onto Negatively Charged Surfaces in the Presence of Anionic Surfactant. *Langmuir* **1994,** *10,* 1093–1100.
170. Pincet, F.; Perez, E.; Belfort, G. Molecular-Interactions between Proteins and Synthetic Membrane Polymer-Films. *Langmuir* **1995,** *11,* 1229–1235.
171. Kontturi, E.; Tammelin, T.; Österberg, M. Cellulose-model films and the fundamental approach. *Chem. Soc. Rev.* **2006,** *35,* 1287–1304.
172. Neuman, R. D.; Berg, J. M.; Claesson, P. M. Direct measurement of surface forces in papermaking and paper coating systems. *Nord. Pulp Pap. Res. J.* **1993,** *8,* 96–104.
173. Schaub, M.; Wenz, G.; Wegner, G.; Stein, A.; Klemm, D. Ultrathin Films of Cellulose on Silicon-Wafers. *Adv. Mater.* **1993,** *5,* 919–922.
174. Buchholz, V.; Wegner, G.; Stemme, S.; Ödberg, L. Regeneration, derivatization and utilization of cellulose in ultrathin films. *Adv. Mater.* **1996,** *8,* 399–402.
175. Tammelin, T.; Saarinen, T.; Österberg, M.; Laine, J. Preparation of Langmuir/Blodgett-cellulose surfaces by using horizontal dipping procedure. Application for polyelectrolyte adsorption studies performed with QCM-D. *Cellulose* **2006,** *13,* 519–535.
176. Bergström, L.; Stemme, S.; Dahlfors, T.; Arwin, H.; Ödberg, L. Spectroscopic ellipsometry characterisation and estimation of the Hamaker constant of cellulose. *Cellulose* **1999,** *6,* 1–13.
177. Holmberg, M.; Wigren, R.; Erlandsson, R.; Claesson, P. M. Interactions between cellulose and colloidal silica in the presence of polyelectrolytes. *Colloid Surf. A-Physicochem. Eng. Asp.* **1997,** *130,* 175–183.
178. Holmberg, M.; Berg, J.; Stemme, S.; Ödberg, L.; Rasmusson, J.; Claesson, P. Surface force studies of Langmuir-Blodgett cellulose films. *J. Colloid Interface Sci.* **1997,** *186,* 369–381.
179. Österberg, M. The effect of a cationic polyelectrolyte on the forces between two cellulose surfaces and between one cellulose and one mineral surface. *J. Colloid Interface Sci.* **2000,** *229,* 620–627.
180. Österberg, M.; Claesson, P. M. Interactions between cellulose surfaces: effect of solution pH. *J. Adhes. Sci. Technol.* **2000,** *14,* 603–618.

181. Poptoshev, E.; Carambassis, A.; Österberg, M.; Claesson, P. M.; Rutland, M. W. Comparison of model surfaces for cellulose interactions: elevated pH. *Prog. Colloid Polym. Sci.* **2000**, *116*, 79–83.

182. Poptoshev, E.; Rutland, M. W.; Claesson, P. M. Surface forces in aqueous polyvinylamine solutions. 2. Interactions between glass and cellulose. *Langmuir* **2000**, *16*, 1987–1992.

183. Stiernstedt, J.; Nordgren, N.; Wågberg, L.; Brumer, H.; Gray, D. G.; Rutland, M. W. Friction and forces between cellulose model surfaces: A comparison. *J. Colloid Interface Sci.* **2006**, *303*, 117–123.

184. Rehfeldt, F.; Tanaka, M. Hydration forces in ultrathin films of cellulose. *Langmuir* **2003**, *19*, 1467–1473.

185. Rundlöf, M.; Karlsson, M.; Wågberg, L.; Poptoshev, E.; Rutland, M.; Claesson, P. Application of the JKR method to the measurement of adhesion to Langmuir-Blodgett cellulose surfaces. *J. Colloid Interface Sci.* **2000**, *230*, 441–447.

186. Rojas, O. J.; Ernstsson, M.; Neumann, R. D.; Claesson, P. M. X-ray photoelectron spectroscopy in the study of polyelectrolyte adsorption on mica and cellulose. *J. Phys. Chem. B* **2000**, *104*, 10032–10042.

187. Poptoshev, E.; Claesson, P. M. Weakly charged polyelectrolyte adsorption to glass and cellulose studied by surface force technique. *Langmuir* **2002**, *18*, 1184–1189.

188. Ahola, S.; Österberg, M.; Laine, J. Cellulose nanofibrils-adsorption with poly(amideamine) epichlorohydrin studied by QCM-D and application as a paper strength additive. *Cellulose* **2008**, *15*, 303–314.

189. Paananen, A.; Österberg, M.; Rutland, M.; Tammelin, T.; Saarinen, T.; Tappura, K.; Stenius, P. Interaction between cellulose and xylan: An atomic force microscope and quartz crystal microbalance study. In *Hemicelluloses: Science and Technology;* Gatenholm, P., Tenkanen, M., Eds.; ACS Symposium Series 864; American Chemical Society: Washington, DC, 2004; pp 269–290.

190. Esker, A.; Becker, U.; Jamin, S.; Beppu, S.; Renneckar, S.; Glasser, W. Self-assembly behavior of some co- and heteropolysaccharides related to hemicelluloses. In *Hemicelluloses: Science and Technology;* Gatenholm, P., Tenhanen, M., Eds.; ACS Symposium Series 864; American Chemical Society: Washington, DC, 2004; pp 198–219.

191. Gradwell, S. E.; Renneckar, S.; Esker, A. R.; Heinze, T.; Gatenholm, P.; Vaca-Garcia, C.; Glasser, W. Surface modification of cellulose fibers: towards wood composites by biomimetics. *C. R. Biol.* **2004**, *327*, 945–953.

192. Kontturi, K. S.; Tammelin, T.; Johansson, L. S.; Stenius, P. Adsorption of cationic starch on cellulose studied by QCM-D. *Langmuir* **2008**, *24*, 4743–4749.

193. Penfold, J.; Tucker, I.; Petkov, J.; Thomas, R. K. Surfactant adsorption onto cellulose surfaces. *Langmuir* **2007**, *23*, 8357–8364.

194. Saarinen, T.; Österberg, M.; Laine, J. Adsorption of polyelectrolyte multilayers and complexes on silica and cellulose surfaces studied by QCM-D. *Colloid Surf. A-Physicochem. Eng. Asp.* **2008**, *330*, 134–142.

195. Geffroy, C.; Labeau, M. P.; Wong, K.; Cabane, B.; Cohen Stuart, M. A. Kinetics of adsorption of polyvinylamine onto cellulose. *Colloid Surf. A-Physicochem. Eng. Asp.* **2000**, *172*, 47–56.

196. Kontturi, E.; Thüne, P. C.; Niemantsverdriet, J. W. Novel method for preparing cellulose model surfaces by spin coating. *Polymer* **2003**, *44*, 3621–3625.

197. Kontturi, E.; Thüne, P. C.; Niemantsverdriet, J. W. Cellulose model surfaces-simplified preparation by spin coating and characterization by X-ray photoelectron spectroscopy, infrared spectroscopy, and atomic force microscopy. *Langmuir* **2003**, *19*, 5735–5741.

198. Rossetti, F. F.; Panagiotou, P.; Rehfeldt, F.; Schneck, E.; Dommach, M.; Funari, S. S.; Timmann, A.; Muller-Buschbaum, P.; Tanaka, M. Structures of regenerated cellulose films revealed by grazing incidence small-angle x-ray scattering. *Biointerphases* **2008**, *3*, 117–127.

199. Gunnars, S.; Wågberg, L.; Cohen Stuart, M. A. Model films of cellulose: I. Method development and initial results. *Cellulose* **2002**, *9*, 239–249.

200. Fält, S.; Wågberg, L.; Vesterlind, E. L.; Larsson, P. T. Model films of cellulose II – Improved preparation method and characterization of the cellulose film. *Cellulose* **2004**, *11*, 151–162.

201. Yokota, S.; Kitaoka, T.; Wariishi, H. Surface morphology of cellulose films prepared by spin coating on silicon oxide substrates pretreated with cationic polyelectrolyte. *Appl. Surf. Sci.* **2007**, *253*, 4208–4214.

202. Eriksson, J.; Malmsten, M.; Tiberg, F.; Callisen, T. H.; Damhus, T.; Johansen, K. S. Enzymatic degradation of model cellulose films. *J. Colloid Interface Sci.* **2005**, *284*, 99–106.

203. Eriksson, J.; Malmsten, M.; Tiberg, F.; Callisen, T. H.; Damhus, T.; Johansen, K. S. Model cellulose films exposed to H-insolens glucoside hydrolase family 45 endo-cellulase – The effect of the carbohydrate-binding module. *J. Colloid Interface Sci.* **2005**, *285*, 94–99.

204. Freudenberg, U.; Zschoche, S.; Simon, F.; Janke, A.; Schmidt, K.; Behrens, S. H.; Auweter, H.; Werner, C. Covalent immobilization of cellulose layers onto maleic anhydride copolymer thin films. *Biomacromolecules* **2005**, *6*, 1628–1634.

205. Notley, S. M.; Pettersson, B.; Wågberg, L. Direct measurement of attractive van der Waals' forces between regenerated cellulose surfaces in an aqueous environment. *J. Am. Chem. Soc.* **2004**, *126*, 13930–13931.

206. Radtchenko, I. L.; Papastavrou, G.; Borkovec, M. Direct force measurements between cellulose surfaces and colloidal silica particles. *Biomacromolecules* **2005**, *6*, 3057–3066.

207. Notley, S. M.; Wågberg, L. Morphology of modified regenerated model cellulose II surfaces studied by atomic force microscopy: Effect of carboxymethylation and heat treatment. *Biomacromolecules* **2005**, *6*, 1586–1591.

208. Notley, S. M.; Eriksson, M.; Wågberg, L.; Beck, S.; Gray, D. G. Surface forces measurements of spin-coated cellulose thin films with different crystallinity. *Langmuir* **2006**, *22*, 3154–3160.

209. Leporatti, S.; Sczech, R.; Riegler, H.; Bruzzano, S.; Storsberg, J.; Loth, F.; Jaeger, W.; Laschewsky, A.; Eichhorn, S.; Donath, E. Interaction forces

between cellulose microspheres and ultrathin cellulose films monitored by colloidal probe microscopy – Effect of wet strength agents. *J. Colloid Interface Sci.* **2005**, *281*, 101–111.

210. Sczech, R.; Riegler, H. Molecularly smooth cellulose surfaces for adhesion studies. *J. Colloid Interface Sci.* **2006**, *301*, 376–385.

211. Notley, S. M. Effect of introduced charge in cellulose gels on surface interactions and the adsorption of highly charged cationic polyelectrolytes. *Phys. Chem. Chem. Phys.* **2008**, *10*, 1819–1825.

212. Fält, S.; Wågberg, L.; Vesterlind, E. L. Swelling of model films of cellulose having different charge densities and comparison to the swelling behavior of corresponding fibers. *Langmuir* **2003**, *19*, 7895–7903.

213. Freudenberg, U.; Zimmermann, R.; Schmidt, K.; Behrens, S. H.; Werner, C. Charging and swelling of cellulose films. *J. Colloid Interface Sci.* **2007**, *309*, 360–365.

214. Eriksson, M.; Notley, S. M.; Wågberg, L. Cellulose thin films: Degree of cellulose ordering and its influence on adhesion. *Biomacromolecules* **2007**, *8*, 912–919.

215. Forsström, J.; Eriksson, M.; Wågberg, L. A new technique for evaluating ink-cellulose interactions: Initial studies of the influence of surface energy and surface roughness. *J. Adhes. Sci. Technol.* **2005**, *19*, 783–798.

216. Enarsson, L. E.; Wågberg, L. Polyelectrolyte Adsorption on Thin Cellulose Films Studied with Reflectometry and Quartz Crystal Microgravimetry with Dissipation. *Biomacromolecules* **2009**, *10*, 134–141.

217. Torn, L. H.; Koopal, L. K.; de Keizer, A.; Lyklema, J. Adsorption of nonionic surfactants on cellulose surfaces: Adsorbed amounts and kinetics. *Langmuir* **2005**, *21*, 7768–7775.

218. Aulin, C.; Shchukarev, A.; Lindqvist, J.; Malmström, E.; Wågberg, L.; Lindström, T. Wetting kinetics of oil mixtures on fluorinated model cellulose surfaces. *J. Colloid Interface Sci.* **2008**, *317*, 556–567.

219. Kleimann, J.; Lecoultre, G.; Papastavrou, G.; Jeanneret, S.; Galletto, P.; Koper, G. J. M.; Borkovec, M. Deposition of nanosized latex particles onto silica and cellulose surfaces studied by optical reflectometry. *J. Colloid Interface Sci.* **2006**, *303*, 460–471.

220. Ahola, S.; Turon, X.; Österberg, M.; Laine, J.; Rojas, O. J. Enzymatic Hydrolysis of Native Cellulose Nanofibrils and Other Cellulose Model Films: Effect of Surface Structure. *Langmuir* **2008**, *24*, 11592–11599.

221. Jausovec, D.; Angelescu, D.; Voncina, B.; Nylander, T.; Lindman, B. The antimicrobial reagent role on the degradation of model cellulose film. *J. Colloid Interface Sci.* **2008**, *327*, 75–83.

222. Josefsson, P.; Henriksson, G.; Wågberg, L. The Physical Action of Cellulases Revealed by a Quartz Crystal Microbalance Study Using Ultrathin Cellulose Films and Pure Cellulases. *Biomacromolecules* **2008**, *9*, 249–254.

223. Rojas, O. J.; Jeong, C.; Turon, X.; Argyropoulos, D. S. Measurement of cellulase activity with piezoelectric resonators. In *Materials, Chemicals, and Energy from Forest Biomass;* Argyropoulos, D. S., Ed.; ACS Symposium Series 954; American Chemical Society: Washington, DC, 2007; pp 478–494.

224. Turon, X.; Rojas, O. J.; Deinhammer, R. S. Enzymatic kinetics of cellulose hydrolysis: A QCM-D study. *Langmuir* **2008**, *24*, 3880–3887.

225. Klash, A.; Ncube, E.; Meincken, M. Localization and attempted quantification of various functional groups on pulpwood fibres. *Appl. Surf. Sci.* **2009**, *255*, 6318–6324.

226. Notley, S. M.; Chen, W.; Pelton, R. Extraordinary Adhesion of Phenylboronic Acid Derivatives of Polyvinylamine to Wet Cellulose: A Colloidal Probe Microscopy Investigation. *Langmuir* **2009**, Article ASAP, DOI:10.1021/la900256s.

227. Bradbury, J. E. AKD sizing reversion: The vapor phase adsorption of the thermal decomposition products of alkyl ketene dimer onto cellulose substrates. Ph.D. Thesis, Georgia Institute of Technology, Atlanta, GA, January 1997. http://etd.gatech.edu/theses/available/ipstetd-292/swanson_re.pdf (accessed March 8, 2009).

228. Garnier, G.; Wright, J.; Godbout, L.; Yu, L. Wetting mechanism of alkyl ketene dimers on cellulose films. *Colloid Surf. A-Physicochem. Eng. Asp.* **1998**, *145*, 153–165.

229. Dickson, L. E.; Berg, J. C. The investigation of sized cellulose surfaces with scanning probe microscopy techniques. *J. Adhes. Sci. Technol.* **2001**, *15*, 171–185.

230. Yang, L.; Pelton, R.; McLellan, F.; Fairbank, M. Factors influencing the treatment of paper with fluorochemicals for oil repellency. *Tappi J.* **1999**, *82*, 128–135.

231. Kurosu, K.; Pelton, R. Simple lysine-containing polypeptide and polyvinylamine adhesives for wet cellulose. *J. Pulp Pap. Sci.* **2004**, *30*, 228–232.

232. Li, X.; Pelton, R. Enhancing wet cellulose adhesion with proteins. *Ind. Eng. Chem. Res.* **2005**, *44*, 7398–7404.

233. Chen, W.; Lu, C.; Pelton, R. Polyvinylamine boronate adhesion to cellulose hydrogel. *Biomacromolecules* **2006**, *7*, 701–702.

234. Su, S. X.; Pelton, R. Bovine Serum Albumin (BSA) as an adhesive for wet cellulose. *Cellulose* **2006**, *13*, 537–545.

235. Zhao, B. X.; Bursztyn, L.; Pelton, R. Simple approach for quantifying the thermodynamic potential of polymer-polymer adhesion. *J. Adhes.* **2006**, *82*, 121–133.

236. DiFlavio, J. L.; Pelton, R.; Leduc, M.; Champ, S.; Essig, M.; Frechen, T. The role of mild TEMPO-NaBr-NaClO oxidation on the wet adhesion of regenerated cellulose membranes with polyvinylamine. *Cellulose* **2007**, *14*, 257–268.

237. Feng, X.; Pouw, K.; Leung, V.; Pelton, R. Adhesion of colloidal polyelectrolyte complexes to wet cellulose. *Biomacromolecules* **2007**, *8*, 2161–2166.

238. Miao, C. W.; Pelton, R.; Chen, X. N.; Leduc, M. Microgels versus linear polymers for paper wet strength – Size does matter. *Appita J.* **2007**, *60*, 465–468.

239. Miao, C. W.; Chen, X. N.; Pelton, R. Adhesion of poly(vinylamine) microgels to wet cellulose. *Ind. Eng. Chem. Res.* **2007**, *46*, 6486–6493.

240. Feng, X. H.; Zhang, D.; Pelton, R. Adhesion to wet cellulose – Comparing adhesive layer-by-layer assembly to coating polyelectrolyte complex suspensions. *Holzforschung* **2009**, *63*, 28–32.

241. Zauscher, S.; Klingenberg, D. J. Normal forces between cellulose surfaces measured with colloidal probe microscopy. *J. Colloid Interface Sci.* **2000**, *229*, 497–510.

242. Zauscher, S.; Klingenberg, D. J. Surface and friction forces between cellulose surfaces measured with colloidal probe microscopy. *Nord. Pulp Pap. Res. J.* **2000**, *15*, 459–468.

243. Zauscher, S.; Klingenberg, D. J. Friction between cellulose surfaces measured with colloidal probe microscopy. *Colloid Surf. A-Physicochem. Eng. Asp.* **2001**, *178*, 213–229.

244. Nigmatullin, R.; Lovitt, R.; Wright, C.; Linder, M.; Nakari-Setälä, T.; Gama, A. Atomic force microscopy study of cellulose surface interaction controlled by cellulose binding domains. *Colloid Surf. B-Biointerfaces* **2004**, *35*, 125–135.

245. Linder, Å.; Gatenholm, P. Effect of Cellulose Substrate on Assembly of Xylan. In *Hemicelluloses: Science and Technology;* Gatenholm, P., Tenhanen, M., Eds.; ACS Symposium Series 864; American Chemical Society: Washington, DC, 2004; pp 236–253.

246. Vaswani, S. Surface modification of paper and cellulose using plasma enhanced chemical vapor deposition employing fluorocarbon precursors. Ph.D. Thesis, Georgia Institute of Technology, Atlanta, GA, January 2005. http://etd.gatech.edu/theses/available/ipstetd-292/swanson_re.pdf (accessed March 8, 2009).

247. Werner, C.; Jacobasch, H. J.; Reichelt, G. Surface Characterization of Hemodialysis Membranes Based on Streaming Potential Measurements. *J. Biomater. Sci.-Polym. Ed.* **1995**, *7*, 61–76.

248. Jacobasch, H. J.; Grundke, K.; Werner, C. Surface characterization of cellulose materials. *Papier* **1995**, *49*, 740–745.

249. Werner, C.; Jacobasch, H. J. Surface characterization of hemodialysis membranes based on electrokinetic measurements. *Macromol. Symp.* **1996**, *103*, 43–54.

250. Werner, C.; König, U.; Augsburg, A.; Arnhold, C.; Körber, H.; Zimmermann, R.; Jacobasch, H. J. Electrokinetic surface characterization of biomedical polymers – A survey. *Colloid Surf. A-Physicochem. Eng. Asp.* **1999**, *159*, 519–529.

251. Grundke, K.; Bogumil, T.; Werner, C.; Janke, A.; Pöschel, K.; Jacobasch, H. J. Liquid-fluid contact angle measurements on hydrophilic cellulosic materials. *Colloid Surf. A-Physicochem. Eng. Asp.* **1996**, *116*, 79–91.

252. Müller, M.; Werner, C.; Grundke, K.; Eichhorn, K. J.; Jacobasch, H. J. Spectroscopic and thermodynamic characterization of the adsorption of plasma proteins onto cellulosic substrates. *Macromol. Symp.* **1996**, *103*, 55–72.

253. Müller, M.; Werner, C.; Grundke, K.; Eichhorn, K. J.; Jacobasch, H. J. ATR-FT-IR spectroscopy of proteins adsorbed on biocompatible cellulose films. *Mikrochim. Acta* **1997**, 671–674.

254. Müller, M.; Grosse, I.; Jacobasch, H. J.; Sams, P. Surfactant adsorption and water desorption on thin cellulose films monitored by in-situ ATR FTIR spectroscopy. *Tenside Surfactants Deterg.* **1998,** *35,* 354–359.

255. Werner, C.; Jacobasch, H. J. Surface characterization of polymers for medical devices. *Int. J. Artif. Organs* **1999,** *22,* 160–176.

256. Yamane, C.; Aoyagi, T.; Ago, M.; Sato, K.; Okajima, K.; Takahashi, T. Two different surface properties of regenerated cellulose due to structural anisotropy. *Polym. J.* **2006,** *38,* 819–826.

257. Rutland, M. W.; Carambassis, A.; Willing, G. A.; Neuman, R. D. Surface force measurements between cellulose surfaces using scanning probe microscopy. *Colloid Surf. A-Physicochem. Eng. Asp.* **1997,** *123,* 369–374.

258. Carambassis, A.; Rutland, M. W. Interactions of cellulose surfaces: Effect of electrolyte. *Langmuir* **1999,** *15,* 5584–5590.

259. Bogdanovic, G.; Tiberg, F.; Rutland, M. W. Sliding friction between cellulose and silica surfaces. *Langmuir* **2001,** *17,* 5911–5916.

260. Poptoshev, E.; Carambassis, A.; Österberg, M.; Claesson, P. M.; Rutland, M. W. Comparison of model surfaces for cellulose interactions: elevated pH. *Surf. Colloid Sci.* **2001,** *116,* 79–83.

261. Plunkett, M. A.; Feiler, A.; Rutland, M. W. Atomic force microscopy measurements of adsorbed polyelectrolyte layers. 2. Effect of composition and substrate on structure, forces, and friction. *Langmuir* **2003,** *19,* 4180–4187.

262. Rutland, M. W.; Tyrrell, J. W. G.; Attard, P. Analysis of atomic force microscopy data for deformable materials. *J. Adhes. Sci. Technol.* **2004,** *18,* 1199–1215.

263. Ralston, J.; Larson, I.; Rutland, M. W.; Feiler, A. A.; Kleijn, M. Atomic force microscopy and direct surface force measurements – IUPAC technical report. *Pure Appl. Chem.* **2005,** *77,* 2149–2170.

264. Theander, K.; Pugh, R. J.; Rutland, M. W. Friction force measurements relevant to de-inking by means of atomic force microscope. *J. Colloid Interface Sci.* **2005,** *291,* 361–368.

265. Stiernstedt, J.; Brumer, H.; Zhou, Q.; Teeri, T. T.; Rutland, M. W. Friction between cellulose surfaces and effect of xyloglucan adsorption. *Biomacromolecules* **2006,** *7,* 2147–2153.

266. Feiler, A. A.; Stiernstedt, J.; Theander, K.; Jenkins, P.; Rutland, M. W. Effect of capillary condensation on friction force and adhesion. *Langmuir* **2007,** *23,* 517–522.

267. Theander, K.; Pugh, R. J.; Rutland, M. W. Forces and friction between hydrophilic and hydrophobic surfaces: Influence of oleate species. *J. Colloid Interface Sci.* **2007,** *313,* 735–746.

268. Zhou, Q.; Rutland, M. W.; Teeri, T. T.; Brumer, H. Xyloglucan in cellulose modification. *Cellulose* **2007,** *14,* 625–641.

269. Nordgren, N.; Eklof, J.; Zhou, Q.; Brumer, H.; Rutland, M. W. Top-down grafting of xyloglucan to gold monitored by QCM-D and AFM: Enzymatic activity and interactions with cellulose. *Biomacromolecules* **2008,** *9,* 942–948.

270. Nordgren, N.; Eronen, P.; Österberg, M.; Laine, J.; Rutland, M. W. Mediation of the Nanotribological Properties of Cellulose by Chitosan Adsorption. *Biomacromolecules* **2009**, *10*, 645–650.

271. Zauscher, S.; Klingenberg, D. J. Surface forces and friction between cellulose surfaces in aqueous media. *Nanotribology: Critical Assessment and Research Needs* **2003**, 411–440.

272. Notley, S. M.; Norgren, M. Measurement of interaction forces between lignin and cellulose as a function of aqueous electrolyte solution conditions. *Langmuir* **2006**, *22*, 11199–11204.

273. Salmi, J.; Österberg, M.; Laine, J. The effect of cationic polyelectrolyte complexes on interactions between cellulose surfaces. *Colloid Surf. A-Physicochem. Eng. Asp.* **2007**, *297*, 122–130.

274. Salmi, J.; Österberg, M.; Stenius, P.; Laine, J. Surface forces between cellulose surfaces in cationic polyelectrolyte solutions: The effect of polymer molecular weight and charge density. *Nord. Pulp Paper Res. J.* **2007**, *22*, 249–257.

275. Horvath, A. E.; Lindström, T. The influence of colloidal interactions on fiber network strength. *J. Colloid Interface Sci.* **2007**, *309*, 511–517.

276. Linder, Å.; Bergman, R.; Bodin, A.; Gatenholm, P. Mechanism of assembly of xylan onto cellulose surfaces. *Langmuir* **2003**, *19*, 5072–5077.

277. Edgar, C. D.; Gray, D. G. Smooth model cellulose I surfaces from nanocrystal suspensions. *Cellulose* **2003**, *10*, 299–306.

278. Fleming, K.; Gray, D. G.; Matthews, S. Cellulose crystallites. *Chem.-Eur. J.* **2001**, *7*, 1831–1835.

279. de Souza Lima, M. M.; Borsali, R. Rodlike cellulose microcrystals: Structure, properties, and applications. *Macromol. Rapid Commun.* **2004**, *25*, 771–787.

280. Lefebvre, J.; Gray, D. G. AFM of adsorbed polyelectrolytes on cellulose I surfaces spin-coated on silicon wafers. *Cellulose* **2005**, *12*, 127–134.

281. Habibi, Y.; Foulon, L.; Aguié-Béghin, V.; Molinari, M.; Douillard, R. Langmuir-Blodgett films of cellulose nanocrystals: Preparation and characterization. *J. Colloid Interface Sci.* **2007**, *316*, 388–397.

282. Habibi, Y.; Heim, T.; Douillard, R. AC electric field-assisted assembly and alignment of cellulose nanocrystals. *J. Polym. Sci. Pt. B-Polym. Phys.* **2008**, *46*, 1430–1436.

283. Ahola, S.; Salmi, J.; Johansson, L. S.; Laine, J.; Österberg, M. Model films from native cellulose nanofibrils. Preparation, swelling, and surface interactions. *Biomacromolecules* **2008**, *9*, 1273–1282.

284. Cranston, E. D.; Gray, D. G. Birefringence in spin-coated films containing cellulose nanocrystals. *Colloid Surf. A-Physicochem. Eng. Asp.* **2008**, *325*, 44–51.

285. Schaffner, L.; Brügger, G.; Nyffenegger, R.; Walter, R.; Rička, J.; Kleimann, J.; Hotz, J.; Quellet, C. Surfactant mediated adsorption of negatively charged latex particles to a cellulose surface. *Colloid Surf. A-Physicochem. Eng. Asp.* **2006**, *286*, 39–50.

286. Zhang, L. N.; Ruan, D.; Gao, S. J. Dissolution and regeneration of cellulose in NaOH/thiourea aqueous solution. *J. Polym. Sci. Pt. B-Polym. Phys.* **2002**, *40*, 1521–1529.

287. Cai, J.; Zhang, L. Rapid dissolution of cellulose in LiOH/Urea and NaOH/Urea aqueous solutions. *Macromol. Biosci.* **2005,** *5,* 539–548.

288. Cai, J.; Zhang, L. Unique gelation behavior of cellulose in NaOH/Urea aqueous solution. *Biomacromolecules* **2006,** *7,* 183–189.

289. Liu, S. L.; Zhang, L. N. Effects of polymer concentration and coagulation temperature on the properties of regenerated cellulose films prepared from LiOH/urea solution. *Cellulose* **2009,** *16,* 189–198.

290. Yan, L. F.; Wang, Y. Q.; Chen, J. Fabrication of a model cellulose surface from straw with an aqueous sodium hydroxide/thiourea solution. *J. Appl. Polym. Sci.* **2008,** *110,* 1330–1335.

291. Yokota, S.; Kitaoka, T.; Sugiyama, J.; Wariishi, H. Cellulose I nanolayers designed by self-assembly of its thiosemicarbazone on a gold Substrate. *Adv. Mater.* **2007,** *19,* 3368–3370.

292. Wiegand, G.; Jaworek, T.; Wegner, G.; Sackmann, E. Heterogeneous surfaces of structured hairy-rod polymer films: Preparation and methods of functionalization. *Langmuir* **1997,** *13,* 3563–3569.

293. Tanaka, M.; Wong, A. P.; Rehfeldt, F.; Tutus, M.; Kaufmann, S. Selective deposition of native cell membranes on biocompatible micropatterns. *J. Am. Chem. Soc.* **2004,** *126,* 3257–3260.

294. Park, M. S.; Kim, J. K. Breath figure patterns prepared by spin coating in a dry environment. *Langmuir* **2004,** *20,* 5347–5352.

295. Kasai, W.; Kondo, T. Fabrication of honeycomb-patterned cellulose films. *Macromol. Biosci.* **2004,** *4,* 17–21.

296. Baker, T. J. Breath Figures. *Philos. Mag.* **1922,** *44,* 752–768.

297. Kontturi, E.; Thüne, P. C.; Alexeev, A.; Niemantsverdriet, J. W. Introducing open films of nanosized cellulose – Atomic force microscopy and quantification of morphology. *Polymer* **2005,** *46,* 3307–3317.

298. Kontturi, E.; Johansson, L. S.; Kontturi, K. S.; Ahonen, P.; Thüne, P. C.; Laine, J. Cellulose nanocrystal submonolayers by spin coating. *Langmuir* **2007,** *23,* 9674–9680.

299. Iler, R. K. Multilayers of colloidal particles. *J. Colloid Interface Sci.* **1966,** *21,* 569–594.

300. Ariga, K.; Hill, J. P.; Ji, Q. M. Layer-by-layer assembly as a versatile bottom-up nanofabrication technique for exploratory research and realistic application. *Phys. Chem. Chem. Phys.* **2007,** *9,* 2319–2340.

301. Jean, B.; Dubreuil, F.; Heux, L.; Cousin, F. Structural details of cellulose nanocrystals/polyelectrolytes multilayers probed by neutron reflectivity and AFM. *Langmuir* **2008,** *24,* 3452–3458.

302. Jean, B.; Heux, L.; Dubreuil, F.; Chambat, G.; Cousin, F. Non-Electrostatic Building of Biomimetic Cellulose-Xyloglucan Multilayers. *Langmuir* **2009,** *25,* 3920–3923.

Cellulose Surfaces

Cellulose Model Films: Challenges in Preparation

Eero Kontturi and Monika Österberg

Laboratory of Forest Products Chemistry, Helsinki University of Technology, P.O. Box 6300, FIN-02015 TKK, Finland

The peculiar supramolecular architecture of cellulose, responsible for its poor solubility, provides challenges when devising preparation methods for cellulose model films because ultrathin film deposition usually requires dissolving the material first. The poor solubility of cellulose is encountered in three different model film preparation methods: (i) the use of cellulose solvents, (ii) the use of colloidal dispersion of cellulose nanostructures, and (iii) preparation of a dissolving derivative that can be regenerated to cellulose after film deposition. A comprehensive literature review on all present methods for cellulose model film preparation is given in this chapter.

Introduction

Model films are used to track down changes in chemistry and morphology of complex materials as well as monitoring the physicochemical interactions within complicated systems. A suspension of pulp fibers in papermaking, for example, is a highly complex system of a multitude of various components. Moreover, each wood fiber has a distinct morphology and it is, therefore, hard to draw a morphologically representative fiber sample. If one is able to take the main component of pulp fibers, cellulose, and reduce its morphology to a smooth film, fundamental research becomes easier and less ambiguous.

Since model films are morphologically well-defined, ultrathin films of chemically pure materials, cellulose is not the most straightforward substance for model film preparation. Especially in its native form, cellulose possesses an exceptional supramolecular architecture (*1*, *2*). The supramolecular character is the reason for the poor solubility of cellulose and solubility, in turn, is an essential factor in preparation of model films. In general, the main requirement for film preparation from any material is to dissolve the material and cast the film from solution by removing the solvent or regenerating the material in one way or another. To tackle the difficult solubility of cellulose, researchers have devised methods that can be roughly divided into three major categories: (i) face the difficulty and use direct cellulose solvents, which may seem rather "exotic" to many scientists, (ii) prepare a dispersion of cellulose nanomaterials (e.g., cellulose nanocrystals), and (iii) prepare a dissolving derivative which can be regenerated to cellulose after film deposition. This chapter is a literature review which consists of a short section on the required instrumentation, and a lengthier passage on these three main categories of preparation methods. The subject partially overlaps with our recent coverage on model films of cellulose and their applications (*3*), but this chapter does not intend to repeat that text. We intend to give a comprehensive and up-to-date review on the current status of the *preparation* of cellulose model films.

Instrumentation

The thickness of model films is usually in the ultrathin regime (<100 nm), which already poses experimental challenges for film deposition. There are two established instrumentation methods, with which smooth cellulose model surfaces have successfully been prepared: Langmuir–Blodgett (LB) deposition and spin coating. In the following, a short overview of both techniques is presented.

Langmuir–Blodgett Deposition

The preparation of an LB film involves the following stages (Figure 1). First, the film-forming compound is dissolved in a suitable solvent. The solution is then spread on a clean water surface in a Langmuir trough and the solvent is allowed to evaporate. After that, the surface area of the film is compressed, using one or two barriers, until the desired surface pressure is reached. Deposition of the film onto the substrate is carried out by moving the substrate through the water surface, either upwards or downwards, while keeping the surface pressure constant. If only one monolayer is desired, the substrate (if hydrophilic) is immersed into the water prior to spreading the monolayer and then one monolayer is deposited upon lifting the substrate through the air–water interface. By re-immersing the substrate through the monolayer, a second layer can be deposited. In this way, LB multilayers can be formed. This technique is named after Irving Langmuir and Katherine Blodgett. In 1920, Langmuir published the first paper on monolayer deposition (*4*), and in 1934, Blodgett extended this technique to multilayer deposition (*5*, *6*).

Commonly, the substrate is dipped vertically and one monolayer is transferred onto the substrate every time the substrate passes the air–water interface, both on lowering and lifting the substrate (Figure 1a). Sometimes it is desired that only one side of the substrate is covered with the film, hence another dipping procedure was developed. In this approach, the substrate is dipped horizontally (Figure 1b). The substrate is brought horizontally into contact with the monolayer on the water surface. Upon withdrawing, two layers are deposited onto the substrate (*7*).

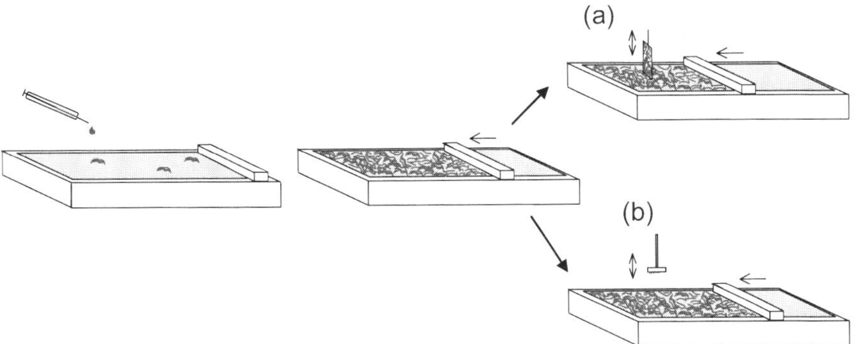

Figure 1. Langmuir–Blodgett deposition. The dissolved film-forming compound is spread on a water surface in a Langmuir trough and the solvent is allowed to evaporate. The surface area of the film is then compressed until the desired surface pressure is reached. Deposition of the film on the substrate is carried out by moving the substrate through the air–water interface, keeping the surface pressure constant. The substrate may be dipped vertically (a) or horizontally (b).

Spin Coating

Spin coating is, in short, a method to create films of dissolved substances by removing the solvent with the aid of high speed spinning. It has been widely used for thin film deposition since the 1950s, and the equipment is ubiquitous in laboratories of physical, inorganic, organic, and polymer chemistry worldwide, not to mention its wide-spread applications in the advanced electronics industry (*8, 9*). The setup for spin coating is depicted in Figure 2. A substrate is attached, often by a means of a suction pump, to a chuck that is spun at a desired rate. Spinning velocity, acceleration, and time of spinning are the adjustable parameters related to the equipment. Concentration of the coating solution and choice of solvent, on the other hand, are parameters related to sample preparation.

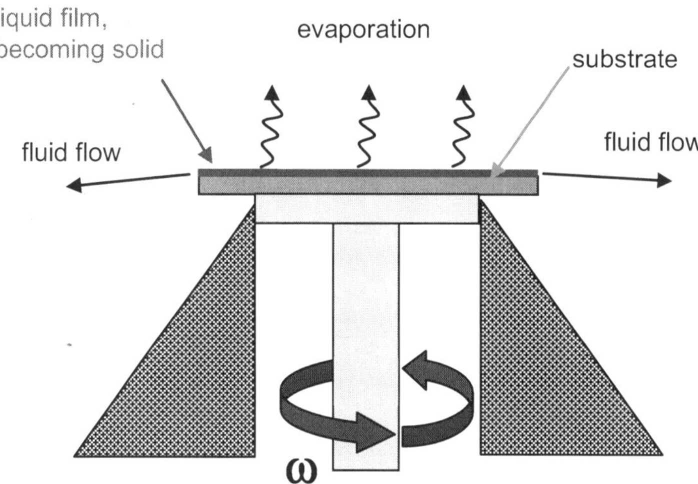

Figure 2. Schematic representation of a spin coating setup. The substrate is attached to a chuck that is spun at a desired rate (normally 1000–6000 rpm). The excess solvent from the initial spin-up stage is collected on the side (exhaust).

There are several theoretical surveys of spin coating (*10–12*). The consensus of these studies is that, after thinning of the spin-induced film starts, there are two stages: first, when the liquid film is thick (several micrometers), fluid flow dominates the thinning, but later, when the film is much thinner and the viscosity higher, fluid flow becomes very slow and film thinning occurs because of evaporation. This transition from flow-control to evaporation-control is abrupt. Initially, with thicker films, the film becomes non-uniform because of the greater velocity and, thus, greater shear rate near the edge of the film. When the film has thinned enough, however, the shear rate becomes low enough throughout the whole film so that the coating solution can behave as a Newtonian fluid. For Newtonian fluids, the film thickness is uniform, according to Emslie et al. (*10*).

Although equations exist to calculate the thickness of the resulting spin coated film (*9, 12*), they are rarely used in practice. The unpredictable nature of, for instance, polymer solutions advocates the experimental method. For a particular film thickness, the experimentalist prepares a set of samples with parameters of concentration, spinning speed, and different solvents, and extrapolates the desired thickness (and smoothness) from the results.

Preparation Methods

Be it LB deposition or spin coating, both techniques involve dissolving the coating material before its deposition on the substrate. This requirement exposes the seminal difficulty in preparation of cellulose model surfaces: its insolubility in common solvents. Solvents for cellulose are somewhat exotic and not straightforward to use concerning thin film deposition. Nonetheless, some of the solvents for cellulose have been successfully exploited in the preparation of model surfaces. A means to circumvent the cumbersome cellulose solvents is to utilize a dispersion of cellulose nanocrystals or synthesize a readily soluble derivative of cellulose that can later be transformed back to cellulose after film deposition. In the following, all three methods will receive a separate treatise. An additional section is devoted to so-called open films, an emerging subject in polymer science but poorly explored with cellulose surfaces.

Deposition of Dissolved Cellulose

The solvents for cellulose, used in model surface preparation, include tri(ethylenediamine) cadmium hydroxide (Cadoxen), lithium chloride containing dimethylacetamide (DMAc/LiCl), and N-methylmorpholine-N-oxide (NMMO). Cadoxen was successfully applied to create monolayers of cellulose on water in the first paper on cellulose model surfaces in 1967 (*13*). The concept of a liquid substrate is unique in the literature concerning cellulose model surfaces, and it is probably something modern surface chemists should look into when generating yet another method of preparing or applying cellulose films. Model surfaces cast directly from solution surfaced again in 1993 in a paper by Neuman et al. (*14*) who used trifluoroacetic acid (TFA) as a solvent to prepare spin cast films on mica. TFA, however, has a problem of reacting with cellulose, which violates the purity of the model substance (*15*). If efforts of casting thick films by evaporation from NMMO (*16*) or by coagulation from sodium hydroxide/urea solution (*17*) are omitted, the first model film to be directly cast from dissolved cellulose and having been adequately characterized was described by Wågberg and co-workers in 2002 (*18*). Their spin coated films (20–270 nm thick) from NMMO, containing small amounts of dimethyl sulfoxide, were deposited onto SiO_2 wafers with glyoxalated polyacrylamide as an anchoring polymer to improve the adhesion between the substrate and the coating. The films were characterized by atomic force microscopy (AFM), X-ray photoelectron spectroscopy (XPS), and size exclusion chromatography. The method was further optimized in a sequel paper in which the effect of numerous parameters of spin coating on film roughness and thickness were investigated (*19*). Unsurprisingly, the concentration of the spin coating solution predominantly determined film thickness but no single factor was found to be associated with surface roughness. The authors' speculation on the crystalline state of the coated cellulose, however, was slightly dubious: the authors' conclusion was that the cellulose films spin coated from NMMO had a similar crystallinity to Lyocell fibers (mixture of cellulose II and amorphous) since Lyocell is prepared by re-crystallization of cellulose from NMMO solution under high shear (*19*). However, the presented evidence was too indirect for more than speculation. The presence of the substrate and the geometrical constraints, for instance, may affect crystallization of cellulose in a thin film. Recent analysis by AFM phase imaging (*20*) points out that a cellulose II/amorphous character of such films is plausible, but hard evidence is nonetheless missing. Perhaps techniques like grazing incidence diffraction could elucidate the crystallinity of thin cellulose films. Furthermore, cellulose films spin coated from NMMO solution received attention in a recent paper by Yokota et al. who studied the effect of anchoring polymer on the morphological orientation of the films (*21*). Overall, the method presented in refs 18 and 19 is one of the best to prepare cellulose model surfaces. The films are smooth and well defined both chemically and physically and their applicability to, for example, surface force and adhesion studies is well established (*22–24*). A slight deficiency lies in the reproducibility of the method: the ranges of film thickness are rather large, such as 20–40 nm or 50–60 nm, but this is probably due to the rather high roughness of the films (*18*). An AFM image of cellulose spin coated from NMMO is presented in Figure 3a.

When exposed to, for instance, aqueous media or heat, it is important that the cellulose films do not detach from the substrate. As in refs 17 and 18, an "anchor" is often used between the substrate and the coating to enhance the physisorption of the coating to the carrier surface. A more efficient means of irreversible attachment was introduced recently by Freudenberg et al. who suggested covalent bonding of cellulose to the substrate to prevent detachment of the film in subsequent treatments (*25*). They utilized the known reaction (*26*) of cellulose with graft copolymers of polypropylene and maleic anhydride to chemisorb cellulose, and eventually came up with 20–300 nm thick layers of cellulose with a thin covalently bonded layer between the film and the substrate. The initial deposition was done by spin coating. These films might prove necessary for studies in which cellulose model surfaces are subjected to harsher conditions, such as the elevated pressure and temperature ranges of pulping or pulp bleaching.

An interesting method for preparing crystalline cellulose films from solution was recently introduced by Yokota et al. (*27*). The authors prepared an end-functionalized cellulose derivative by selectively modifying the reducing end of dissolved cellulose by thiosemicarbazide (TSC) in NMMO solution. When a gold substrate was exposed to this NMMO solution, the TSC end of the TSC-cellulose attached covalently to the gold. As the cellulose chains were forced to a parallel alignment by the attachment from their functionalized reducing ends, electron diffraction analysis of the resulting film showed that crystallization to cellulose I type crystallinity had taken place. Cellulose II is the crystalline form that normally forms when cellulose is precipitated from solution (*28*). However, cellulose II possesses the thermodynamically more favorable anti-parallel chain alignment (*28*) which is not possible when the chains are forced to a parallel alignment by end-grafting them to the substrate, as in the work by Yokota et al. (*27*). This study is not only groundbreaking by introducing a method to regenerate a cellulose I structure; it is also the first proper surface of crystalline native cellulose I with high chemical purity.

Although one of the most common contemporary laboratory solvents for cellulose, DMAc/LiCl appears to be a particularly difficult solvent for the preparation of model surfaces because of the high boiling point of DMAc and the large amount of lithium chloride in the mixture. DMAc/LiCl has nevertheless been exploited as a spin coating solvent for cellulose films (*29*). Elevated temperature (100 °C) was necessary during spin coating and lithium chloride had to be removed by rinsing with water after spin coating. The resulting films were relatively pure according to the XPS data, but morphologically rather rough with a root mean square roughness of almost 5 nm for 28 nm thick films (z-scale variation of ca. 20 nm). Although the films were rough, their suitability for studies on enzyme adsorption and activity was demonstrated in the paper (*29*).

A more detailed survey on spin coating cellulose films from DMAc/LiCl was conducted recently by Sczech and Riegler (*30*). By optimizing the spin coating parameters, they managed to reduce the roughness variation down to only a few nanometers with very thin (<10 nm) films. This roughness variation is actually comparable to the smoothest cellulose films, prepared by regeneration of a hydrophobic cellulose derivative (*31, 32*). Moreover, the study presents extensive characterization by X-ray reflectivity, ellipsometry, FTIR spectroscopy, AFM, and contact angle measurements (*30*). The applicability of the films was demonstrated by studying polyelectrolyte adsorption and subsequent adhesion properties. An AFM image of a cellulose film prepared according to ref 30 is presented in Figure 3b.

Ordinary dialysis membranes from regenerated cellulose were used after thorough washing as model surfaces for surface force studies (*33, 34*). The problem with these surfaces is that they are rather rough and, according to the authors' experiences, the roughness varies throughout the film. To our knowledge, cellulose membranes have not been utilized as model surfaces since.

(a) spin coated from NMMO   (b) Spin coated from DMAc/LiCl

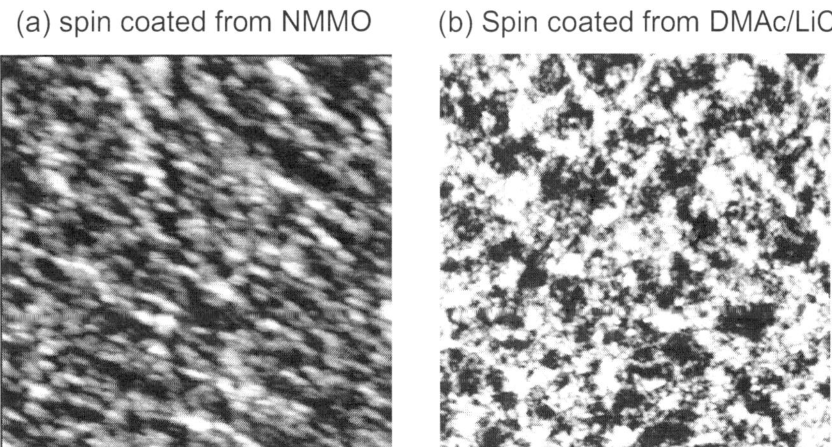

Figure 3. AFM images of model cellulose films: (a) 1×1 μm² image of a film spin coated from NMMO solution (Reproduced from ref 22. Copyright 2004 American Chemical Society), (b) 5×5 μm² image of a film spin coated from DMAc/LiCl solution, prepared according to ref 30.

Deposition of Dispersed Nanocellulose

Spin coating is, in fact, also applicable to colloidal suspensions, not only molecularly dissolved systems (*9*). The group of Gray and co-workers took advantage of this and spin coated colloidal (or "nanocrystal") water suspensions of nanocrystalline cellulose (NCC) on mica substrates (*35*). NCC is prepared from native cellulose samples by hydrolyzing the amorphous cellulose regions using concentrated acid solutions (*36*). The resulting structure is NCC, often also called nanorods, consisting of crystallites of cellulose I with dimensions of ca. (3–20) × (100–2000) nm, depending on the cellulose source. If the NCC is hydrolyzed using sulfuric acid, a sulfate ester is introduced to approximately every tenth anhydroglucose unit on the surface of the crystallites. The electrostatic repulsion of the anionic sulfate groups on the surface yields a stable aqueous suspension of NCC. Spin coating NCC suspensions has recently been further refined, using silicon wafers as substrates, yielding films smooth enough for surface force studies (*37*). Although films from NCC have been characterized by XPS (*35*), X-ray diffraction (XRD) (*35*), and AFM (*35, 37*), they unfortunately lack thickness analysis. On the other hand, the determination of crystallinity was much more reliable in the case of NCC than with the alleged films of cellulose II (*18*): XRD was performed for the film, not the starting material. Besides, spin coating is very unlikely to affect the crystallinity since NCC is a colloidal suspension and it is not re-crystallized in the process.

The nanocrystal suspensions from NCC have also been exploited in a study which applies layer-by-layer (LbL) assembly to prepare nanostructured composites of NCC and poly(dimethyldiallylammonium chloride) (*38*). LbL assembly overcomes the problem of poor matrix connectivity of NCC with the host polymers. The authors elaborate that immobilizing NCC inside a polymer matrix opens a new route for nanoscale organization of the material which has potential for new developments.

A more detailed study on LbL deposition of NCC was performed recently by Cranston and Gray (*39, 40*). The LbL assembly was achieved by both conventional means and spin coating, and the obtained films were characterized by AFM, optical microscopy, and ellipsometry. In addition, the deficiencies of ellipsometry for characterizing the thickness of NCC films were discussed.

Recently, also LB deposition of NCC has been achieved. Habibi et al. took advantage of a complex formation between NCC and an amphiphilic molecule, dioctadecyldimethylammonium (DODA), on the air–water interface of an NCC dispersion (*41*). First, the authors characterized Langmuir layers (i.e., monolayers on water) of the NCC–DODA complex by tensiometry, ellipsometry, and Brewster angle microscopy. After transferring the NCC–DODA complex by LB deposition onto a silicon substrate, the resulting films were washed with chloroform and sodium hydroxide solution, thus removing DODA and leaving NCC intact, as demonstrated by XPS analysis.

Overall, from the point of view of model films, all NCC films presented in the literature so far possess a fundamental disadvantage: the sulfate groups on the crystallite surface, introduced by sulfuric acid hydrolysis when preparing the NCC suspension (*36*). Although NCC films are genuinely of the native cellulose I structure, the sulfate groups on the surface result in a polyelectrolyte like nature (*23*) that is certainly not an intrinsic physicochemical feature of pure cellulose I. It may be possible to remove the sulfate groups after NCC film deposition, but such work has not been reported to date. Furthermore, it is feasible to prepare NCC by, for instance, hydrolysis with hydrochloric acid, which does not introduce sulfate groups on the NCC surface. However, NCC prepared by hydrochloric acid hydrolysis has not been utilized in model film preparation—probably because the lack of anionic sulfate groups and, thus, lack of electrostatic repulsion between the crystallites results in poor dispersion of such NCC. Spin coating a suspension that has a tendency to flocculate can be a tedious task. According to our experience, homogeneity, i.e., full coverage over the substrate and proper smoothness of the films is difficult to achieve by spin coating poorly dispersed NCC.

Whatever the hindrances for NCC usage as model films may be, the sulfate groups are surely not an obstacle to utilize NCC films in materials science and technology where various functionalized biological structures are in high demand at the moment.

Besides the efforts to prepare films from NCC, Ahola et al. recently utilized nanofibrillar cellulose (NFC), prepared by mechanical treatment of chemical pulp fibers (*42*), for model film preparation (*43*). NFC consists of bundles of native cellulose microfibrils with dimensions of 5–10 nm in width and several micrometers in length. The films were prepared by spin coating onto pretreated silica substrates. The advantage of NFC model films is the close resemblance of NFC to cellulose in its native form. In the same paper, the authors demonstrated the applicability of NFC films for swelling and surface force studies.

Deposition of Cellulose Derivatives and Subsequent Conversion to Cellulose

Partial substitution of the hydroxyl groups of cellulose leads to cellulose derivatives that often dissolve in common solvents. To generalize, a charged substituent, such as a sulfate group, yields water soluble derivatives whereas an organosoluble derivative is achieved with a hydrophobic substituent, a silyl ether, for instance. Cellulose derivatization has received extensive reviews, e.g. in ref 44.

If a model surface is cast using a cellulose derivative, the reversibility of the initial reaction is important, i.e. converting the derivative back to cellulose must be relatively effortless. An early paper explains the use of cellulose xanthogenate and cellulose acetate to cast thick films on glass plates (*45*). The films were regenerated to cellulose by sulfuric acid and methanol, respectively. Unfortunately, the characterization was poor and the authors lacked modern morphological characterization methods like AFM.

Trimethylsilyl cellulose (TMSC) has been used as a model cellulose surface medium since the early 1990s. TMSC is a hydrophobic derivative, soluble in common non-polar solvents, such as chloroform or toluene (*46, 47*). Schaub et al. used TMSC to prepare well characterized cellulose films in a seminal publication in the field of cellulose model surfaces (*48*). Their method introduced ultrathin cellulose films of <10 nm thickness on silicon wafers, glass slides, and gold surfaces. The elegance of the technique lies in the vapor phase transition of TMSC to cellulose after LB deposition of TMSC (Figure 4). Exposure to a liquid substance is detrimental to the smooth morphology created by LB deposition. Therefore, the easy acid hydrolysis of TMS groups, carried out in the vapor phase, is ideal. The films were credibly characterized by IR-spectroscopy and X-ray reflectivity (*48*). A follow-up publication covered more elaborate characterization (surface plasmon resonance spectroscopy) and explored the influence of varying LB parameters on film thickness (*49*).

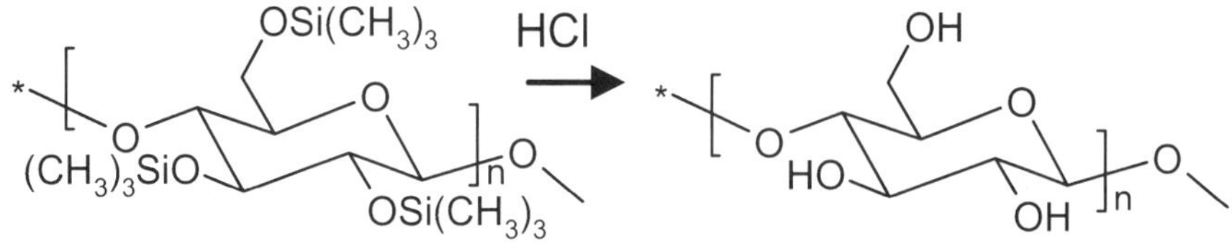

Figure 4. Hydrolysis of TMSC to cellulose. The bulky, "hairy-rod" TMSC structure is compressed to a compact, tightly hydrogen bonded structure of a cellulose network by the removal of TMS groups.

LB deposition of TMSC and the subsequent hydrolysis to cellulose were further refined by Holmberg et al. (*32*). TMSC was cast onto a mica substrate, hydrophobized by a surfactant mixture, and the subsequent hydrolysis to cellulose took place in 10% hydrochloric acid for 1 minute. The characterization of these films is extensive: XPS, ellipsometry, surface force measurements, contact angle measurements, and AFM were applied. Furthermore, some fundamental issues were investigated, such as the swelling of the films in a humid atmosphere. The layer thickness of cellulose was determined as 0.5 nm in dry air. Figure 5a shows an AFM image of LB-deposited TMSC that has been subsequently converted back to cellulose by hydrochloric acid hydrolysis.

The aforementioned publications (*32, 48, 49*) established LB deposition of TMSC as a viable and reproducible method for preparing smooth, ultra-thin films of cellulose by hydrolyzing the TMSC. The thickness of the films could be varied between ca. 5 and 50 nm. The pivotal issues were that hydrolysis of TMSC to cellulose is complete and that the morphology remains smooth after hydrolysis. (In fact, the roughness variation was smaller in the resulting cellulose films than in the corresponding TMSC films because of the bulkiness of the TMSC groups and the tightly bound hydrogen bonding network of cellulose.). The suitability of LB-deposited cellulose films for, e.g., surface force studies has been demonstrated in several occasions (*50–52*). Moreover, Tammelin et al. recently revisited LB deposition of TMSC and updated its use also for horizontal dipping in order to have cellulose films for quartz crystal microbalance studies (*53*).

Despite the fundamental work (*32, 48, 49*), model surfaces of cellulose have started to attract more interest only during the present decade. Geffroy et al. were the first to apply spin coating with TMSC and succeeding hydrolysis, but the characterization of the films was minimal (*54*). In 2003, Rehfeldt and Tanaka published a study comparing LB deposition and spin coating of TMSC and its hydrolysis (*55*). However, the work focused on examining hydration forces (swelling) of the films in humid atmosphere and, consequently, was not methodology oriented. Spin coating received more attention in a survey by Kontturi et al. (*31, 56*) of spin coating TMSC onto untreated silicon and gold substrates. The research included characterization by XPS, IR spectroscopy, ellipsometry, and AFM, confirming the purity of the coated cellulose and smoothness of the morphology. An AFM image of a spin coated TMSC film subsequently hydrolyzed to cellulose is depicted in Figure 5b. The advantages of this method are fast preparation and high degree of reproducibility. On the other hand, smooth films (<10% roughness/thickness) are only produced with <20 nm film thickness, whereas in LB deposition the smoothness is independent of the number of monolayers as long as the film has full coverage over the substrate (*49*). In any case, the influence of spin coating parameters—solution concentration, spinning speed, choice of solvent—on the resulting cellulose film is covered extensively (*31, 56*). The hydrolysis of TMSC to cellulose has been shown to proceed from the film surface to the film–substrate interface by a comparison of IR and XPS data (*31*). Furthermore, both Rehfeldt (*55*) and Kontturi (*31*) point out that the transformation of the TMSC film to a cellulose film results in a ca. 60% contraction in thickness with <20 nm thick films. LB films of the same thickness, in contrast, contract 50% or less (*32, 49, 55*). Thus, in the ultra-thin region, LB deposition appears to be able to cast more compact layers than spin coating.

(a) LB-deposited   (b) spin coated

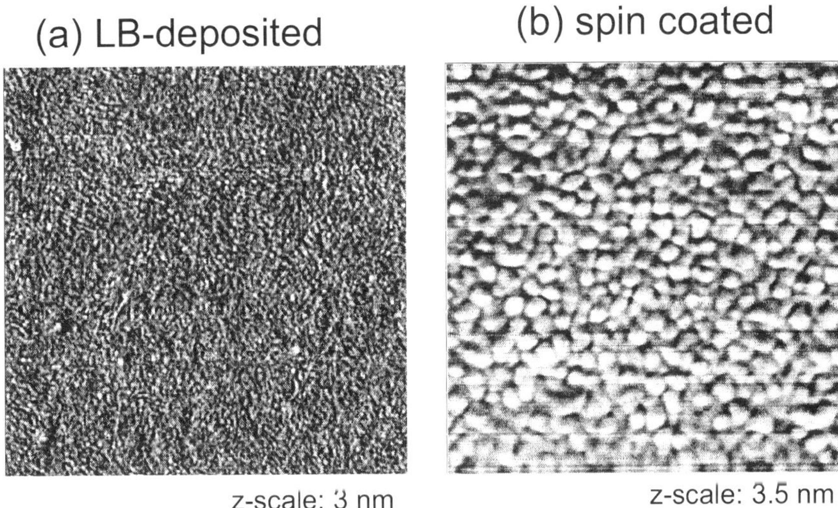

z-scale: 3 nm   z-scale: 3.5 nm

Figure 5. 1×1 μm² AFM images of cellulose films: (a) LB-deposited TMSC film that has subsequently been hydrolyzed to cellulose as described in ref 32; (b) spin coated TMSC film which has subsequently been hydrolyzed to cellulose as described in ref 31.

Open Films of Cellulose

All cellulose surfaces covered in this review so far have been so-called closed films, i.e. the coating substance (cellulose) uniformly and completely covers the substrate. If the concentration of the solution, from which the films are cast, is decreased substantially, there is not enough matter to cover the substrate totally. In this case, an open film occurs. Open films may consist of aggregates ("islands") on a flat substrate (*57*) or—as is often the case with polymers—of evenly spread individual molecules (*58*). The latter type is often used to study polymer molecules by AFM. Applications of open films of individual polymer molecules and AFM include determining the molecular weight distribution (*59*), visualizing conformation and structural diversity (*60*), measuring elasticity (*61*), and visualizing conformational transitions (*62*). Open films are also an indispensable tool in modern supramolecular chemistry (*63*).

By refining the well understood spin coating method (*31*) through a simple decrease in concentration, open films of cellulose on silicon were achieved (*64*). These open films consisted of nanosized cellulose domains that were ca. 50–200 nm long, 20 nm wide, and only 1 nm high (Figure 6a). The reproducibility of the open films was confirmed by quantifying the volume of the cellulose domains in AFM images. The size of these domains suggests that they are conglomerates of a few tens of individual cellulose chains and clearly not individual molecules. As the cellulose domains are conspicuous, yet very small, they provide a novel medium for interpreting supramolecular changes in cellulose networks at the nanoscale. The applicability of these open films of nanosized cellulose was demonstrated by exploring the supramolecular chemistry of cellulose by AFM upon wetting by water and drying (*64*).

The previously discussed paper by Rehfeldt and Tanaka also briefly introduces a method to prepare open cellulose films by micropatterning with UV photolithography (*55*). A grid was placed on the closed cellulose film, regenerated from TMSC, and the film was subsequently illuminated with a mercury lamp, ablating the exposed cellulose but leaving the cellulose underneath the grid intact. The result was a 40 nm wide grid of cellulose with 60 nm rectangles of silica in between. These films were revisited in a follow-up paper which introduced an additional method to prepare open films of cellulose: stamping protein barriers of bovine serum albumin labeled with fluorescein isothiocyanate onto closed cellulose films (*65*).

Another paper on open films of cellulose describes the preparation of micrometer sized cellulose islands of ca. 10–15 nm in height on top of a thin layer (3–5 nm) of cellulose (Figure 6b) (*66*). The films were achieved by exploiting the phase separation of blends of incompatible polymers in thin films (*67*, *68*), in this case polystyrene and TMSC. Hydrolysis transformed the TMSC to completely hydrophilic cellulose, after which the polystyrene could be washed away with a hydrophobic solvent. These surfaces overcome a certain limitation of organic model surfaces on inorganic substrates, namely that the model substance is vastly different—physically and chemically—from the substrate. For instance, large differences in Young's modulus and thermal expansion coefficient can lead to rupture and delamination upon harsh treatments (*69*).

A recent survey on the arrangement of submonolayers of NCC on various substrates included open films of NCC on silica, titania, and amorphous cellulose regenerated from TMSC (*70*). The NCCs on amorphous cellulose (Figure 6c) were the first surfaces to contain conspicuous domains of crystalline and amorphous cellulose but the NCCs suffered from the aforementioned problem of sulfate groups on their surface, as discussed earlier. However, films containing both crystalline and amorphous cellulose in domains, whose quantities can be characterized, will probably be important in the future, especially in studies on enzymatic adsorption and activity where crystallinity is a central issue.

Recently, Yokota et al. managed to prepare open films of single cellulose molecules on a pyrolytic graphite surface (*71*). The deposition was achieved by drop casting with fast (<1 s) evaporation from a dilute cellulose solution of aqueous cupri-ethylenediamine. AFM imaging of single cellulose molecules revealed conformational details about the polysaccharide's alignment on the substrate. The strong tendency of individual cellulose chains to aggregate probably prevents the use of these films in aqueous solutions but they may have potential in, for instance, studies of photodegradation where one could visualize chain scission and/or end-wise degradation on a truly molecular level.

Although the previously discussed films of nanofibrillar cellulose (NFC) exhibit full coverage over the substrate (*43*) and, thus, are not open films by definition, the NFC units within the films do possess conspicuous morphology. Therefore, we can foresee the use of NFC model films also in various morphological studies.

A curious offshoot within cellulose films is the method by Kasai et al. to prepare honeycomb-patterned cellulose films (*72*). The application uses the recently discovered self-organization of a hexagonal array of micropores by casting a polymer emulsion of water in oil onto a substrate (*73*). The authors cast films from cellulose triacetate, dissolved in chloroform, in a water suspension, and convert the acetate back to cellulose with aqueous ammonium hydroxide after deposition. The resulting film is honeycomb shaped cellulose on a layer of cellulose. The pore size of the honeycombs can be altered between 1 and 100 µm. The width of the features was around 10 µm and the height ca. 1–3 µm, which makes traditional AFM imaging difficult. The IR data illustrates convincingly that deacetylation was complete and that the films were thus pure cellulose.

(a) nanosized cellulose on silicon

(b) cellulose on cellulose

(c) crystalline cellulose on cellulose

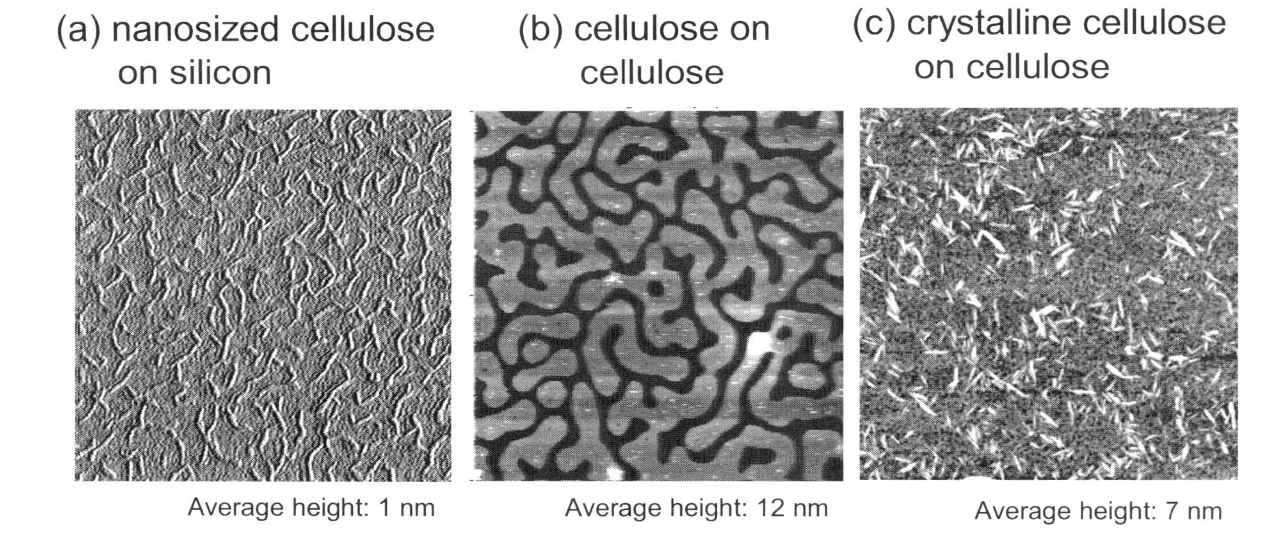

Average height: 1 nm

Average height: 12 nm

Average height: 7 nm

Figure 6. AFM images of open films of cellulose: (a) 1×1 μm^2 height image of nanosized open films of cellulose on untreated silicon, prepared as described in ref 64; the height of the cellulose domains is 1 nm on average; (b) 25×25 μm^2 height image of cellulose islands on cellulose, prepared by spin coating a TMSC/polystyrene mixture onto silicon, hydrolyzing the TMSC to cellulose, and selectively dissolving the polystyrene, as described in ref 66; (c) 5×5 μm^2 height image of nanocrystalline cellulose (NCC) on a thin film of amorphous cellulose, prepared according to ref 70.

Summary and Conclusion

The modern methodology of preparing cellulose model surfaces was initiated by Schaub et al. in their paper about LB deposition of TMSC and its subsequent complete hydrolysis to cellulose in vapor phase acidic conditions (*48*). The method was further refined by Buchholz et al. (*49*) and Holmberg et al. (*32*). The extensive characterization of these LB films in various conditions established LB deposition via TMSC as a reliable and adjustable technique already in the 1990s. LB deposition has recently been established also for cellulose nanocrystals (*41*).

The other route to cellulose films has been developed during the present decade with the use of spin coating. Gunnars et al. demonstrated how ultrathin cellulose films may be spin coated directly from NMMO solution (*18*, *19*). Kontturi et al. explored the possibilities of spin coating TMSC and hydrolyzing it to cellulose (*31*, *56*), precisely as with the LB films (*32*, *48*, *49*) but in a more simplified procedure. Meanwhile, Gray and co-workers focused on spin coating a cellulose nanocrystal suspension, resulting in a smooth film of native crystalline cellulose I (with sulfate groups on the surface) (*35*, *37*).

The advantage of the LB-technique over spin coating is its controlled adjustability. Thickness of the films may be controlled by deposition of one monolayer at a time while the roughness of the films remains constant. Spin coating, on the other hand, provides a faster method. The reproducibility of the films is also reliable with spin coating but there is certain robustness within the control parameters. For instance, a 20 nm thick film may be reproduced by spin coating with as high a degree of smoothness as with LB deposition, but this is not necessarily the case for a 50 nm film (*31*). Moreover, LB deposition and spin coating are not the only methods for successful cellulose model film preparation. A fine example is the recent study by Yokota et al. who demonstrated how crystalline films of native cellulose I can be generated by grafting end-functionalized dissolved cellulose onto gold substrates (*27*).

The newly introduced open films of cellulose have potential for a more morphologically oriented interpretation (*55*, *64*, *66*, *71*, *72*). Films consisting of conspicuous cellulose domains of defined size and shape—or, indeed, individual cellulose chains—are set to be important tools for tracking the behavior of cellulose in diverse conditions that resemble natural or industrial conditions. The hitherto dominant smooth films are excellent for adsorption or surface force studies since they simplify the porous morphology of cellulosic fibers to the extreme. In fact, when totally smooth, the model surfaces reduce the morphology parameter to non-existence. The deliberately "rough" films, however, with defined cellulose shapes may prove to be important in interpreting supramolecular and chemical changes in cellulosic materials during different reaction conditions, such as pulp bleaching, or even in simple wetting and drying processes (swelling/shrinking).

Acknowledgements

Laura Nyfors (M.Sc.) is acknowledged for providing the AFM image for Figure 3b. Ms. Aila Rahkola is acknowledged for preparing the sample that was used for Figure 5a.

References

1. Nishiyama, Y.; Langan, P.; Chanzy, H. *J. Am. Chem. Soc.* **2002**, *124*, 9074.
2. Nishiyama, Y.; Sugiyama, J.; Chanzy, H.; Langan, P. *J. Am. Chem. Soc.* **2003**, *125*, 14300.
3. Kontturi, E.; Tammelin, T.; Österberg, M. *Chem. Soc. Rev.* **2006**, *35*, 1287.
4. Langmuir, I. *Trans. Faraday Soc.* **1920**, *15*, 62.
5. Blodgett, K. B. *J. Am. Chem. Soc.* **1935**, *37*, 1007.
6. Blodgett, K. B. *J. Am. Chem. Soc.* **1934**, *36*, 495.
7. Lee, S.; Virtanen, J. A.; Virtanen, S. A.; Penner, R. M. *Langmuir* **1992**, *8*, 1243.
8. Norman, K.; Ghanbari-Siahkali, A.; Larsen, N. B. *Annu. Rep. Prog. Chem., Sect. C: Phys. Chem.* **2005**, *101*, 174.
9. Larson, R. G.; Rehg, T. in *Liquid Film Coating*, Kistler, S. F.; Schweizer, P. M., Eds.; Chapman & Hall: London, 1997, ch. 14.
10. Emslie, A. G.; Bonner, F. T.; Peck, L. G. *J. Appl. Phys.* **1958**, *29*, 858.
11. Meyerhofer, D. *J. Appl. Phys.* **1978**, *49*, 3993.
12. Bornside, E. B.; Brown, R. A; Ackmann, P. W.; Frank, J. R.; Tryba, A. A.; Geyling, F. T. *J. Appl. Phys.* **1993**, *73*, 585 and refs therein.
13. Giles, C. H.; Agnihotri, V. G. *Chem. Ind. (London, U. K.)* **1967**, *44*, 1874.
14. Neuman, R. D.; Berg, J. M.; Claesson, P. M. *Nord. Pulp Pap. Res. J.* **1993**, *8*, 96.
15. Nehls, I.; Wagenknecht, W.; Philipp, B. *Cellul. Chem. Technol.* **1995**, *29*, 243.
16. Togawa, E.; Kondo, T. *J. Polym. Sci., Part B: Polym. Phys.* **1999**, *37*, 451.
17. Zhang, L.; Ruan, D.; Zhou, J. *Ind. Eng. Chem. Res.* **2001**, *40*, 5923.
18. Gunnars, S.; Wågberg, L.; Cohen Stuart, M. A. *Cellulose* **2002**, *9*, 239.
19. Fält, S.; Wågberg, L.; Vesterlind, E.-L.; Larsson, P. T. *Cellulose* **2004**, *11*, 151.
20. Notley, S. M.; Wågberg, L. *Biomacromolecules* **2005**, *6*, 1586.
21. Yokota, S.; Kitaoka, T.; Wariishi, H. *Appl. Surf. Sci.* **2007**, *253*, 4208.
22. Notley, S. M.; Pettersson, B.; Wågberg, L. *J. Am. Chem. Soc.* **2004**, *126*, 13930.
23. Notley, S. M.; Eriksson, M.; Wågberg, L.; Beck. S.; Gray, D. G. *Langmuir* **2006**, *22*, 3154.
24. Eriksson, M.; Notley, S. M.; Wågberg, L. *Biomacromolecules* **2007**, *8*, 912.
25. Freudenberg, U.; Zschoche, S.; Simon, F.; Janke, A.; Schmidt, K.; Behrens, S. H.; Auweter, H.; Werner, C. *Biomacromolecules* **2005**, *6*, 1628.
26. Qiu, W.; Zhang, F.; Endo, T.; Hirotsu, T. *J. Appl. Polym. Sci.* **2004**, *91*, 1703.

27. Yokota, S.; Kitaoka, T.; Sugiyama; J.; Wariishi, H. *Adv. Mater.* **2007**, *19*, 3368.

28. Langan, P.; Nishiyama, Y.; Chanzy, H. *J. Am. Chem. Soc.* **1999**, *121*, 9940.

29. Eriksson, J.; Malmsten, M.; Tiberg, F.; Callisen, T. H.; Damhus, T.; Johansen, K. S. *J. Colloid Interface Sci.* **2005**, *284*, 99.

30. Sczech, R.; Riegler, H. *J. Colloid Interface Sci.* **2006**, *301*, 376.

31. Kontturi, E.; Thüne, P. C.; Niemantsverdriet, J. W. *Langmuir* **2003**, *19*, 5735.

32. Holmberg, M.; Berg, J.; Stemme, S.; Ödberg, L.; Rasmusson, J.; Claesson, P. *J. Colloid Interface Sci.* **1997**, *186*, 369.

33. Zauscher, S.; Klingenberg, D. J. *J. Colloid Interface Sci.* **2000**, *229*, 497.

34. Zauscher, S.; Klingenberg, D. J. *Colloids Surf. A* **2001**, *178*, 213.

35. Edgar, C. D.; Gray, D. G. *Cellulose* **2003**, *10*, 299.

36. Fleming, K.; Gray, D. G.; Matthews, S. *Chem. – Eur. J.* **2001**, *7*, 1831.

37. Lefebvre, J.; Gray, D. G. *Cellulose* **2005**, *12*, 127.

38. Podsiadlo, P.; Choi, S.-Y.; Shim, B.; Lee, J.; Chuddihy, M.; Kotov, N. A. *Biomacromolecules* **2005**, *6*, 2914.

39. Cranston, E. D.; Gray, D. G. *Sci. Technol. Adv. Mater.* **2006**, *7*, 319.

40. Cranston, E. D.; Gray, D. G. *Biomacromolecules* **2006**, *7*, 2522.

41. Habibi, Y.; Foulon, L.; Aguié-Béghin, V.; Molinari, M.; Douillard, R. *J. Colloid Interface Sci.* **2007**, *316*, 388.

42. Pääkkö, M.; Ankerfors, M.; Kosonen, H.; Nykänen, A.; Ahola, S.; Österberg, M.; Ruokolainen, J.; Laine, J.; Larsson, P. T.; Ikkala, O.; Lindström, T. *Biomacromolecules* **2007**, *8*, 1934.

43. Ahola, S.; Salmi, J.; Johansson, L.-S.; Laine, J.; Österberg, M. *Biomacromolecules* **2008**, *9*, 1273.

44. Klemm, D.; Philipp, B.; Heinze, T.; Heinze, U.; Wagenknecht, W. *Comprehensive Cellulose Chemistry;* Wiley-VCH: Weinheim, 2001; Vol. 2, ch. 4.

45. Luner, P.; Sandell, M. *J. Polym. Sci.: Part C* **1969**, *28*, 115.

46. Greber, G.; Paschinger, O. *Papier* **1981**, *35*, 547.

47. Schempp, W.; Krause, T.; Seifried, U.; Koura, A. *Papier* **1984**, *38*, 607.

48. Schaub, M.; Wenz, G.; Wegner, G.; Stein, A.; Klemm D. *Adv. Mater.* **1993**, *5*, 919.

49. Buchholz, V.; Wegner, G.; Stemme, S.; Ödberg, L. *Adv. Mater.* **1996**, *8*, 399.

50. Holmberg, M.; Wigren, R.; Erlandsson, R.; Claesson, P. M. *Colloids Surf., A* **1997**, *129–130*, 175.

51. Österberg, M. *J. Colloid Interface Sci.* **2000**, *229*, 620.

52. Popotshev, E.; Rutland, M. W.; Claesson, P. M. *Langmuir* **2000**, *16*, 1987.

53. Tammelin, T.; Saarinen, T.; Österberg, M.; Laine, J. *Cellulose* **2006**, *13*, 519.

54. Geffroy, C.; Labeau, M. P.; Wong, K.; Cabane, B.; Cohen Stuart, M. A. *Colloids Surf., A* **2000**, *172*, 47.

55. Rehfeldt, F.; Tanaka, M. *Langmuir* **2003**, *19*, 1467.

56. Kontturi, E.; Thüne, P. C.; Niemantsverdriet, J. W. *Polymer* **2003**, *44*, 3621.

57. Niemantsverdriet, J. W.; Engelen, A. F. P.; de Jong, A. M.; Wieldraaijer, W.; Kramer, G. J. *Appl. Surf. Sci.* **1999**, *144–145*, 366.

58. Sheiko, S. S.; Möller, M. *Chem. Rev.* **2001,** *101,* 4099 and refs therein.
59. Sheiko, S. S.; da Silva, M.; Shirvaniants, D.; LaRue, I.; Prokhorova, S.; Moeller, M.; Beers, K.; Matyjaszewski, K. *J. Am. Chem. Soc.* **2003,** *125,* 6725.
60. Falvo, M. R.; Clary, G. J.; Taylor, R. M., II; Chi, V.; Brooks, F. P., Jr., Washburn, S.; Superfine, R. *Nature* **1997,** *389,* 582.
61. Zhenyu, L.; Novak, W.; Lee, G.; Marszalek, P. E.; Yang, W. *J. Am. Chem. Soc.* **2004,** *126,* 9033.
62. Sheiko, S. S.; Prokhorova, S. A.; Beers, K. L.; Matyjaszewski, K.; Potemkin, I. I.; Khokhlov, A. R.; Möller, M. *Macromolecules* **2001,** *34,* 8354.
63. Samori, P. *Chem. Soc. Rev.* **2005,** *34,* 551 and refs therein.
64. Kontturi, E.; Thüne, P. C.; Alexeev, A.; Niemantsverdriet, J. W. *Polymer* **2005,** *46,* 3307.
65. Tanaka, M.; Wong, A. P.; Rehfeldt, F.; Tutus, M.; Kaufmann, S. *J. Am. Chem. Soc.* **2004,** *126,* 3257.
66. Kontturi, E.; Thüne, P. C.; Niemantsverdriet, J. W. *Macromolecules* **2005,** *38,* 10712.
67. Walheim, S.; Böltau, M.; Mlynek, J.; Krausch, G.; Steiner, U. *Macromolecules* **1997,** *30,* 4995 and refs therein.
68. Budkowski, A. *Adv. Polym. Sci.* **1999,** *148,* 1 and refs therein.
69. Iyer, K. S.; Luzinov, I. *Langmuir* **2003,** *19,* 118 and refs therein.
70. Kontturi, E.; Johansson, L.-S.; Kontturi, K. S.; Ahonen, P.; Thüne, P. C.; Laine, J. *Langmuir* **2007,** *23,* 9674.
71. Yokota, S.; Ueno, T.; Kitaoka, T.; Wariishi, H. *Carbohydr. Res.* **2007,** *342,* 2593.
72. Kasai, W.; Kondo, T. *Macromol. Biosci.* **2004,** *4,* 17.
73. Nishikawa, T.; Nishida, J.; Ookura, R.; Nishimura, S.-I.; Scheumann, V.; Zizlsperger, M.; Lawall, R.; Knoll, W.; Shimomura, M. *Langmuir* **2000,** *16,* 1337 and refs therein.

Chapter 3

Model Cellulose I Surfaces: A Review

Emily D. Cranston and Derek G. Gray

Department of Chemistry, Pulp and Paper Research Centre, McGill University, Montréal, Québec, H3A 2A7, Canada

Past work on model cellulose surfaces has most often used cellulose II or amorphous cellulose regenerated from solvents; natural cellulose I surfaces are generally rough and intractable. However, smooth cellulose I surfaces can be prepared using an aqueous colloidal suspension of cellulose I nanocrystals with suitable substrates. We review a number of ways to make flat, uniform films by solvent-casting, spin-coating, electrostatic adsorption, and Langmuir-Blodgett deposition. The methods to make cellulose I films with internal structure and surface orientation are compared and some applications of these model surfaces are discussed.

Introduction

Cellulose, a long-chain polysaccharide composed of β-1,4-linked D-glucose rings, is the most abundant biomacromolecule in nature (*1*). It is of great industrial importance as a renewable resource, due to its low cost and availability. Cellulose in plants is normally found in close association with other polymers such as lignin and hemicelluloses. Purification and removal of these components, for example by chemical pulping of wood, normally leaves a porous structure with a rough fibrillar surface of cellulose. For many modern characterization techniques (spectroscopies, force measurements, contact angle) smooth surfaces are needed. Model cellulose surfaces can be used to elucidate intermolecular forces, interactions with polymers, metals, and water, bio- and thermal degradation, and photochemistry. These fundamental investigations are significant in papermaking, textiles, and pharmaceutical industries, as well as in the development of new composite materials and biomass gasification.

76

Most naturally occurring cellulose is 'cellulose I', found as metastable partially-crystalline microfibrils, which make up the primary structural component in plant cell walls. These microfibrils are composed of a mixture of two different crystalline polymorphs, designated cellulose I_α and I_β. The I_α crystal structure dominates in cellulose from cotton sources and has a one-chain triclinic unit cell (1–3). The I_β form is defined by a two-chain monoclinic unit cell and interconversion between them is due to shifting of hydrogen bonded cellulose sheets along the chain axis (4).

For many polymers, smooth surfaces can be prepared by solvent or melt casting. However, cellulose does not dissolve in common solvents, and decomposes before melting (1). Once dissolved, cellulose can be regenerated but the resulting film is either amorphous or incorporates the more thermodynamically stable form of cellulose II. Past work on model cellulose surfaces (5–13) has most often used regenerated cellulose which differs in crystallinity, strength, and density from the natural crystalline form. Crystallinity plays an important role when the interactions of cellulose in water are considered. For example, even though cellulose does not dissolve in water, amorphous cellulose swells substantially by taking water into the matrix, which disrupts hydrogen bonding between chains. On the other hand, highly crystalline cellulose I shows little tendency to swell. This review focuses on various methods to prepare chemically defined, reproducible, and smooth surfaces comprised solely of cellulose in its native crystal form.

Nanocrystalline Cellulose as a Source of Cellulose I

Cellulose can be hydrolyzed with sulfuric acid to give an aqueous suspension of cellulose I nanocrystals (14, 15). Nanocrystals can be made from wood pulp (16–18), cotton (19, 20), tunicate (21, 22), bacterial (23, 24) and algal (25, 26) sources and are stable in water due to negative surface charges (sulfate ester or carboxyl groups). The production of nanocrystalline cellulose proceeds in two steps: 1) through controlled chemical hydrolysis to destroy amorphous regions and thus the fibrillar structure and 2) through the use of mechanical energy in the form of ultrasonication to disperse the nanocrystals in water. Suspensions are cleaned of residual acid by extensive centrifugation, rinsing, and dialysis. The resultant highly crystalline, rod-like colloids are hydrophilic, with a surface charge ranging from 0.15 e/nm^2 to 0.4 e/nm^2, corresponding to about 1 in 10 surface glucose units that are substituted with sulfate esters (4). The counterions are most commonly protons or Na^+ ions, with the latter leading to more heat stable films (27, 28).

Cellulose nanocrystals are optically active and rod-shaped and as a result they possess liquid crystalline properties and form a chiral nematic phase above a critical concentration (18). More specifically, a spontaneous phase separation occurs into upper isotropic and lower anisotropic (chiral nematic) phases in the concentration range of 5–7% w/w for nanocrystals from cotton (20).

Cellulose nanocrystals are referred to in the literature by various terms such as whiskers, needles, nanowires, monocrystals, microcrystals, crystallites, depending on their source, axial ratio, and application. Transmission electron microscopy (TEM) and atomic force microscopy (AFM) images of cellulose nanocrystals from pulp and cotton, respectively, are shown in Figure 1. For the majority of the model cellulose surfaces presented here, cellulose nanocrystals are prepared from wood pulp and cotton, and have average dimensions of 5–10 nm wide by 100–350 nm long. The exact physical dimensions and size heterogeneity in cellulose nanocrystals depend on several factors including the cellulose source, reaction time, reaction temperature, acid concentration, and ionic strength. Spherical cellulose nanoparticles have also been made by pre-treating cotton with DMSO and NaOH and hydrolyzing while sonicating with sulfuric and hydrochloric acids. These particles are polydisperse, composed of cellulose I and II, and as of yet, have not been used to make solid cellulose materials (*28–30*). Cellulose nanocrystal properties from different sources and processes are summarized in Table I.

Recently, microfibrillated cellulose (MFC) has been introduced as a high aspect ratio cellulose I alternative. The nanofibrils are prepared though a mild enzymatic hydrolysis (with a monocomponent endoglucanase) combined with mechanical shearing and high-pressure homogenization. The process is well controlled giving two fractions with consistent fibril diameters of 5–7 nm and 10–20 nm. This results in a network of highly entangled, micrometer long cellulose I elements that form a strong gel. The method of Pääkkö et al. (*31*) differs from the usual route to MFC which gives a wide size distribution of cellulose particles with low crystallinity (*32, 33*). Spin-coated films of microfibrillated cellulose have been studied by colloid-probe AFM and quartz crystal microbalance with dissipation. MFC is promising as a reinforcing agent in nanocomposites but contains hemicelluloses. Solid MFC films are mesh-like and therefore not smooth, giving more of a model *fiber* surface than a model *cellulose* surface.

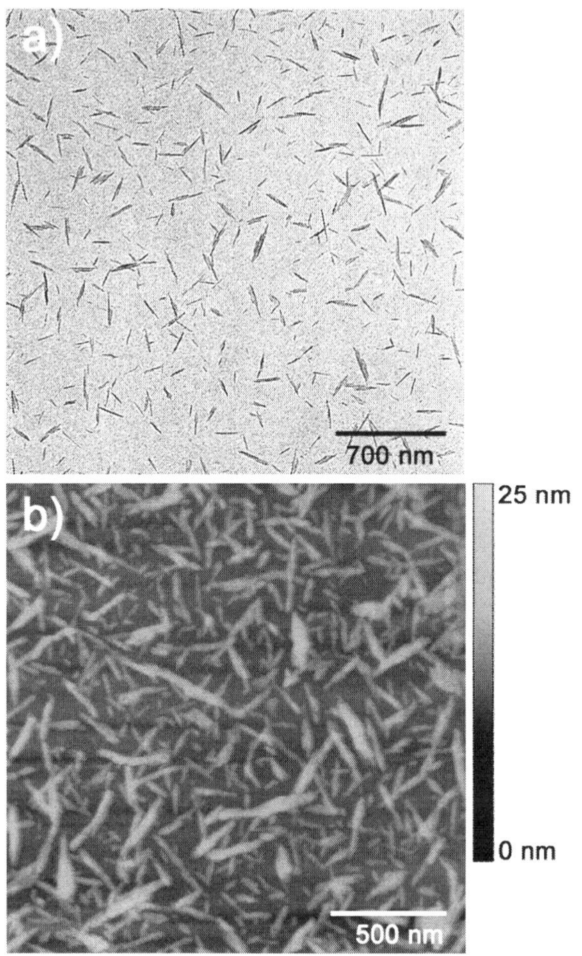

Figure 1. Cellulose nanocrystals imaged by (a) TEM (courtesy of J.-F. Revol and Paprican) and (b) tapping-mode AFM (height image).

Table I. Cellulose nanocrystal dimensions and surface charge from various sources.

Cellulose source	Length	Cross section	Charge	Ref.
bleached softwood	100–300 nm	3–5 nm	0.33–0.38 e/nm^2	(4, 16)
cotton	100–350 nm	7 nm	0.15–0.4 e/nm^2	(4, 16, 20, 25)
bleached hardwood	147 ± 7 nm	3–5 nm	0.29 ± 0.1 e/nm^2	(4, 16)
ramie	150–250 nm	6–8 nm	0.63 e/nm^2	(44)
tunicin	100 nm–several μm	8–15 nm	0.31 e/nm^2	(44)
bacteria	100 nm–several μm	5–10 × 30–50 nm	–	(16, 24)
algae (*Valonia* & *Cladophora*)	100 nm–several μm	20–40 nm	–	(4, 20, 35, 49)
microfibrillated cellulose from bleached softwood	several μm	5 nm	44.2 μeq/g (from hemicelluloses)	(31)
spherical cellulose from cotton	ca. 60 nm (radius)	–	0.16 e/nm^2	(28–30)

Surface Preparation Techniques

The film preparation methods used in the studies discussed below are solvent-casting, spin-coating, electrostatic adsorption, and Langmuir-Blodgett (LB) deposition. Solvent-casting involves choosing a suitable volatile solvent or suspension medium, and a suitable container or mold. The wettability of the container surface is an important variable; for aqueous suspensions of cellulose nanocrystals, the surface should be sufficiently hydrophilic to avoid beading up of the fluid during evaporation, but not so polar that removal of the cellulose film is impossible without damaging the film. During rapid evaporation, the equilibrium liquid crystalline structures expected at the high nanocrystal concentrations are often not achieved, leading to metastable film morphologies with surface defects and undulations that are strongly dependant on the speed of drying. The film thickness is often variable and while the equipment needed for solvent-casting is simple, films are often rough and thick (*8*). Spin-coating is used to apply uniform thin films (generally less than a few 100 nm) to flat surfaces. An excess amount of the solution is placed on the substrate, which is then rotated at high speed in order to spread the fluid. Film properties can be tailored by adjusting experimental parameters such as spin velocity and acceleration, solution concentration, and molecular weight. Optical and mechanical properties can therefore be easily varied by changing the film thickness and density. Creating thick films (necessary if freestanding films are desired) can take many deposition steps with spin-coating and are therefore easier to make using solvent-casting techniques.

Electrostatic adsorption is commonly used to make charged monolayers (or multilayers) on flat substrates. The negative surface charges on cellulose nanocrystals makes them amenable to such preparation methods on positively charged surfaces. Either the substrate is dipped into a dilute aqueous suspension of cellulose nanocrystals or some suspension is poured onto the substrate and then rinsed. In addition to naturally cationic surfaces, positively charged substrates may be prepared by functionalization (i.e. silylation) or by adsorbing cationic polymers. Anchoring polymers used in cellulose film preparations have included chitosan (*8*), polyvinylamine (*8, 34*), glyoxalated-polyacrylamide (*8*), polyallylamine hydrochloride (*22, 35, 36*), polydiallyldimethlyammonium chloride (*37, 38*), and polyethlyeneimine (*22*). Finally, Langmuir-Blodgett films are made by immersing a solid substrate into a monolayer of molecules floating on a liquid. This results in a well-defined surface prepared at a specific surface pressure. The LB technique can flawlessly coat a surface with a single layer of particles but also can be time consuming and is very sensitive to contaminants (*8*).

Model Cellulose I Surfaces

The first reported preparation of a flat model cellulose I surface used cellulose nanocrystals (*5*). A number of preparation methods were tried, and the films were characterized by AFM, X-ray photoelectron spectroscopy (XPS), and X-ray diffraction. Freestanding films were made by evaporating 5 ml of a 2% w/w suspension of cellulose nanocrystals in polystyrene dishes. The root mean square (RMS) roughness of these films determined by AFM was 1.2 nm and 1.0 nm for cellulose from cotton and pulp sources, respectively. The surface originally contacting the polystyrene substrate was usually flatter than the upper free surface. X-ray diffraction confirmed that all of the surfaces prepared from cellulose nanocrystals were solely cellulose I (*5*). XPS indicated that the difference in chemical composition between the two kinds of nanocrystals was minimal; cellulose nanocrystals from cotton had a slightly larger O/C ratio than for the wood fibre nanocrystals and both types had a O/C ratio substantially lower than theoretically predicted (*39*). This is generally attributed to hydrocarbon impurities. Cellulose films were heat-treated to enhance stability and make them less susceptible to swelling and re-dispersion in water. Depending on the study and cellulose suspension counterions, various treatments were found to be effective, 105 °C overnight (*5*), 35 °C for 24 hours in vacuum (*27*), 90 °C for 4 hours (*40*), 80 °C for 15 minutes (*37*), with milder heating preferred for films made from acid-form suspensions.

Mica was also used as a substrate for model cellulose films (*5*). Atomically flat mica is freshly cleaved before use, ensuring a clean smooth surface. It is however, negatively charged leading to electrostatic repulsion between the surface and the cellulose nanocrystals. Films prepared on mica were made by solvent-casting and spin-coating (2700 rpm for 60 s) 2% w/w suspensions. Solvent-cast films could be peeled off the substrate and the surface roughness was measured on the top and the bottom of the film. The top surface was slightly smoother than the bottom and the RMS roughness values for films from cotton and pulp nanocrystals were 2.3 nm and 1.5 nm, respectively. Film properties for these and other model cellulose I surfaces are summarized in Table II.

Similarly, Lefebvre and Gray (*37*) prepared spin-coated cellulose nanocrystal films on silicon substrates. Nanocrystal suspensions with concentrations ranging from 3–11% w/w were used to create films of varying thickness. Although silicon wafers are atomically flat, thorough cleaning can be challenging. Cleaning procedures for silicon and glass are numerous (*41*) and include piranha (H_2SO_4 + 30% H_2O_2 in 3:1 ratio), Chromerge® (chromium(III) oxide/sulfuric acid) (*42*), UV-ozonator (*43*), HF (*41*), boiling chloroform, and ethanol (*37*). When exposed to air, silicon surfaces oxidize resulting in a silica layer which is negatively charged under most solution conditions. The average "native" SiO_2 layer is ca. 20 Å as measured by ellipsometry (*35*, *44*). Films spin-coated on silicon were smooth (RMS roughness 5.3 nm dry and 5.4 nm wet) and iridescent (*37*). The colors vary with film thickness and are a result of thin-film interference and not chiral nematic ordering (*35*). In these cases, the electrostatic repulsion between cellulose nanocrystals and silica or mica surfaces does not appear to be an inhibiting factor in achieving smooth uniform films but this electrostatic effect will be discussed further in a later section.

Ordered Model Cellulose I Surfaces

It is also possible to prepare oriented films of cellulose nanocrystals due to their rod-like shape and liquid crystalline properties. Films with high degrees of crystallinity and orientation are useful for studying crystal structure (by X-ray, electron, and neutron diffraction) as well as various physical properties of cellulose. The concentration at which chiral nematic organization occurs in cellulose nanocrystal suspensions is dependent on nanocrystal length, size polydispersity, surface charge, ionic strength, and the presence of other macromolecules in the suspension (*19, 20, 27, 35, 45–47*). When preparing solvent-cast films, ordered polydomains grow as the water evaporates resulting in a solid cellulose sample with oriented regions. This is visible in polarized-light microscopy as light and dark bands or "fingerprint textures" due to the helicoidal arrangement of the nanocrystals (Figure 2). However, alignment is not uniform over the entire film. Additionally, internal structure, such as chiral nematic ordering, does not always manifest itself as surface orientation of the nanocrystals which may be desired for surface force measurements. Complex microstructures, such as parabolic focal conic defect textures, have been observed in solvent-cast films of cellulose nanocrystals (*48*). Polarized-light microscopy shows this bulk ordering (Figure 3a), but no apparent orientation of the nanocrystals is seen when the top of the films are imaged by AFM (Figure 3b, c). In fact, in this case the internal structure leads to a non-planar surface topology, which is undesirable for model surfaces. The concentration and preparation method used to make solid films must therefore be carefully chosen depending on the desired film properties.

Edgar and Gray (*5*) observed long-range orientation induced by spin-coating and shearing of nanocrystalline suspensions. Spin-coating on mica (*5*) and silicon (*37*) gave radially oriented films as determined by polarized-light microscopy and surface roughness was minimal (i.e. 2.6 nm). Two methods of shearing were compared: a concentrated drop of suspension was sheared across a long strip of mica and a dilute suspension was allowed to concentrate in a polystyrene dish and then the dish was tilted during the final drying stage. Both methods gave linearly aligned cellulose nanocrystal films with RMS roughness values of 2.4 nm.

Figure 2. Polarized-light microscopy image of a drying chiral nematic cellulose nanocrystal suspension on glass showing the characteristic fingerprint texture. Ordered polydomains grow as the film dries, resulting in a solid cellulose sample with oriented regions.

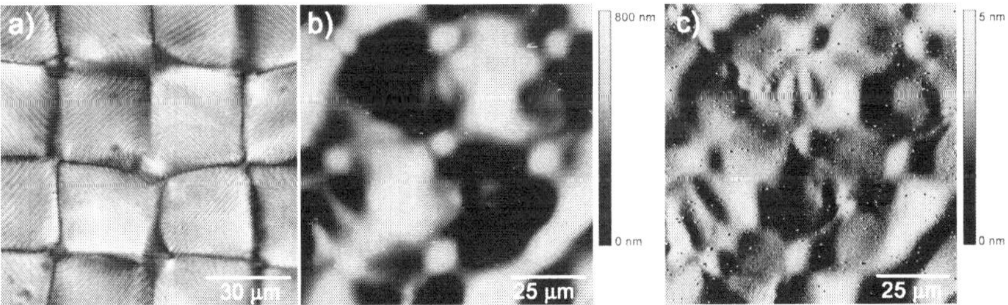

Figure 3. Solvent-cast parabolic focal conic textured cellulose films imaged by (a) polarized-light microscopy, (b) contact-mode AFM (height image) and (c) contact-mode AFM (deflection image). (Adapted with permission from ref 46. Copyright 2005 American Chemical Society.)

Table II. Summary of some model cellulose I surface preparation methods and reported properties.

Nanocrystalline cellulose source	Substrate	Preparation method	RMS roughness	Film thickness	Oriented	Ref.
Pulp fibres	freestanding	solvent-cast on PS[a]	1.0 nm	~ μm	No	(5)
Whatman 1 filter paper (cotton)	freestanding	solvent-cast on PS[a]	1.2 nm	~ μm	No	(5)
Pulp fibres	mica	solvent-cast	1.5 nm (top)	~ μm	No	(5)
Whatman 1 filter paper (cotton)	mica	solvent-cast	2.3 nm (top) 2.5 nm (underside)	~ μm	No	(5)
Whatman 1 filter paper (cotton)	mica	sheared concentrated suspension	2.3 nm	~ μm	Yes (linearly)	(5)
Whatman 1 filter paper (cotton)	polystyrene	concentrated sus-pension in PS[a] dish and then tilted to dry	Not reported	~ μm	Yes (linearly)	(5)
Whatman 1 filter paper (cotton)	mica	spin-coated	2.6 nm	~ nm	Yes (radially)	(5, 40)
Whatman 1 filter paper (cotton)	silicon/sili ca	spin-coated	5.3 nm (dry) 5.4 nm (wet)	ca. 100 nm	Yes (radially)	(37)

Table II. (continued)

Cladophora	freestanding	solvent-cast sheared gel in rotating glass vial	Not reported	~ µm	Yes (linearly)	(49)
Tunicin	freestanding	solvent-cast on glass in magnetic field	Not reported	0.05–0.15 µm (for TEM) 0.5–1 µm (for X-ray)	Yes (linearly)	(50)
Pulp fibres	freestanding	solvent-cast on Teflon and PS[a] in magnetic field	Not reported	~ µm	Yes (chiral nematic)	(51, 52)
Whatman ashless cotton cellulose powder (cotton)	silicon/silica	adsorbed on PAH in magnetic field	3 nm	~ nm	Yes (linearly)	(36)
Tunicin	silicon/silica	LB with DODA[b]	3.1 nm	7 nm (monolayer)	Yes (linearly)	(44)
Ramie	silicon/silica	LB with DODA[b]	1.8 nm	9.3 nm (monolayer)	Yes (linearly)	(44)
Whatman 541 ashless filter paper (cotton)	silica, titania, amorphous cellulose	spin-coated	Not reported	Sub-monolayer to 50 nm	No	(43)

[a] PS = polystyrene

[b] DODA= dioctadecyldimethylammonium bromide

To create freestanding films with a surprisingly high uniaxial orientation, Nishiyama et al. (*49*) first created a sheared gel. A 1% w/w cellulose nanocrystal suspension from the green alga *Cladophora* sp. (with residual sulfuric acid) was rotated at 500 rpm in a 50 mL glass vial for 12 hours. The silky, ordered gel was then rinsed with ethanol and dried under warm air and removed from the vial wall with forceps. The large axial ratio of these nanocrystals (40 nm × 4 μm) and high ionic strength made the formation of an anisotropic gel possible at low concentrations where spontaneous liquid crystal phase separation was not observable. In this case, the suspension was never chiral nematic and the nanocrystals were linearly aligned in the bulk and at the surface, as shown by scanning electron microscopy. Film orientation was quantified by X-ray and electron diffraction and found to be more ordered than a ramie fibre, one of the highest oriented samples of native cellulose known.

Alternatively, aligned cellulose nanocrystal films can be prepared in a magnetic field. Sugiyama et al. (*50*) solvent-cast 0.18% w/w suspensions of tunicate cellulose nanocrystals on glass slides in a 7 T magnetic field. The suspension was dry within 3 hours and the films were removed from the substrate and examined by polarized-light microscopy, TEM and X-ray/electron diffraction. When a film prepared in the magnetic field was compared to one dried outside of the magnet, it was obvious that the effect of the magnetic field was to align the nanocrystals perpendicular to the field (indicating a negative diamagnetic anisotropy) and that the glucose monomers were oriented perpendicular to the film surface. It was concluded that the anisotropic magnetic susceptibility of the individual C–C, C–O, C–H and O–H bonds led to the orientation of the nanocrystals and that the surface sulfate ester groups did not interfere with this effect. As in the work of Nishiyama et al. (*49*) the films were linearly aligned throughout because of the long nanocrystals and low concentrations used.

The effect of an induced magnetic dipole is amplified in liquid crystals. In the case of cellulose nanocrystal suspensions, the chiral nematic axis orients parallel to the field, meaning that the individual cellulose nanocrystals are aligned perpendicular to the field. Phases composed of species with positive diamagnetic anisotropy, untwist in a magnetic field, but here the ordered cellulose nanocrystal phase retains its chiral nematic structure, and simply reorients to give a more uniform planar texture. Revol et al. (*51*) and Edgar and Gray (*52*) made solvent-cast freestanding films by drying a biphasic suspension of nanocrystals in a 2.1 T and 7 T magnetic field. The films were iridescent, uniformly colored, and reflected one hand of circularly polarized light. The reflected wavelength was found to be adjustable by changing the ionic strength and the chiral nematic ordering was over a large surface area (>5 cm) and locked into the film upon drying.

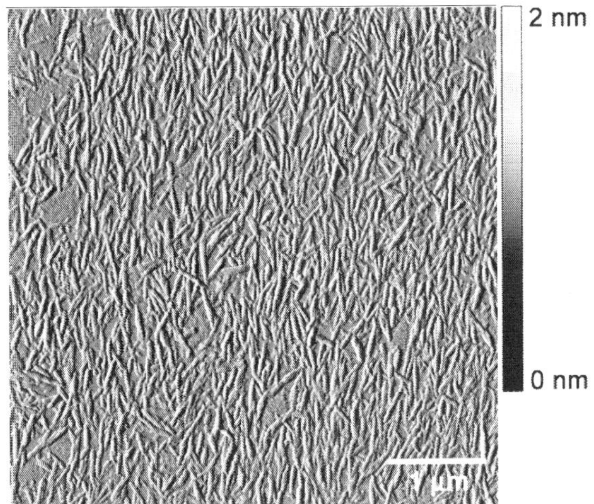

Figure 4. AFM contact-mode image (deflection) of ordered thin films of cellulose nanocrystals prepared by electrostatic adsorption in a 7 T magnetic field.

This negative diamagnetic anisotropy was also used to create an ordered monolayer of cellulose nanocrystals (*36*). A concentrated suspension of cellulose nanocrystals from cotton (9% w/w and completely anisotropic) was placed into the 7 T magnetic field of an NMR spectrometer. A silicon wafer was rendered cationic by the adsorption of polyallylamine hydrochloride and rinsed thoroughly. This coated substrate was then placed in the cellulose suspension in the magnetic field. At long adsorption times (24 hours) a linearly ordered monolayer was seen by AFM (Figure 4) whereas after 30 minutes only small nanocrystals had adsorbed and no orientation was visible. The films were 5–10 nm thick, relatively smooth, and retained their order after thorough rinsing. The long time for aligned adsorption is attributed to the twist elastic energy necessary for nanocrystals that are not aligned parallel to the substrate, and that need to twist out of the chiral nematic structure for efficient adsorption. Albeit slow, it is surprising nonetheless that liquid crystalline order could be transferred to a solid substrate though the electrostatic adsorption process by dip-coating in a magnetic field.

There is a noteworthy cellulose surface preparation method that does not use cellulose nanocrystals and gives ordered "films" of cellulose I (*25*). This treatment was used by Sugiyama et al. (*53*) for detailed examination of the cellulose crystal structure in algal *Valonia* cell walls. The *Valonia* was pre-treated, first by rinsing, drying, and scraping to remove non-cellulosic components, followed by boiling in both water and 0.1 N NaOH. The resulting cell wall specimen was neutralized and washed. An individual layer of microfibrils was delaminated using sharp needles and placed on a carbon grid for electron microscopy. Multiple layers of microfibrils were used for X-ray diffraction studies. The samples were found to be perfectly crystalline, composed of both the I_α and I_β crystal structures. The molecular roughness of these surfaces was not indicated.

Monolayers and Sub-monolayers of Cellulose Nanocrystals

Monolayers of cellulose nanocrystals on atomically flat surfaces have been prepared by a number of authors (*36, 43, 44*). Habibi et al. (*44*) made Langmuir-Blodgett films of cellulose nanocrystals from ramie and tunicin on silicon wafers. The LB methodology was possible because they were able to make a stable layer of cellulose at the air–water interface by floating the nanocrystals on a cationic amphiphilic layer of dioctadecyldimethylammonium bromide (DODA). When the surface pressure exceeded 40 mN/m and the weight ratio of cellulose nanocrystal: DODA was above a critical value of 125 for ramie and 250 for tunicin, the surface layer was homogeneous, stable and amenable to LB deposition. Various trials indicated that transfer was most efficient on the upward stroke at a maximum speed of 2 mm/min (*44*). The films were rinsed with chloroform and dilute NaOH to remove residual DODA and found to be smooth and stable in aqueous and organic solutions. XPS confirmed that the monolayers were relatively pure cellulose (*44*). The LB transfer resulted in preferential alignment of nanocrystals in the dipping direction and gave the smoothest films to date, 31 Å and 18 Å for tunicin and ramie, respectively.

A recent study by Kontturi et al. (*43*) has looked at sub-monolayers (or open-films) prepared by spin-coating cellulose nanocrystals on various substrates. Dilute suspensions of nanocrystals from cotton were deposited on silica (anionic), titania (cationic), and amorphous cellulose (neutral, polymeric). The nanocrystals were found to aggregate on the anionic surface and disperse uniformly on the cationic one. The distribution of cellulose nanocrystals on the amorphous cellulose substrate was uniform but the deposited amount was low and the resultant films were rough. These cellulose-on-cellulose films present an interesting addition to previously prepared model cellulose surfaces: distinct amorphous and crystalline (cellulose I) regions are present which means one can measure surface forces of both segments and the interactions between them using only one material. Open-films of cellulose nanocrystals (prepared on appropriate surfaces and from different cellulose sources) are useful for fundamental studies as well. For example, being able to image individual nanocrystals by AFM can provide structural detail about the microfibrils, which may give insight into biosynthetic differences among the source organisms.

In the same study, closed-film thickness was also found to be dependent on the substrate, where the combination of spin-coating deposition and electrostatic adsorption led to thicker films in the case of nanocrystals on titania (*43*). Surface coverage as a function of cellulose nanocrystal concentration is either linear (which is predicted by basic spin-coating theory (*54*)) for silica and amorphous cellulose substrates or resembles an adsorption isotherm in the case of titania. Very uniform monolayers/ultrathin films can therefore be made by spin-coating cellulose nanocrystals (at 4000 rpm at concentrations above 0.0125% w/w) on titania substrates as determined by AFM and XPS (*43*). Ellipsometry indicates that films of cellulose nanocrystals on titania are consistently 10 nm thicker than films prepared at identical conditions on silica and amorphous cellulose. This is due to the initial electrostatically adsorbed monolayer of cellulose nanocrystals which are ca. 10 nm in diameter; further film growth is mainly deposition due to solvent removal (*43*). Perfect monolayer films of cellulose nanocrystals would be ideal for some model cellulose studies, however, with such thin films, substrate effects are non-negligible.

Fundamental Studies Using Model Cellulose I Surfaces

While surface force studies are an obvious application of smooth model surfaces, adsorption, swelling, adhesion, friction, and structure-determining experiments have also been reported. The first direct imaging of the crystallographic surfaces of cellulose was performed by AFM on cellulose nanocrystal surfaces by Baker et al. (*3, 55*). Their goal was to image the crystal structure of cellulose I and determine the location of I_α and I_β in microfibrils. Cellulose nanocrystals from *Valonia* were solvent-cast on aminopropyltrioxysilanated mica for imaging under water and propanol. The large aspect-ratio nanocrystals provided an ideal surface to measure the structure and crystallographic spacings. They could image the 0.52 nm glucose sub-unit pitch and, by optimizing the AFM scan angle, they could see even more structural detail corresponding to the 1.04 nm cellobiose repeat unit which indicates two-fold screw symmetry. This was observed in high-resolution AFM because of the bulky O6 groups, which were detected topographically and would not have been visible on a rougher surface. Overall, the triclinic (or I_α crystal structure) was identified as the primary crystal form on the outer surface of the nanocrystals.

Surface force measurements on cellulose films have been made using AFM and colloid-probe AFM (*7, 9, 10, 37, 40, 56–67*), the surface force apparatus (SFA)(*11, 12*) and direct adhesion measurements (*34, 68*). The first attempt at measuring interactions between cellulose and hemicelluloses used spin-coated regenerated cellulose surfaces and SFA (*11*). Film instability, roughness, and extended polymer chains were a few of the problems encountered. Further investigations using SFA on regenerated cellulose films prepared by the LB method were reported (*12*).

Recently, many friction, force, and adhesion studies have been carried out on the model cellulose I surfaces described above. The model surfaces of spin-coated cotton cellulose nanocrystals on Si/SiO$_2$ were investigated by AFM with a standard silicon nitride tip (*37*). Polyelectrolytes were deposited on the cellulose nanocrystal film by spin-coating (monolayers, bilayers, or multilayers of polydiallyldimethyl ammonium chloride and carboxymethylcellulose). Cantilever deflection vs. distance curves were collected on the films and then on clean Si wafers to determine whether material was transferred from the surface to the AFM tip. No assumptions were made as to the contact point or cantilever spring constant. In water, deflection–distance curves indicated that the cationic polymer deposited on a cellulose nanocrystal film was only weakly attached and tended to desorb except when anchored to the nanocrystal films by the strongly anionic polyelectrolyte. Negatively charged carboxymethylcellulose was stable on the nanocrystal films in water but screening of the charge with salt appeared to facilitate its transfer to the tip. The multilayers were found to swell in dilute salt solutions, although no clear double-layer or ionic-strength effects were observed (*37*). In general, electrostatic and steric effects were not directly detectable in this work.

Stiernstedt et al. (*56*) used colloid-probe AFM to compare model cellulose surface pre-contact forces, friction, and the effect of xyloglucan adsorption. The colloid-probe used was a relatively non-swelling and non-deformable amorphous cellulose sphere made by the viscose process. Interactions were measured between the regenerated sphere and 1) another cellulose sphere, 2) regenerated cellulose films from N-methylmorpholine oxide, pre- and post-annealing, and 3) a cellulose nanocrystal film cast from aqueous suspension. The nanocrystal film was found to be the smoothest with the smallest friction coefficient. (The friction coefficient was actually found to increase monotonically with surface roughness.) As expected, xyloglucan selectively adsorbed onto all cellulose surfaces and greatly reduced the friction coefficient. DLVO-type forces (mostly short-range electrosteric repulsion and long-range double-layer interactions), surface potential and adhesion were measured and fit to theory for the various model surfaces. The observed onset of steric repulsion for the nanocrystalline films was comparatively large at 18 nm.

Instead of measuring pre-contact forces for *suspension-cast* model cellulose I surfaces, Notley et al. (*40*) examined *spin-coated* films of cellulose nanocrystals. Again, colloid-probe AFM was used to look at surfaces of varying crystallinity. Amorphous cellulose spheres were used to measure 1) spin-coated films of amorphous cellulose, 2) cellulose II, and 3) cellulose I nanocrystals. The study agreed with Stiernstedt et al. (*56*) in that the forces are dependent on the way the surfaces are prepared. The interactions for cellulose nanocrystal films were found to be monotonically repulsive over the pH and ionic strength range investigated. (The cellulose II and amorphous sample force profiles were dominated by van der Waals and steric forces, respectively.) Some swelling of the cellulose I film was also indicated by the decreasing surface potential that was measured with increasing ionic strength. Generally, the data was well described by DLVO theory.

Spin-coated cellulose nanocrystals on polyvinyl amine coated silicon was used by Eriksson et al. (*34*) for adhesion studies. The contact angle of this film was 19.5° for water and 34° for iodide. The microadhesion measurement apparatus was used with polydimethylsiloxane caps and loading and unloading curves were fit to Johnson-Kendall-Roberts theory. The work of adhesion for cellulose I, II, and amorphous cellulose was found to be similar and dominated by dispersive interactions, however, the adhesion *hysteresis* was strongly dependent on the degree of crystallinity, with cellulose nanocrystal surfaces showing the least hysteresis.

Conclusion

It is apparent that much progress has been made towards creating suitable model cellulose I surfaces. The studies outlined here focus on cellulose nanocrystals, not only because they are highly crystalline and in the native cellulose form but also because they possess desirable material properties. These include low density, high aspect ratio, high specific strength and modulus, as well as the the possibility of surface chemical modification, all of which may prove useful in finding applications for these cellulose I-based materials.

References

1. Klemm, D.; Philipp, B.; Heinze, T.; Heinze, U.; Wagenknecht, W. *Comprehensive Cellulose Chemistry, Volume 1: Fundamental and Analytical Methods.* 1st ed; Wiley-VCH: Weinheim, 1998.
2. Nishiyama, Y.; Sugiyama, J.; Chanzy, H.; Langan, P. *J. Am. Chem. Soc.* **2003,** *125,* 14300–14306.
3. Baker, A. A.; Helbert, W.; Sugiyama, J.; Miles, M. J. *Biophys. J.* **2000,** *79,* 1139–1145.
4. Fleming, K.; Gray, D. G.; Mathews, S. *Chem. – Eur. J.* **2001,** *7,* 1831–1835.
5. Edgar, C. D.; Gray, D. G. *Cellulose* **2003,** *10,* 299–306.
6. Kontturi, E.; Thüne, P. C.; Niemantsverdriet, J. W. H. *Langmuir* **2003,** *19,* 5735–5741.
7. Notley, S. M.; Wågberg, L. *Biomacromolecules* **2005,** *6,* 1586–1591.
8. Gunnars, S.; Wågberg, L.; Cohen Stuart, M. A. *Cellulose* **2002,** *9,* 239–249.
9. Carambassis, A.; Rutland, M. W. *Langmuir* **1999,** *15,* 5584–5590.
10. Notley, S. M.; Pettersson, B.; Wågberg, L. *J. Am. Chem. Soc.* **2004,** *126,* 13930–13931.
11. Neuman, R. D.; Berg, J. M.; Claesson, P. M. *Nord. Pulp Pap. Res. J.* **1993,** *8,* 96.
12. Holmberg, M.; Berg, J.; Stemme, S.; Ödberg, L.; Rasmusson, J.; Claesson, P. M. *J. Colloid Interface Sci.* **1997,** *186,* 369–381.
13. Kontturi, E.; Tammelin, T.; Österberg, M. *Chem. Soc. Rev.* **2006,** *35,* 1287–1304.
14. Rånby, B. G. *Discuss. Faraday Soc.* **1951,** *11,* 158–164.
15. Battista, O. A.; Coppick, S.; Howsmon, J. A.; Morehead, F. F.; Sisson, W. A. *J. Ind. Eng. Chem.* **1956,** *48,* 333–335.
16. Beck-Candanedo, S.; Roman, M.; Gray, D. G. *Biomacromolecules* **2005,** *6,* 1048–1054.
17. Araki, J.; Wada, M.; Kuga, S.; Okano, T. *Colloids Surf., A* **1998,** *142,* 75–82.
18. Revol, J. F.; Bradford, H.; Giasson, J.; Marchessault, R. H.; Gray, D. G. *Int. J. Biol. Macromol.* **1992,** *14,* 170–172.
19. Dong, X. M.; Kimura, T.; Revol, J.-F.; Gray, D. G. *Langmuir* **1996,** *12,* 2076–2082.
20. Dong, X. M.; Revol, J. F.; Gray, D. G. *Cellulose* **1998,** *5,* 19–32.

21. Favier, V.; Chanzy, H.; Cavaille, J. Y. *Macromolecules* **1995**, *28*, 6365–6367.
22. Podsiadlo, P.; Sui, L.; Elkasabi, Y.; Burgardt, P.; Lee, J.; Miryala, A.; Kusumaatmaja, W.; Carman, M. R.; Shtein, M.; Kieffer, J.; Lahann, J.; Kotov, N. A. *Langmuir* **2007**, *23*, 7901–7906.
23. Roman, M.; Winter, W. T. *Biomacromolecules* **2004**, *5*, 1671–1677.
24. Tokoh, C.; Takabe, K.; Fujita, M.; Saiki, H. *Cellulose* **1998**, *5*, 249–261.
25. Revol, J. F. *Carbohydr. Polym.* **1982**, *2*, 123–134.
26. Hanley, S. J.; Giasson, J.; Revol, J. F.; Gray, D. G. *Polymer* **1992**, *33*, 4639–4642.
27. Dong, X. M.; Gray, D. G. *Langmuir* **1997**, *13*, 2404–2409.
28. Wang, N.; Ding, E.; Cheng, R. *Polymer* **2007**, *48*, 3486–3493.
29. Li, X.-F.; Ding, E.; Li, G.-K. *Chin. J. Polym. Sci.* **2001**, *19*, 291–296.
30. Zhang, J.; Elder, T. J.; Pu, Y.; Ragauskas, A. J. *Carbohydr. Polym.* **2007**, *69*, 607–611.
31. Pääkkö, M.; Ankerfors, M.; Kosonen, H.; Nykänen, A.; Ahola, S.; Österberg, M.; Ruokolainen, J.; Laine, J.; Larsson, P. T.; Ikkala, O.; Lindström, T. *Biomacromolecules* **2007**, *8*, 1934–1941.
32. Turbak, A. F.; Snyder, F. W.; Sandberg, K. R. *J. Appl. Polym. Sci.* **1983**, *37*, 815.
33. Herrick, F. W.; Casebier, R. L.; Hamilton, J. K.; Sandberg, K. R. J. *J. Appl. Polym. Sci.* **1983**, *37*, 797–813.
34. Eriksson, M.; Notley, S. M.; Wågberg, L. *Biomacromolecules* **2007**, *8*, 912–920.
35. Cranston, E. D.; Gray, D. G. *Biomacromolecules* **2006**, *7*, 2522–2530.
36. Cranston, E. D.; Gray, D. G. *Sci. Technol. Adv. Mater.* **2006**, *7*, 319–321.
37. Lefebvre, J.; Gray, D. G. *Cellulose* **2005**, *12*, 127–134.
38. Podsiadlo, P.; Choi, S.-Y.; Shim, B.; Lee, J.; Cuddihy, M.; Kotov, N. A. *Biomacromolecules* **2005**, *6*, 2914–2918.
39. Dorris, G. M.; Gray, D. G. *Cellul. Chem. Technol.* **1978**, *12*, 9–23.
40. Notley, S. M.; Eriksson, M.; Wågberg, L.; Beck-Candanedo, S.; Gray, D. G. *Langmuir* **2006**, *22*, 3154–3160.
41. Hattori, T. *Ultraclean Surface Processing of Silicon Wafers*; Springer-Verlag: Berlin, 1998.
42. Kang, J.; Rowntree, P. A. *Langmuir* **2007**, *23*, 509–516.
43. Kontturi, E.; Johansson, L.-S.; Kontturi, K. S.; Pa"ivi, A.; Thu"ne, P. C.; Laine, J. *Langmuir* **2007**, *23*, 9674–9680.
44. Habibi, Y.; Foulon, L.; Aguie-Beghin, V.; Molinari, M.; Douillard, R. *J. Colloid Interface Sci.* **2007**, *316*, 388–397.
45. Edgar, C. D.; Gray, D. G. *Macromolecules* **2002**, *35*, 7400–7406.
46. Beck-Candanedo, S.; Viet, D.; Gray, D. G. *Langmuir* **2006**, *22*, 8690–8695.
47. Beck-Candanedo, S.; Viet, D.; Gray, D. G. *Cellulose* **2006**, *13*, 629–635.
48. Roman, M.; Gray, D. G. *Langmuir* **2005**, *21*, 5555–5561.
49. Nishiyama, Y.; Kuga, S.; Wada, M.; Okano, T. *Macromolecules* **1997**, *30*, 6395–6397.
50. Sugiyama, J.; Chanzy, H.; Maret, G. *Macromolecules* **1992**, *25*, 4232–4234.
51. Revol, J. F.; Godbout, L.; Gray, D. G. *J. Pulp Pap. Sci.* **1998**, *24*, 146–149.
52. Edgar, C. D.; Gray, D. G. *Cellulose* **2001**, *8*, 5–12.

53. Sugiyama, J.; Chanzy, H.; Revol, J. F. *Planta* **1994,** *193,* 260–265.

54. Schubert, D. W. *Polym. Bull.* **1997,** *38,* 177–184.

55. Baker, A. A.; Helbert, W.; Sugiyama, J.; Miles, M. J. *Appl. Phys. A: Mater. Sci. Process.* **1998,** *66,* S559–S563.

56. Stiernstedt, J.; Nordgren, N.; Wågberg, L.; Brummer, H. I.; Gray, D. G.; Rutland, M. W. *J. Colloid Interface Sci.* **2006,** *303,* 117–123.

57. Leporatti, S.; Sczech, R.; Riegler, H.; Bruzzano, S.; Storsberg, J.; Loth, F.; Jaeger, W.; Laschewsky, A.; Eichhorn, S. J.; Donath, E. *J. Colloid Interface Sci.* **2005,** *281,* 101–111.

58. Radtchenko, I. L.; Papastavrou, G.; Borkovec, M. *Biomacromolecules* **2005,** *6,* 3057–3066.

59. Notley, S. M.; Norgren, M. *Langmuir* **2006,** *22,* 11199–11204.

60. Stiernstedt, J.; Brummer, H. I.; Zhou, Q.; Teeri, T. T.; Rutland, M. W. *Biomacromolecules* **2006,** *7,* 2147–2153.

61. Rutland, M. W.; Carambassis, A.; Willing, G. A.; Neuman, R. D. *Colloids Surf., A* **1997,** *123–124,* 369–374.

62. Theander, K.; Pugh, R. J.; Rutland, M. W. *J. Colloid Interface Sci.* **2005,** *291,* 361–368.

63. Zauscher, S.; Klingenberg, D. J. *J. Colloid Interface Sci.* **2000,** *229,* 497–510.

64. Zauscher, S.; Klingenberg, D. J. *Colloids Surf., A* **2001,** *178,* 213–229.

65. Bogdanovic, G.; Tiberg, F.; Rutland, M. W. *Langmuir* **2001,** *17,* 5911–5916.

66. Schaub, M.; Wenz, G.; Wegner, G.; Stein, A.; Klemm, D. *Adv.Mater.***1993,** *5,* 919.

67. Ahola, S.; Salmi, J.; Johansson, L.-S.; Laine, J.; Österberg, M. *Biomacromolecules* **2008,** *9,* 1273–1282.

68. Rundlof, M.; Karlsson, M.; Wagberg, L.; Poptoshev, E.; Rutland, M. W.; Claesson, P. M. *J. Colloid Interface Sci.* **2000,** *230,* 441–447.

Polyelectrolyte Multilayer Films Containing Cellulose: A Review

Emily D. Cranston and Derek G. Gray

Department of Chemistry, Pulp and Paper Research Centre, McGill University, Montréal, Québec, H3A 2A7, Canada

In the past decade, electrostatic layer-by-layer (LBL) self-assembly has gained attention because it is a facile and robust method to prepare thin polymer films. The low-cost technique is ideally suited to create chemically defined, reproducible, and smooth films with tailor-made properties. Due to the industrial importance and natural abundance of cellulose, its incorporation into LBL films has been widespread. Here we review research into multilayered composite materials containing cellulose and cellulose derivatives with favourable properties including high strength, flexibility, and biocompatibility. Preparation and characterization of polyelectrolyte multilayer films containing (1) cellulose derivatives, (2) cellulose nanocrystals, and (3) using cellulose fibers as substrates are presented. The applications and advantages of these films and their potential as model cellulose surfaces are discussed.

Introduction

The biosynthesis of cellulose proceeds by building up glucose units into long crystalline microfibrils which form an organized lattice in plant cell walls. Algae, bacteria, and some marine animals also perform this biosynthesis, making cellulose the most ubiquitous natural resource in the biosphere. The cellulose polymer, β-1,4-linked D-glucose, is often used as a construction material, either as wood, natural textile fibers (cotton and flax), or in the form of paper and board. Furthermore, cellulose is a versatile starting material for chemical modification in the extensive field of carbohydrate chemistry. Artificial and natural cellulose-based materials and soluble cellulose derivatives are used in many areas of industry and everyday life.

In a method analogous to the biosynthesis of cellulose, which happens in nature in close association with hemicellulose, one can create multilayered films containing cellulose and other chemical components. The layer-by-layer (LBL) electrostatic self assembly technique is ideally suited to create chemically defined, reproducible, and smooth thin films with controlled internal architecture. These model cellulose surfaces can then be used to study cellulose force interactions and adsorption phenomena. This methodology presents a unique way to understand how cellulose and cellulose derivatives interact with other polymers and colloids in a stratified material. As well, industrial realization of cellulose products is dependent on cellulose interfacial properties and surface morphology, which can be systematically varied through polymer adsorption processes.

The demand for films and coatings with tunable properties has lead to much research in the field of LBL assembly. The technique, first introduced by Decher (*1*, *2*), is simple and adaptable yet has been exploited for the fabrication of sophisticated nanocomposites. In this method, a charged or hydrophilic substrate is exposed to a solution of charged polymer (polyelectrolyte) followed by rinsing. Polyelectrolyte adsorption is driven by electrostatics and entropy, is irreversible, and is self-limiting as a result of rinsing. During multilayer film build-up, the surface charge is reversed because the polymeric material adhering to the surface has more than the required number of charges needed to compensate the previous layer. This allows for easy adsorption of the next oppositely charged polyelectrolyte, leading to step-wise film growth. The deposition and rinsing steps can be repeated to give an arbitrary number of alternating polycation–polyanion layers. Although generally performed with linear polyelectrolytes, the procedure is suitable to a large range of poly-charged nano-objects with no inherent restrictions on substrate geometry (*2*).

While the processing window is quite large in LBL assembly, multilayer films are uniform and reproducible under constant deposition conditions (*3*). The film properties can be tailored by changing the solution pH, ionic strength, polyelectrolyte molar mass, concentration, temperature, stirring/shear forces, and deposition time. Weak polyelectrolyte systems are more adaptable than strong systems because the polymer charge density can be controlled through pH and salt concentration. This in turn adjusts the layer structure and film swelling properties because the polyelectrolyte chain conformation is modified. Rinsing and drying steps (as well as ambient humidity) can also affect the film build-up. Films with variable thickness, density, permeability, morphology, and roughness are thus easily made through the appropriate choice of assembly parameters (*2*, *4*). Additionally, films prepared by electrostatic multilayering are generally free of defects and show long-life stability and self-healing characteristics (*5*, *6*). The robustness of these films is attributed to ionic-crosslinking and extensive interpenetration of layers.

Conventionally, LBL assembly is done by solution dipping solid substrates in beakers containing dilute aqueous polymer solutions, making it a low cost and environmentally friendly procedure. Alternatively, spin-coating, spray-coating, and electrically-driven assemblies can be used to make smooth multilayer films which require less volume of polyelectrolyte solution and minimize the assembly time. Multilayer films with cellulose derivatives have also been prepared by Langmuir–Blodgett deposition (*7–13*) and solvent casting (*14, 15*) but these techniques are not strictly electrostatic assembly and will not be described in this review.

The LBL methodology is a type of template-assisted or directed assembly; films are often prepared on flat substrates such as silicon, glass, quartz, and most metals, which facilitate characterization. However, colloids have also been coated with polyelectrolyte multilayers and in some cases the core can be dissolved leaving hollow microcapsules. In other studies, metal nanorods, inorganic fibers, polymer microspheres, enzymes, and biological cells have been susceptible to LBL assembly. Coating of colloidal particles is done by adding dilute polyelectrolyte solution to a colloidal suspension, generally under sonication. The coated particles are then rinsed multiple times by centrifugation and the process is repeated with the oppositely charged polymer. In general, any charged surface which can be immersed in a polyelectrolyte solution or have solution flowed through it to coat the interior, can be subjected to polyelectrolyte multilayering.

Industrial applications of LBL films include light-emitting diodes (*16*), electrochromic devices (*16*), non-linear optical devices (*16*), antireflective coatings (*17, 18*), separation technologies (*1*), dielectric mirrors (*19*), and optical sensors (*1, 5*). The technique also presents a novel way to design nanocomposites where interfacial properties between the binding matrix and strengthening component are crucial. In biological uses, multilayer films have been investigated for drug-delivery systems (*19*), bio-fouling agents, and bioinert/biocompatible coatings (i.e. in contact lenses) (*2*).

Understanding polyelectrolyte film assembly and properties may well lead to achieving desired macroscopic properties from carefully designed nano-architectures. Secondary function is easily incorporated into films with great positional precision. Components in LBL films have included synthetic and bio-polymers, organic molecules, nanotubes, and various colloids of biological, metallic, and inorganic nature. Here we present an overview of polyelectrolyte multilayer films containing cellulose and cellulose derivatives. These films are currently being used as controlled drug-release systems, biofunctional coatings, high performance nanocomposites, renewable and non-toxic alternatives for pigments and antireflective coatings, and as model cellulose surfaces for research on cellulose surface forces and interactions.

This review summarizes recent research in the field of polyelectrolyte multilayer films containing cellulose divided into three categories:

1. Polyelectrolyte multilayer films containing cellulose derivatives (Figure 1a)

2. Polyelectrolyte multilayer films containing cellulose nanocrystals (Figure 1b)

3. Polyelectrolyte multilayers prepared on cellulose fiber substrates (Figure 1c)

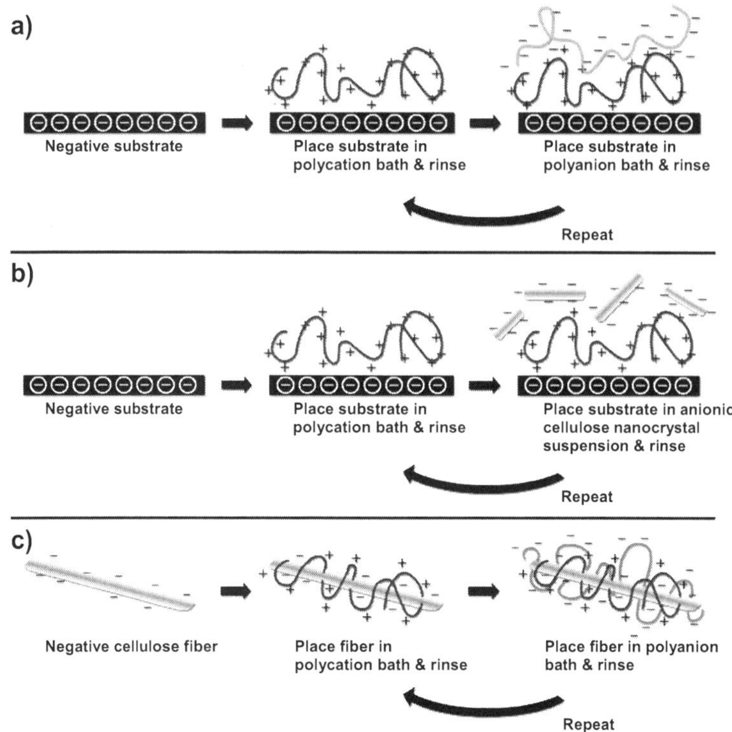

Figure 1. Schematic representation of the build-up of LBL films divided into 3 categories: a) polyelectrolyte multilayer films containing cellulose derivatives, b) polyelectrolyte multilayer films containing cellulose nanocrystals, and c) polyelectrolyte multilayers prepared on cellulose fiber substrates. Polycations are shown in black, polyanions in gray, cellulose nanocrystals are depicted as short straight rods. Counterions have been omitted for clarity.

Polyelectrolyte Multilayer Films Containing Cellulose Derivatives

Relatively little work has been done in the field of LBL with linear polymers of cellulose. Solvents that dissolve pure cellulose include N-methylmorpholine N-oxide, dimethylacetamide/LiCl, trifluoroacetic acid/chlorinated alkanes, ferric sodium tartrate, NH_4SCN/NH_3, Cadoxen, Cuam, Cuen, etc., which are not compatible with the aqueous multilayer assembly technique. As a result, there have been no multilayer films prepared with pure polymeric cellulose. Functionalization of cellulose at the hydroxyl groups with hydrophilic substituents results in derivatives with varying degrees of solubility in water. These include, but are not limited to, carboxymethylcellulose (CMC), cellulose acetate, methylcellulose, and hydroxypropylcellulose. Charged cellulose derivatives are amenable to the electrostatic multilayering technique; the sodium salt form of CMC is the most commonly used polyanion in this regard (*20–23*). LBL films containing cellulose derivatives have been prepared on silicon (*21*), gold (*20*), polyethyleneimine (PEI) coated SiO_2 (*24*), cellulose surfaces (*21, 25*), β-ferric hydrous oxide particles (*22*), and ibuprofen microparticles (*23*). Polycations used include poly(diallyldimethylammonium chloride) (PDDA) (*20, 21, 26*), chitosan (*22*), PEI (*20*), poly(allylamine hydrochloride) (PAH) (*27, 28*), and poly(methylene-co-guanidine) (*24*).

Well-controlled and characterized films of biocompatible polymers are of practical importance for preparing drug delivery or drug encapsulation systems. Polysaccharide multilayers of CMC and chitosan were found to have linear film build-up that was strongly dependent on the solution pH (*22*). Radeva et al. described the film growth as a series of adsorption–desorption steps attributed to partial desorption of chitosan due to a stripping effect by the long chain CMC. Polyelectrolyte complexes of CMC and chitosan were detectable in solution, however more chitosan was adsorbed in subsequent steps than was removed during desorption (*22*). At pH values above the pK_a of CMC, bilayer thickness dropped substantially because of intra-chain electrostatic repulsion which lead to a more extended polymer conformation (fewer loops) and facilitated the desorption because of stronger electrostatic attraction between components. Overall, the bilayer thickness was easily adjusted, ranging from 3 nm to 14 nm, by changing the solution-dipping conditions (*22*).

The same polyelectrolyte system of CMC and chitosan was used by Qiu et al. to coat ibuprofen microparticles with up to 5 bilayers (*23*). The microcapsules obtained were tough and homogeneous and release properties were studied by UV–vis spectroscopy and determined to be a function of microcapsule size, capsule thickness, and pH value of the dispersing bulk solutions. Results were compared to microcapsules of chitosan/sodium alginate and chitosan/dextran sulfate and indicated that this system was reliable, reproducible, and ideal for controlled drug-release (*23*).

The potential of cellulose as functional coatings in biosensors, bioreactors, and biofunctional electrodes was evaluated by Anzai and coworkers (*20, 27, 28*). Gold electrodes were LBL coated with CMC and the effect of the polycation was examined using PAH, PEI, and PDDA. Quartz-crystal microbalance (QCM) studies showed that multilayer film growth was exponential (*27, 28*). LBL films were permeable to $[Fe(CN)_6]^{3-}$ ions, and redox reactions proceeded smoothly as measured by cyclic voltammetry. High charge density polyelectrolytes bound the most $[Fe(CN)_6]^{3-}$ ions (*20*). Loading was also enhanced when the outermost layer was cationic and increased with film thickness but high permeability and smooth diffusion within the membranes were observed even with an anionic capping layer. Multilayers prepared *without* polysaccharide components were less permeable, and redox reactions were only detected when the outer layer was cationic (*27*). The redox potential of $[Fe(CN)_6]^{3-}$ shifted slightly negative for PDDA/CMC films and positive for PEI/CMC films (*20*). In contrast, the permeability of $[Ru(NH3)_6]^{3+}$ cation was very limited in the LBL films examined (*27*). It was concluded that the $[Fe(CN)_6]^{3-}$ ions were strongly immobilized due to electrostatic binding sites within the multilayer films, and that these films could be used as electrocatalysts to oxidize ascorbic acid in solution (*20*).

Polyelectrolyte multilayer films are typically smooth and ideal for fundamental adsorption, adhesion, and force measurements. Muller et al. made anionic polyelectrolyte complexes between cellulose sulfate and sodium alginate which were assembled into multilayer films with poly(methylene-co-guanidine) as the polycation (*24*). The films were then employed as a model system to adsorb bovine serum albumin (BSA) which has a negative charge at neutral pH (*24*). Multilayer growth and protein adsorption was monitored by in situ attenuated total reflection Fourier transform infrared spectroscopy (ATR-FTIR). The films were successfully assembled and the amount of BSA adsorbed was found to be enhanced when the cationic polymer, poly(methylene-co-guanidine), was the outermost layer (*24*).

The effect that LBL film growth has on adhesion was studied by Rundlöf and Wågberg (*26*) with respect to papermaking, e.g. the build-up of deposits on processing equipment and fiber–fiber adhesion. The polyanion used was a complex of wood extractives referred to as "dissolved and colloidal substances" (DCS) in the pulping industry. The components included glyceride esters, terpenes, terpenoids, polysacharrides, low molecular weight lignin, lignans, and pectins. Film growth was monitored by stagnation point adsorption reflectometry (SPAR) with 10 minute deposition times for the DCS and cationic PDDA layers. The multilayer films were irregular with incomplete surface coverage as seen by scanning electron microscopy (SEM). A microadhesion apparatus was used to measure elastic deformation under load, and curves were fitted to Johnson–Kendall–Roberts theory. When PDDA was adsorbed on silica, the adhesion decreased by blocking the specific interactions between the polydimethylsiloxane cap of the device and the silica surface. However, depositing multiple layers of DCS and PDDA on the silica increased adhesion due to the build-up of a thick soft layer which led to additional energy dissipation upon separation. Annealing of the film at 105 °C also greatly decreased the adhesion (*26*).

Adhesion between polyelectrolyte multilayers and silicon nitride atomic force microscopy (AFM) probes was also observed by Lefebvre and Gray (*21*). LBL films were prepared with 1–5 bilayers of CMC and PDDA on spin-coated cellulose nanocrystal substrates. CMC was found to give a compact bilayer structure due to its relative stiffness. The diffusive nature of the multilayers was found to decrease with increasing deposition steps; repulsive forces decreased as measured by AFM force–distances curves. Generally, salt facilitated the desorption of CMC and above 0.1 M NaCl, the films were found to dissolve as aggregates into solution (*21*).

In summary, charged cellulose derivatives are readily incorporated into polyelectrolyte multilayer films. Biocompatibility, film robustness, and relative stiffness are a few characteristics routinely seen in these films. Linear vs. exponential film build-up, film thickness, and porosity can be tailored by choosing a suitable polycation and appropriate deposition conditions. Moreover, the pH adjustable charge density (and thus chain configuration) as well as the tendency of CMC to desorb at high pH and high salt concentration, may make LBL films with cellulose derivatives useful in drug delivery applications.

Polyelectrolyte Multilayer Films Containing Cellulose Nanocrystals

To circumvent the problem of cellulose solubility, cellulose can be acid hydrolyzed to give an aqueous suspension of nanocrystals (*29, 30*). The rod-shaped particles are composed of highly crystalline cellulose in its native "cellulose I" form. The suspensions are polydisperse and nanocrystal dimensions are strongly dependent on the cellulose source and hydrolysis conditions (*31, 32*). The reinforcing properties are tailorable through modifying the aspect ratio and the materials have been used in various polymer matrices for applications ranging from transportation vehicle paneling to low thickness polymer electrolytes in lithium batteries (*33–39*). For more information on cellulose nanocrystal preparation and properties, please see Chapter 3, *Model Cellulose I Surfaces: A Review*. In the studies discussed below, cellulose nanocrystals come from cotton and tunicate sources having average dimensions of 10 nm × 129 nm and 4 nm × many microns, respectively. The nanocrystal surface is charged with sulfate ester groups, a product of the sulfuric acid hydrolysis, making them amenable to LBL assembly. Cellulose nanocrystals are strong polyelectrolytes; the surface charge varies from 0.15 e/nm^2 to 0.4 e/nm^2 (depending on the hydrolysis conditions) and they are fully charged at all pH values. (More precisely, the pK_a is 1.9 implying full dissociation at regular solution conditions (*40*).) The suspension counterions can be varied but are normally Na$^+$ or H$^+$.

LBL films have often included one or more types of spherical nanoparticles. Cellulose nanocrystals however, are one of few anisometric colloids which have been successfully incorporated into multilayer films: clay platelets, inorganic sheets, the tobacco mosaic virus, DNA, proteins, rod-shaped dye particles, nanotubes, Fe$_2$O$_3$, and semiconductor rods are others. Along with the low cost and natural origins, the mechanical properties of cellulose nanocrystals are impressive, including a bending strength of ~10 GPa and an elastic modulus of 150 GPa (*41*), which can be attributed to the high crystallinity of cellulose polymer chains in the nanocrystals.

Recently, anionic cellulose nanocrystals have been used in LBL assembly to create thin films ranging in thickness from 10 nm to ~1 μm (*18, 41, 42*). These materials have been prepared on silicon wafers, glass slides and as freestanding films (removed from glass with HF). The film growth is dependent on the assembly conditions: firstly, solution-dipping versus spin-coating makes films with different thicknesses but the assembly pH, polycation, and counterion in the cellulose nanocrystal suspension also affects the final film properties.

Cranston and Gray prepared multilayer films with PAH and Na^+-form cellulose nanocrystals at neutral pH (*42*). Solution-dipped and spin-coated films were found to have linear growth although the solution-dipped assembly was believed to be diffusion (transport) limited. In the past, LBL films made with polyelectrolytes and colloids have often shown slow or irregular film growth (*2*), but this was not the case for multilayer films containing cellulose nanocrystals. Complete surface coverage was observed after 1 bilayer for spin-coating deposition but required 2.5 bilayers for the films made by solution-dipping. AFM was used to examine surface morphology (Figure 2) and no detectable difference was seen, based on whether cellulose nanocrystals or PAH was the outermost layer. Interestingly, some orientation of the cellulose nanocrystals was visible in the spin-coated LBL films as more layers were deposited. The alignment is radial and arises from viscous shear as the suspension of cellulose nanocrystals flows outward during spin-coating. Linear orientation of cellulose nanocrystals in multilayer films was also achieved by LBL solution-dipping in a strong magnetic field (*43*).

Figure 2. AFM (height image) showing surface morphology of films prepared by LBL with cellulose nanocrystals and PAH (25 bilayers prepared by spin-coating).

Films with up to 25 bilayers, (PAH/cellulose)$_{25}$, were assembled and the film thickness (measured by ellipsometry and optical reflectometry) is shown in Figure 3a. (It should be noted that the abbreviation (PAH/cellulose)$_n$ describes the deposition process, rather than the actual sequence of species in the film. Integer bilayers have cellulose as the outermost layer and half-integer bilayers end with PAH. Distinct layered structure is rarely seen, although is sometimes more evident in spin-coated films (*2*).)

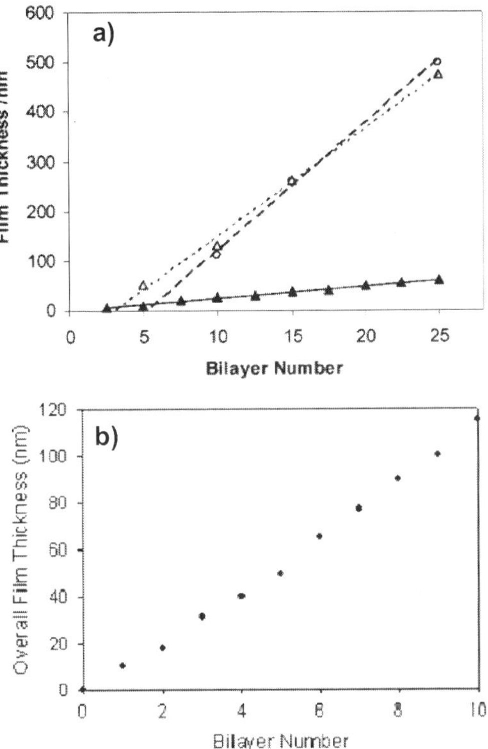

Figure 3. Variation in film thickness of multilayers containing nanocrystalline cellulose as a function of bilayer number. a) (PAH/cellulose)$_n$ films measured by ellipsometry (triangles) and reflectivity (circles) for solution-dipped films (closed symbols) and spin-coated films (open symbols). Lines are linear least-squares fits of the data. b) (PDDA/cellulose)$_n$ films measured by ellipsometry. (Adapted with permission from refs 41 (a) and 42 (b). Copyright 2005 and 2006, respectively, American Chemical Society.)

Strong interpenetration between layers leads to blurred boundaries, which was observed by SEM in both spin-coated and solution-dipped films (*42*). Spin-coated films were substantially thicker, up to 7 times thicker for (PAH/cellulose)$_{25}$ than for solution-dipped films. Spin-coated LBL films also displayed angle dependent colors (Figure 4) as a result of interference between light reflected from the air–film interface and the film–substrate interface. Because the film thickness was in the UV to visible range, optical reflectometry was used to measure thin film interference peaks (Fabry–Perot fringes). The reflected light intensity is a function of the film thickness and the mean refractive index. This was a convenient way to determine these parameters, which were unreliable when obtained by ellipsometry due to the anisotropic nature of the cellulose nanocrystals. Reflectivity was modeled using the transfer matrix method (44) and the mean refractive index was determined to be 1.50 ± 0.01 for films prepared by both methods.

Figure 4. Digital photographs of iridescent colors seen in spin-coated LBL films of cellulose nanocrystals and PAH on Si. (See color insert)

Multilayer films of PAH and cellulose nanocrystals were found to be relatively smooth as determined by AFM; root-mean-square (RMS) roughness was always below 10 nm. The films prepared by solution-dipping show a slight linear increase in roughness per deposition step whereas the spin-coated films are more uniform (ca. 5 nm) with a negligible change upon the addition of more layers. As far as film stability is concerned, films prepared by both solution-dipping and spin-coating were found to be unaffected after 5 days immersion in water at temperatures ranging from 10–100 °C. AFM images of the surfaces before and after exposure to water did not indicate any change in the surface morphology, roughness, or nanocrystal alignment. The studies showed enhanced resistance to swelling and redispersion, which is often a problem in pure nanocrystalline cellulose materials (45), and these results are consistent with previous reports on multilayer stability (46).

Very recently, Jean et al. (47) confirmed the work of Cranston and Gray (42) using neutron reflectivity and AFM on PAH/cellulose nanocrystal LBL films. Smooth films with very few defects and linear growth were observed leading to an average bilayer thickness of 15 nm. The adsorption was described as being in two layers where the surface coverage of the bottom layer was 50% and the top was only 25%. Clear Bragg peaks in neutron reflectivity curves indicated that the films were structured from the electrostatic adsorption of cellulose nanocrystals and the smoothening effect of the flexible PAH. AFM images showed that films prepared by solution-dipping in the anisotropic phase of a chiral nematic suspension were aligned and that such structured materials are close to mimicking the organization of cellulose microfibrils in living organisms (47).

In work by Podsiadlo et al., LBL films were prepared using cellulose nanocrystals in the acid form (H^+ counterions on the sulfate ester groups) and the strong polycation PDDA (41). Films were solution-dipped, with rinse baths adjusted to pH 2–3 to be consistent with the acidic cellulose nanocrystal suspension pH. Films were dried with compressed air between deposition steps. The average thickness increment was 11 nm per bilayer, intermediate between the solution-dipping value of 2 nm and spin-coating value of 16 nm, observed by Cranston and Gray (42). This indicates that the cellulose nanocrystal adsorption from solution is fast and may be due to the concentrated dipping solutions, high surface charge because of the acidic pH, and high ionic strength, which leads to less repulsion between charged species. Surface morphologies of the films, up to (PDDA/cellulose)$_{10}$, were consistent with other cellulose nanocrystal LBL films and film thickness by ellipsometry is shown for comparison in Figure 3b. Film thickness can also be monitored by UV–vis absorbance at 360 nm when the films are thin enough to not show interference patterns. Podsiadlo et al. commented that the high density and uniformity in LBL films with cellulose nanocrystals was superior to similar studies with carbon nanotubes (*41*).

Subsequently, antireflective coatings were prepared using cellulose nanocrystals from tunicate in LBL films (*18*). The prepared films were extremely porous with a mesh-like morphology as a result of the high aspect-ratio "nanowires" of cellulose. Random orientation and overlapping of nanocrystals led to porosity and optical properties which varied with film thickness. To make an ideal homogenous antireflective coating on glass in air, the film thickness should be $\lambda/4$ (λ being the maximum transmittance wavelength) with a refractive index of ~1.22. This was achieved for films with cellulose nanocrystals and PEI after 12 bilayers were adsorbed. Light transmittance was measured as 100%, with a thickness of 85 nm and refractive index of 1.28. The maximum transmittance peak red-shifted as more layers were added due to increased film thickness and light scattering. The antireflective properties appear after one bilayer is deposited and increase until 20 bilayers at which point the porosity and transparency starts to decrease because of thickly stacked nanocrystals. The average bilayer increment was 7 nm which corresponds to two tunicate nanocrystals thick. Similar results were seen with PAH, PDDA, and chitosan polycations. One minute adsorption times were used in the dipping procedure, resulting in fast coating preparation when compared to 10 and 25 min deposition times used in other studies (*41, 42*). Fast film growth is attributed to the factors mentioned above and the length of the tunicate nanocrystals.

The use of cellulose here is novel because in the past, LBL antireflective coatings have either used nanoporous polyelectrolyte assemblies (*17*) or spherical nanoparticles (*2*). This type of structure was only possible because of the long rigid nanocrystals, indicating that a minimum critical length is necessary (*18*). Freestanding films of PAH and tunicate cellulose nanocrystals were found to have tensile strengths as high as 110 MPa, Young's modulus of ~6 GPa, and an in-plane modulus between 20 and 30 GPa (*18*).

Disadvantages of the LBL technique have also recently been addressed by Shim et al. These include, long adsorption times, laborious rinsing steps which generate large amounts of waste, and lack of lateral structure control during film build-up (*48*). To remedy these problems a "dewetting method" of LBL was employed for a variety of films including multilayered polyvinyl alcohol and cellulose nanocrystals (from tunicate) (*48*). The dewetting effect was achieved by adding dimethylformamide to the aqueous dipping solutions to produce a high contact angle at the solid–liquid interface. This results in self-cleaning/rinsing and alignment of axial colloids, conserving the general applicability of LBL and accelerating the overall process. Concentration, advancing/receding angles and speed of fluid movement are key factors for controlling surface morphology in dewetting LBL. The tunicate nanocrystals were found to preferentially align when the substrates were held vertically and air was blown to guide the dewetting lines (*48*).

We can conclude from these studies that the LBL technique applied to cellulose nanocrystals and various polycations is robust and sufficient to create densely packed uniform films with a choice of tunable properties. Changing the cellulose source and hydrolysis conditions varies the aspect-ratio and polydispersity which can be adjusted for optimal reinforcing properties. Additionally, the iridescent colors can be controlled in a minimal number of steps when compared to linear polymer systems requiring over 1000 deposition steps to achieve similar optical properties (*19*). Orientational ordering of the cellulose component increases the possibility of preparing materials with anisotropic mechanical properties. The stability of LBL films with cellulose and polycations has made them amenable to further studies that are not always possible for films containing only cellulose nanocrystals. Quantifying cellulose birefringence in ordered films and measuring surface forces using colloid-probe atomic force microscopy are two such measurements (*49*).

Polyelectrolyte Multilayers Prepared on Cellulose Fiber Substrates

Polyelectrolyte adsorption on wood fibers is important in papermaking. Flocculating agents, which are normally polyelectrolytes, are used to retain fillers and fiber fragments in paper sheets (*50*). Fiber–fiber interactions can be controlled through adsorbed polymers and have a significant influence on paper strength. Polyelectrolytes are also added to control drainage and recover fibers from effluent water in the papermaking process (*51*). Both wet and dry adhesion are affected, and the LBL method shows promise as a way of improving the physical properties of paper (*52*). Factors that affect individual polymer adsorption on fibers include the fiber porosity, polyelectrolyte charge, molecular weight, and specific interactions; these factors have been looked at in detail elsewhere (*50, 51, 53*) and are not the focus of this review.

Wågberg and coworkers have extensively studied the effect of polyelectrolyte multilayers on fiber–fiber bond/paper strength (*52, 54–63*). Wood fibers[1] coated by LBL with PAH and poly(acrylic acid) (PAA) were found to have linear or exponential growth, as monitored by SPAR (*54*) and QCM with dissipation monitoring (QCM-D) (*59*). The fiber surface carboxyl groups (anionic) are fully charged above pH 7, PAA (anionic) charge increases between pH 3.5 and 7.5, and PAH (cationic) charge decreases with increasing pH. Due to this ionizability, LBL formation at pH 7.5 was linear (where all components were fully charged and tended to adsorb in a flat conformation), while at pH 5.0 the growth was described as exponential because of the less charged, loopy conformation of PAA (*54*). The multilayered fibers were made into hand sheets and characterized by light microscopy, mechanical testing,

---

[1] Unless otherwise noted, wood fibers are dried, chlorine free (TCF) bleached, softwood Kraft fibers.

elemental analysis, light scattering, and ATR-FTIR. Film properties of (PAH/PAA)$_n$ were strongly dependent on the outermost polyelectrolyte layer; the coating was more rigid and elastic (less viscous) with a PAA cap, leading to smaller pull-off forces when adhesion was measured by AFM (*59*). Conversely, a softer, more water rich LBL film on wood fibers was seen when the PAA charge density was lowered and contributed to a stronger sheet (*59*). The LBL treatment increased hand sheet density. However, the increase in paper strength was attributed to three factors: (1) an increased number of fiber–fiber joints, (2) an increased degree of contact in the fiber–fiber joint, and (3) an increased covalent nature of the fiber–fiber bond (*56*).

Fiber wettability was studied by coating wood fibers with up to 10 layers of various polymer combinations: PAH/PAA (*52*), polyethylene oxide (PEO)/PAA (*52*), and PDDA/polystyrene sulfonate (PSS) (*55*). A dynamic contact angle analyser was used by Lingström et al. to look at individual coated fiber wettability according to the Wilhelmy technique (*52, 55*). Wettability was pH dependent for PAH/PAA coatings and lowest with PAH as the outer layer. PEO/PAA coatings (held together by hydrogen bonds and not electrostatic interactions) were independent of the outer layer and had a generally higher wettability (lower contact angle). The PDDA/PSS films were of intermediate wettability as shown in Figure 5. Colloid-probe AFM also indicated a correlation between a higher pull-off force for coated surfaces and lower wettability (*52*). Treated fibers were made into sheets and the tensile index was found to be improved (and increased linearly with adsorbed amount) by 90% for (PEO/PAA)$_{4.5}$ (*52*), 60–200% (depending on pH and outermost polymer) for (PAH/PAA)$_n$ (*54*) and 90% for (PDDA/PSS)$_{5.5}$ (*55*).

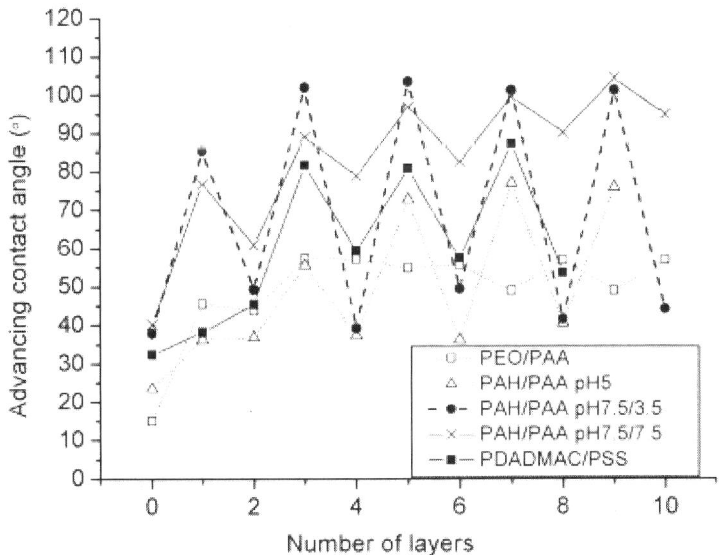

Figure 5. Advancing contact angle as a function of layer number measured on individual fibers treated with PAA/PEO, PAH/PAA (treated at pH 5 and pH 7.5/3.5) and PDDA/PSS (called PDADMAC/PSS in graph). (Reprinted with permission from ref 52. Copyright 2007 Elsevier.)

Often printing grade papers contain silica, titania, clay, and other filler particles to adjust brightness, opaqueness, and wettability but can decrease sheet strength. Lu et al. found that paper brightness was enhanced without diminishing tensile strength when fibers were LBL coated with polyelectrolytes and nanoparticles (*64*). Deposition was controlled by LBL processing with 2–4 bilayers of either TiO_2/PSS, PDDA/SiO_2 or PDDA/halloysite clay nanotubes on a precursor bilayer of PDDA/PSS (*64*). The cationic TiO_2 particles were 20–80 nm in diameter, anionic SiO_2 particles were 80 ± 10 nm in diameter and slightly anionic clay nanotubes were 50×500–700 nm. All of the coatings led to linear film growth and complete surface coverage of the fibers was seen by fluorescence and scanning electron microscopy. Brightness was the most enhanced for $(TiO_2/PSS)_2$ where the nanoparticle loading was 1 wt% and the brightness was increased by 4%. The SiO_2 and clay films with 2 bilayers reduced the brightness by 4.5% and 2.6%, respectively. When hand sheets were prepared with the nanoparticle coated fibers, tensile strength was claimed to remain close that of the control sample, but the porosity was found to increase by 30–50% (*64*). Highly porous paper sheets have promise as controlled drug release reservoirs and other pharmaceutical and biomedical applications.

Instead of making paper sheets from LBL coated fibers, sheets in the form of fibrous mats (*16*) and fabrics (*65*) were also subjected to multilayering through the dipping procedure. Polymer nanofiber mats were prepared by electrospinning cellulose acetate (*16*). Partial hydrolysis of the surface ester groups makes the fibers anionic and the films are insoluble in water with a high surface area making them ideal for LBL coating. Multilayer films of PAH/PAA were prepared on the fibrous mats and found to have the same surface morphology as uncoated sheets. However, the cellulose acetate fibers also had an increased surface roughness. The film composition and surface properties were studied by FTIR, field-emission-SEM, and AFM. The pH dependence of the weak polyelectrolytes resulted in thicker films being prepared at pH 7.5 (for PAH) and pH 3.5 (for PAA) than when both polymers were deposited at pH 5. This is shown in Figure 6a–c. The thicker layers at pH 7.5/3.5 are attributed to a thick loopy conformation of the partially ionized PAA that adsorbs in large amounts because of the previously deposited and highly charged PAH (*16*). If it is desired to keep the original fiber morphology then one can obviously get high loading and thin films by adsorbing these polyelectrolytes at pH 5.

Functional textiles can be prepared by LBL assembly for use in protective clothing and selective filtration fabrics. Cellophane film and cotton fabric surfaces were rendered cationic by reacting with 2,3-epoxypropyltrimethylammonium chloride in base. Multilayer build-up with PSS/PAH was successful as monitored by ATR-FTIR, X-ray photoelectron spectroscopy, and transmission electron microscopy, and gave uniform coatings (*65*).

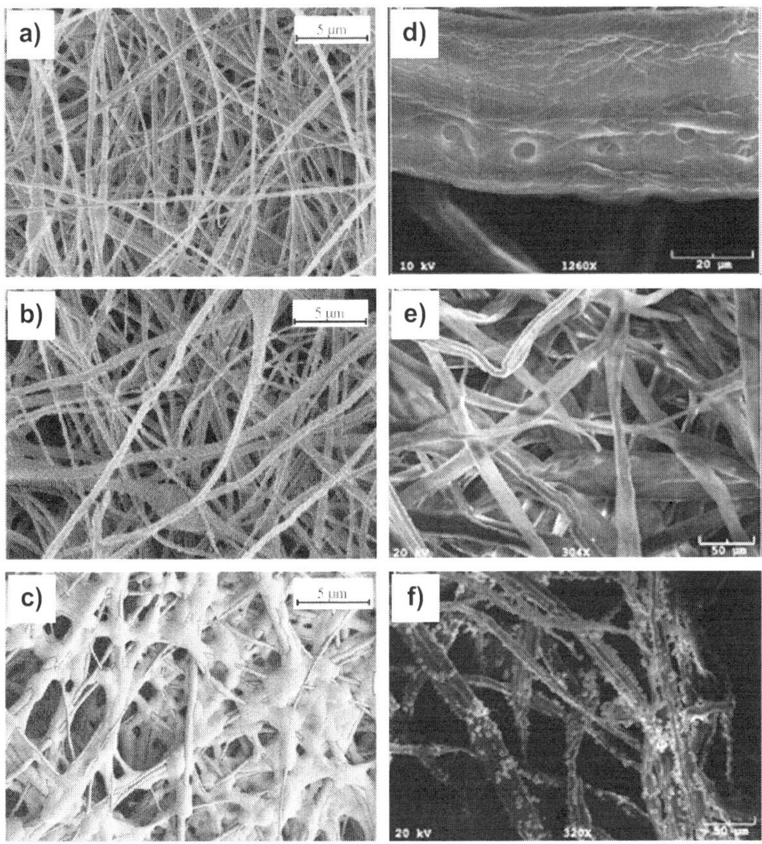

Figure 6. SEM images of multilayers on fiber surfaces. Electrospun cellulose acetate mats (a–c, scale bar 5 μm), (a) uncoated, (b) (PAH/PAA)₅ coated fibers at pH 5, and (c) (PAH/PAA)₅ coated fibers at pH 7.5/3.5. (c). Softwood fibers (d–f, scale bar 20 μm, 50 μm, 50 μm, respectively), (d) (PDDA/PSS)₁.₅ (PDDA/urease)₃ coated fiber, (e) (PDDA/PSS)₃ coated fibers with Ca²⁺ (no microparticle formation observed), and (f) (PDDA/PSS)₁.₅ (PDDA/urease)₃ coated fibers after biocatalysis reaction (CaCO₃ microparticles are visible growing on the fibers.) (Reprinted with permission from refs 16 (a–c) and 66 (d–f). Copyright 2005 Elsevier and 2007 American Chemical Society, respectively.)

110

Two recent applications of LBL coating of cellulose fibers have resulted in composites with novel biological function (*66*) and electrical conducting properties (*63*). Xing et al. incorporated anionic enzymes, laccase, and urease into multilayered films on bleached Kraft softwood fibers (*66*). These modified fibers can be used to decompose urea or lignin and synthesize inorganic particles and polyphenols. The fibers were first coated with $(PDDA/PSS)_{1.5}$ to avoid penetration of the enzyme into the fibers and then $(PDDA/laccase)_3$ or $(PDDA/urease)_3$ was assembled on top. Multilayer build-up was monitored by QCM and zeta-potential and films were found to be 15–20 nm thick. The enzymatic activity was monitored in the film and increased with the number of enzyme layers. These coatings were stable and retained 50% enzymatic activity after 14 days storage in water at 4 °C. The fiber–$(PDDA/urease)_n$ multilayers were used to grow calcium carbonate microparticles of 1–7 μm in diameter upon the addition of urea and $CaCl_2$. This is practical because $CaCO_3$ is often used to increase paper brightness either in the form of precipitated calcium carbonate (particle size: 0.1–2.5 μm) or ground calcium carbonate (particle size: 0.1–0.4 μm) (*67*). The SEM images of coated fibers without enzyme and with enzyme after biocatalysis (i.e. with $CaCO_3$ particles) are shown in Figure 6d–f (*66*). Figure 6 also emphasizes the differences in surface morphology and fiber size between electrospun cellulose acetate fibers and softwood pulp fibers; however both are amenable to LBL coating.

Wistrand et al. prepared electrically conducting cellulose fibers using a conjugated polymer complex of poly(2,3-ethylenedioxythiophene) and PSS (PEDOT:PSS), LBL assembled with PAH (*63*). The softwood fibers were carboxymethylated to enhance the substrate charge and give more uniform films. Multilayer build-up was found to be linear and resulted in a smoother fiber surface after polyelectrolyte deposition. Conductivity increased substantially (2–3 orders of magnitude) for the first layer of PEDOT:PSS adsorbed but only increased slightly for the next bilayer. When PAH was the outermost layer, the conductivity was lower and again, fiber wettability was also decreased. Overall, the hand sheet conductivity and tensile index was higher for fibers coated with $(PAH/PEDOT:PSS)_n$ when the fibers were carboxymethylated. When 5 bilayers are deposited, however, the tensile index decreases by 25%. This suggests that creating a cellulose material with strong fiber contacts and high contact area is more significant than adsorbing large amounts of conducting polymer (*63*).

Not to be overlooked is the recent use of microfibrillated cellulose (MFC) in polyelectrolyte multilayer films (*68*, *69*). Nanofibrils, 5–20 nm in diameter and 1 μm long, are prepared though a mild enzymatic hydrolysis (with a monocomponent endoglucanase) combined with mechanical shearing and high-pressure homogenization (*69–71*). Carboxymethylated MFC was used with PEI, PDDA, and PAH polycations to assemble well-defined, stable LBL films. These flexible cellulose I elements with crystalline and amorphous regions were amenable to the multilayering technique and film growth was strongly pH and ionic strength dependent. Under highly charged conditions, the MFC layers tended to be thinner (*68*, *69*). The resultant films showed randomly oriented fibers with good surface coverage, smooth surfaces, and a water content of 41% (*68*). This type of nanocellulose shows promise as a model cellulose system

(*71, 72*) but also as a reinforcing agent in nanocomposites and in the preparation of optical sensor materials (*69*).

The LBL method has thus presented a new way of engineering fiber surfaces to achieve desired properties and function. Various kinds of smart paper and drug release reservoirs can be prepared by depositing multilayer coatings on fibers. Specifically, the ability of adsorbed polymers to increase wet and dry adhesion between fibers and improve overall paper properties are noteworthy. The studies of LBL on fiber surfaces have helped to show that fiber–fiber joint strength is dependent on both chemical properties and contact area; lower wettability (high contact angle) and having many bilayers with the polyanion component as the outermost layer, all contribute to making stronger paper. High loading of polymeric material can be achieved because of the increased surface area of both fibrous mats and individual fibers, when compared to flat substrates. Incorporating large amounts of polymer into these coatings not only improves the paper tensile index but can be particularly beneficial when the adsorbed polyelectrolyte has a secondary function, such as being conductive, biologically active or possessing sensors capabilities.

Conclusion

Much research has gone into preparing multilayered films containing polyelectrolytes and cellulose using the electrostatic layer-by-layer methodology. The technique generally gives rise to smooth and stable thin films, or uniformly coated fibers, as confirmed by various surface morphology measurements, elemental analyses, and spectroscopies. Contrary to other nanomaterial preparation methods, LBL assembly is relatively insensitive to defects which are quickly hidden by adding further layers. Additionally, LBL can be used to modify fiber surfaces to obtain special functionality without losing any basic structure or properties. It should be evident from the studies presented here that multilayered cellulose films can easily be used to tailor surface interactions and create novel and useful materials.

References

1. Decher, G. *Science* **1997**, *277*, 1232–1237.
2. Decher, G.; Schlenoff, J. B. *Multilayer Thin Films*, 1 ed; Wiley-VCH: 2002.
3. Johal, M. S.; Casson, J. L.; Chiarelli, P. A.; Liu, D.-G.; Shaw, J. A.; Robinson, J. M.; Wang, H.-L. *Langmuir* **2003**, *19*, 8876–8881.
4. Lee, S.-S.; Lee, K.-B.; Hong, J.-D. *Langmuir* **2003**, *19*, 7592–7596.
5. Jiang, C.; Markutsa, S.; Pikus, Y.; Tsukruk, V., V. *Nat. Mater.* **2004**, *3*, 721–728.
6. Jiang, C.; Markutsa, S.; Tsukruk, V., V. *Adv. Mater.* **2004**, *16*, 157–161.
7. Tammelin, T.; Saarinen, T.; Österberg, M.; Laine, J. *Cellulose* **2006**, *13*, 519–535.
8. Itoh, T.; Tsujii, Y.; Suzuki, H.; Fukuda, T.; Miyamoto, T. *Polym. J. (Tokyo, Jpn.)* **1992**, *24*, 641–652.

112

9. Schaub, M.; Wenz, G.; Wegner, G.; Stein, A.; Klemm, D. *Adv. Mater.* **1993,** *5,* 919–922.

10. Ifuku, S.; Kamitakahara, H.; Takano, T.; Tsujii, Y.; Nakatsubo, F. *Cellulose* **2005,** *12,* 361–369.

11. Schaub, M.; Fakirov, C.; Schmidt, A.; Lieser, G.; Wenz, G.; Wegner, G.; Albouy, P.-A.; Wu, H.; Foster, M. D.; Majrkzak, C.; Satija, S. *Macromolecules* **1995,** *28,* 1221–1228.

12. Cohen-Atiya, M.; Vadgama, P.; Mandler, D. *Soft Matter* **2007,** *3,* 1053–1063.

13. Buchholz, V.; Wegner, G.; Stemme, S.; Ödberg, L. *Adv. Mater.* **1996,** *8,* 399–402.

14. Li, X.-G.; Huang, M.-R. *J. Appl. Polym. Sci.* **1997,** *66,* 2139–2147.

15. Huang, H.; He, P.; Hu, N.; Zeng, Y. *Bioelectrochemistry* **2003,** *61,* 29–38.

16. Ding, B.; Fujimoto, K.; Shiratori, S. *Thin Solid Films* **2005,** *491,* 23–28.

17. Hiller, J. A.; Mendelsohn, J. D.; Rubner, M. F. *Nat. Mater.* **2002,** *1,* 59–63.

18. Podsiadlo, P.; Sui, L.; Elkasabi, Y.; Burgardt, P.; Lee, J.; Miryala, A.; Kusumaatmaja, W.; Carman, M. R.; Shtein, M.; Kieffer, J.; Lahann, J.; Kotov, N. A. *Langmuir* **2007,** *23,* 7901–7906.

19. Zhai, L.; Nolte, A. J.; Cohen, R. E.; Rubner, M. F. *Macromolecules* **2004,** *37,* 6113–6123.

20. Wang, B.; Anzai, J.-I. *Langmuir* **2007,** *23,* 7378–7384.

21. Lefebvre, J.; Gray, D. G. *Cellulose* **2005,** *12,* 127–134.

22. Radeva, T.; Kamburova, K.; Petkanchin, I. *J. Colloid Interface Sci.* **2006,** *298,* 59–65.

23. Qiu, X.; Leporatti, S.; Donath, E.; Moehwald, H. *Langmuir* **2001,** *17,* 5375–5380.

24. Müller, M.; Briššova, M.; Rieser, T.; Powers, A. C.; Lunkwitz, K. *Mater. Sci. Eng., C* **1999,** *C8–C9,* 163–169.

25. Geffroy, C.; Labeau, M. P.; Wong, K.; Cabane, B.; Cohen Stuart, M. A. *Colloids Surf., A* **2000,** *172,* 47–56.

26. Rundlöf, M.; Wågberg, L. *Colloids Surf., A* **2004,** *237,* 33–47.

27. Noguchi, T.; Anzai, J.-i. *Langmuir* **2006,** *22,* 2870–2875.

28. Noguchi, T.; Anzai, J.-I. *Electrochemistry* **2006,** *74,* 125–127.

29. Battista, O. A.; Coppick, S.; Howsmon, J. A.; Morehead, F. F.; Sisson, W. A. *J. Ind. Eng. Chem.* **1956,** *48,* 333–335.

30. Ranby, B. G. *Discuss. Faraday Soc.* **1951,** *11,* 158–164.

31. Beck-Candanedo, S.; Roman, M.; Gray, D. G. *Biomacromolecules* **2005,** *6,* 1048–1054.

32. Dong, X. M.; Revol, J. F.; Gray, D. G. *Cellulose* **1998,** *5,* 19–32.

33. Azizi Samir, M. A. S.; Chazeau, L.; Alloin, F.; Cavaille, J. Y.; Dufresne, A.; Sanchez, J. Y. *Electrochim. Acta* **2005,** *50,* 3897–3903.

34. Favier, V.; Chanzy, H.; Cavaille, J. Y. *Macromolecules* **1995,** *28,* 6365–6367.

35. Ljungberg, N.; Bonini, C.; Bortolussi, F.; Boisson, C.; Heux, L.; Cavaille, J. Y. *Biomacromolecules* **2005,** *6,* 2732–2739.

36. Orts, W.; Shey, J.; Imam, s.; Glenn, G.; Guttman, M.; Revol, J. F. *J. Polym. Environ.* **2005,** *13,* 301–306.

Figure 4.4. Digital photographs of iridescent colors seen in spin-coated LBL films of cellulose nanocrystals and PAH on Si.

37. Azizi Samir, M. A. S.; Alloin, F.; Sanchez, J.-Y.; Dufresne, A. *Polymer* **2004**, *45*, 4149–4157.

38. Samir, M. A. S. A.; Alloin, F.; Dufresne, A. *Biomacromolecules* **2005**, *6*, 612–626.

39. Mathew, A. P.; Oksman, K.; Sain, M. *J. Appl. Polym. Sci.* **2005**, *97*, 2014–2025.

40. Notley, S. M.; Eriksson, M.; Wågberg, L.; Beck-Candanedo, S.; Gray, D. G. *Langmuir* **2006**, *22*, 3154–3160.

41. Podsiadlo, P.; Choi, S.-Y.; Shim, B.; Lee, J.; Cuddihy, M.; Kotov, N. A. *Biomacromolecules* **2005**, *6*, 2914–2918.

42. Cranston, E. D.; Gray, D. G. *Biomacromolecules* **2006**, *7*, 2522–2530.

43. Cranston, E. D.; Gray, D. G. *Sci. Technol. Adv. Mater.* **2006**, *7*, 319–321.

44. Hecht, E. *Optics,* 4th ed.; Addison–Wesley: Reading, MA, 2001.

45. Edgar, C. D.; Gray, D. G. *Cellulose* **2003**, *10*, 299–306.

46. Mermut, O.; Barrett, C. J. *Analyst* **2001**, *126*, 1861–1865.

47. Jean, B.; Dubreuil, F.; Heux, L.; Cousin, F. *Langmuir* **2008**, *24*, 3452–3458.

48. Shim, B.; Podsiadlo, P.; Lilly, D. G.; Agarwal, A.; Lee, J.; Tang, Z.; Ho, S.; Ingle, P.; Paterson, D.; Lu, W.; Kotov, N. A. *Nano Lett.* **2007**, *7*, 3266–3273.

49. Cranston, E. D.; Gray, D. G.; Barrett, C. J. *Abstracts,* 32nd Northeast Regional Meeting of the American Chemical Society, Rochester, NY, United States, Oct 31–Nov 3, 2004; GEN-332.

50. Winter, L.; Wågberg, L.; Odberg, L.; Lindström, T. *J. Colloid Interface Sci.* **1986**, *111*, 537–543.

51. Onabe, F. *J. Appl. Polym. Sci.* **1978**, *22*, 3495–3510.

52. Lingström, R.; Notley, S. M.; Wågberg, L. *J. Colloid Interface Sci.* **2007**, *314*, 1–9.

53. Wågberg, L. *Nord. Pulp Pap. Res. J.* **2000**, *15*, 586–597.

54. Eriksson, M.; Notley, S. M.; Wågberg, L. *J. Colloid Interface Sci.* **2005**, *292*, 38–45.

55. Lingström, R.; Wågberg, L.; Larsson, P. T. *J. Colloid Interface Sci.* **2006**, *296*, 396–408.

56. Eriksson, M.; Torgnysdotter, A.; Wågberg, L. *Ind. Eng. Chem. Res.* **2006**, *45*, 5279–5286.

57. Notley, S. M.; Biggs, S.; Craig, V. S. J.; Wågberg, L. *Phys. Chem. Chem. Phys.* **2004**, *6*, 2379–2386.

58. Notley, S. M.; Wågberg, L. *Biomacromolecules* **2005**, *6*, 1586–1591.

59. Notley, S. M.; Eriksson, M.; Wågberg, L. *J. Colloid Interface Sci.* **2005**, *292*, 29–37.

60. Wågberg, L.; Forsberg, S.; Johansson, A.; Juntti, P. *J. Pulp Pap. Sci.* **2002**, *28*, 222–228.

61. Wågberg, L.; Pettersson, G.; Notley, S. *J. Colloid Interface Sci.* **2004**, *274*, 480–488.

62. Winter, L.; Wågberg, L.; Ödberg, L.; Lindström, T. *J. Colloid Interface Sci.* **1986**, *111*, 537–543.

63. Wistrand, I.; Lingström, R.; Wågberg, L. *Eur. Polym. J.* **2007**, *43*, 4075–4091.

64. Lu, Z.; Eadula, S.; Zheng, Z.; Xu, K.; Grozdits, G.; Lvov, Y. *Colloids Surf., A* **2007**, *292*, 56–62.
65. Hyde, K.; Rusa, M.; Hinestroza, J. *Nanotechnology* **2005**, S422.
66. Xing, Q.; Eadula, S. R.; Lvov, Y. *Biomacromolecules* **2007**, *8*, 1987–1991.
67. Biermann, C. J. *Handbook of Pulping and Papermaking*, 2 ed.; Academic Press, Inc.: San Diego, 1996.
68. Aulin, C.; Varga, I.; Claesson, P. M.; Wågberg, L.; Lindström, T. *Langmuir* **2008**, *24*, 2509–2518.
69. Wågberg, L.; Decher, G.; Norgren, M.; Lindström, T.; Ankerfors, M.; Axnas, K. *Langmuir* **2008**, *24*, 784–795.
70. Pääkkö, M.; Ankerfors, M.; Kosonen, H.; Nykänen, A.; Ahola, S.; Österberg, M.; Ruokolainen, J.; Laine, J.; Larsson, P. T.; Ikkala, O.; Lindström, T. *Biomacromolecules* **2007**, *8*, 1934–1941.
71. Ahola, S.; Salmi, J.; Johansson, L.-S.; Laine, J.; Österberg, M. *Biomacromolecules* **2008**, *9*, 1273–1282.
72. Ahola, S.; Turon, X.; Osterberg, M.; Laine, J.; Rojas, O. J. *Langmuir* **2008**, *24*, 11592–11599.

Chapter 5

Preparation of Ordered Films of Cellulose Nanocrystals

Véronique Aguié-Béghin[1,2], Michaël Molinari[3], Arayik Hambardzumyan[4], Laurence Foulon[1,2], Youssef Habibi[1,5], Thomas Heim[6], Ralph Blossey[6], and Roger Douillard[1,2]

[1]INRA, UMR FARE 614, Fractionation of Agricultural Resources and Environment, CREA, 2 Espl. R. Garros, F-51686 Reims, France
[2]Université Reims-Champagne-Ardennes (URCA), UMR FARE 614, F-51686 Reims, France
[3]Laboratoire de Microscopies et d'Etude de Nanostructures, Université Champagne-Ardennes, 21 Rue Clément Ader, BP 138, F-51685 Reims, France
[4]Laboratory of Chemistry and Physics, Erevan State University, 1 A. Manoukian Street, 0025 Erevan, Armenia
[5]Present address: Ecole Française de Papeterie et des Industries Graphiques, Institut National Polytechnique de Grenoble, BP 65, F-38402 St Martin d'Hères Cedex, France
[6]Institute of Electronics, Micro-electronics and Nanotechnology, IEMN, Cité Scientifique, Avenue Poincaré, BP 60069, F-59652 Villeneuve d'Ascq Cedex, France

To understand the structure and reactivity of lignified plant cell walls, ordered cellulose films were developed from nanocrystals (whiskers) to mimic successive native cellulose layers in plant cell walls. The films were prepared by deposition of whisker suspensions in an electric field or by transfer of whisker monolayers onto a solid substrate by the Langmuir–Blodgett technique. In a controlled electric field (strength, frequency), completely ordered nanocrystal films were obtained between the electrodes. With the Langmuir–Blodgett technique, ordered domains of several square millimeters in area were obtained by controlling the surface concentration of the whiskers. These results should allow defining the experimental conditions for the preparation of new model systems from cellulose nanocrystal suspensions for the investigation of surface properties of plant cell walls.

Introduction

When plant cells have completed the stages of growth and expansion, their walls undergo significant structural and chemical modifications that impart functional specificity. The secondary plant cell wall is of high rigidity. It contains a large proportion of highly crystalline cellulose microfibrils, forming a compact network. These microfibrils form concentric layers, within which they are aligned parallel (*1–3*). The microfibrils are surrounded by a complex network made of different amorphous polymers, including hemicelluloses, lignins, and proteins, depending on species (*4*). The intricate architectures of plant cell walls still raise multiple questions, such as, for example, the role of hemicelluloses and lignins in the cohesion between the successive layers of cellulose. Since the role of the individual components in the structure and properties of plant cell walls is currently not well understood, a number of studies have been carried out to develop model surfaces of cellulose by spin coating or Langmuir–Blodgett techniques, primarily using cellulose derivatives (trimethylsilylcellulose) (*5–7*), or cellulose dissolved in N-methyl-morpholine-N-oxide (NMMO) (*8, 9*) or dimethylacetamid/lithium chloride (*10*). The advantage of using dissolved cellulose is that some crystallinity (such as cellulose II in the NMMO case) is retained in the model films. The majority of these surfaces have been used to study their adhesion properties (*10*), the adsorption of polyelectrolytes and surfactants (*11–13*), interaction forces between cellulose surfaces (*5, 14, 15*) or between cellulose and other macromolecules (lignins, xylans) (*5, 14–19*), or their reactivity towards enzymes (*20, 21*).

More recently, other model cellulose surfaces have been prepared from cellulose monocrystals (*19, 22–24*). These monocrystalline particles display interesting properties since the native crystallinity of cellulose is preserved. Spin coating and electrostatic layer-by-layer processing were used to obtain polyelectrolyte multilayer nanocomposites with enhanced mechanical, optical, and electrical properties. Both techniques gave rise to smooth and stable thin films. A radial orientation of the nanocrystals was noticed in the films formed by the spin coating technique (*19, 24*), yet areas with a parallel orientation were limited. Additionally, experiments have been performed to control the orientation of cellulose nanocrystals in colloidal suspensions. The suspensions were subjected to external fields, such as magnetic fields (*25, 26–28*), shear flow (*29, 30*), and more recently electric fields (*31*). Depending on the technique used, the orientation of the nanocrystals' long axes was perpendicular to the magnetic field, parallel or perpendicular to the shear direction, depending on shear rate, or parallel to the electric field direction. However, these different techniques were carried out starting from relatively concentrated suspensions (> 0.5% (w/v)) and could not produce nanocrystal monolayers, even after drying.

The aim of the present paper is to describe new results on the preparation and characterization of ordered monolayers of ramie and tunicin cellulose whiskers obtained from aqueous dilute suspensions, by spreading in an electric field or by the Langmuir–Blodgett technique. The density of the nanocrystals in the monolayer was controlled by the volume concentration of the suspension deposited onto a silicon wafer in an electric field or by the surface pressure at the air–liquid interface before transfer onto the wafer by the Langmuir–Blodgett method. The extent of alignment of the nanocrystals in the monolayer was controlled by the strength and frequency of the electric field for the first method, and by the conditions of the transfer step for the second. The composition, structure (thickness, orientation), and properties of the monolayers were characterized by ellipsometry and atomic force microscopy (AFM).

Experimental Section

Preparation of Cellulose Nanocrystals from Tunicin and Ramie

Tunicin whiskers were prepared from the tunic of *Microscosmus fulcatus*. Small fragments of the external wall of the tunicates were first treated overnight with a solution of KOH (5% (w/v)). The mantles were then washed and submitted to three successive bleaching treatments according to Wise et al. (*32*). Finally, the bleached fragments were disintegrated in water with a Waring blender. Ramie fibers were cut into small pieces and treated with 2% NaOH at 80 °C for 2 h in order to remove residual additives.

The homogeneous suspensions obtained from tunicin or purified ramie fibers were submitted to overnight hydrolysis (~16 h) with 65% (w/w) H_2SO_4 at room temperature under stirring. The suspensions were washed with water until neutrality and dialyzed against a concentrated solution of polyethylene glycol (35000 g/mol) in order to reduce their volume. The resulting concentrated suspensions of nanocrystals from tunicin and ramie were stored at 4 °C. Cellulose nanocrystal suspensions, with appropriate concentrations, were sonicated with a Branson sonifier for a few minutes before use. The average crystal dimensions (19 nm × 9 nm × 1000 nm for tunicin and 7 nm × 7 nm × 250 nm for ramie) were estimated with ImageJ from TEM (or AFM) images of dilute nanocrystal suspensions.

Preparation of Silicon Substrates

Silicon wafers with (1 0 0) surface orientation and an oxide layer thickness of 20 Å were cut to the desired dimensions (squares of about 1–1.5 cm^2) and cleaned with a mixture of H_2SO_4 and H_2O_2 (7/3) for 30 min at 60 °C, before continuous rinsing with purified water. Deposition of the electrodes, the whisker suspension, or the whisker monolayer was performed immediately after cleaning and drying of the silicon substrate. The metal electrodes were manufactured by a standard photolithographic technique. A photosensitive resist (AZ 1505 provided by Hoechst) was first spin coated onto the silicon substrate, leading to a homogenous polymer film of uniform thickness (~ 1 μm). This step was followed by thermal annealing at 110 °C for 1 min in order to remove excess solvent. Electrode patterns with gaps of 20 μm, 50 μm, or 100 μm were photo printed onto the resist by UV exposure (10 mW/cm^2) for 3 s through a quartz/chromium shadow mask. The substrates were then dipped into a developer (MIF 726 provided by Hoechst) for 10–15 s, which removed the UV light-exposed resist. Titanium (~ 100 Å thick) and then platinum (~ 700 Å thick) were subsequently deposited onto the substrate by evaporation of the metal in ultra high vacuum. The excess metal on top of the resist was removed by dissolution of the polymer in acetone. This step was followed by extensive sonication in acetone, isopropanol, and deionized water. Finally, chemical modification of the surface was performed by grafting aminopropyl trimethoxysilane (APTMS) molecules, which promote the adhesion of cellulose nanocrystals. For this purpose, the substrates were subjected to oxygen plasma etch (100 W, 30 s, 100 mbar), leading to –OH rich surfaces, which are strongly reactive towards silane groups ($-Si(O-CH_3)_3$). A few milliliters of APTMS in a small glass flask were positioned near the silicon substrates for 2 h under a glass bell jar filled with dry nitrogen for gas phase transfer.

Preparation of Cellulose Films in an Electric Field

The experimental set up comprised an EG&G galvanostat generator (model 5208) and an oscilloscope connected with wires and alligator clips to the electrodes on the substrate (Figure 1a). After connecting the substrate to the generator, cellulose nanocrystal suspension was carefully deposited between the electrodes. After a few seconds, excess suspension was removed and the surface dried in a stream of air. The experiments were done with voltage ranging from 250 V/cm to 20 kV/cm at frequencies varying from 1 kHz to 2 MHz.

Preparation of Cellulose Films by the Langmuir–Blodgett Technique

Stable layers of whiskers, formed at the air–liquid interface in the presence of a cationic amphiphilic molecule (dioctadecyldimethylammonium bromide, DODA), were transferred onto silicon wafers by the Langmuir–Blodgett procedure (Figure 1b). The pressure–area isotherms of the Langmuir films were determined with a Langmuir–Blodgett trough (KSV Technology, minitrough 75 × 330 mm equipped with a Wilhelmy-type film balance). One drop (20 μL) of a DODA chloroform solution (1 mg/mL) was spread with a microsyringe on the surface of an aqueous cellulose nanocrystal suspension. The concentration increase of molecules at the air–liquid interface was evidenced by the increase in surface pressure during the compression of the film. Transfer of the nanocrystal monolayer onto a silicon substrate was performed by vertical deposition at controlled speed. The silicon substrate, in a vertical orientation, was moved downward and then upward through the air–liquid interface. The nanocrystal monolayer adhered to the silicon substrate during the upward stroke. The transferred layer was washed with chloroform and dilute NaOH in order to remove adsorbed DODA on the whisker film surface.

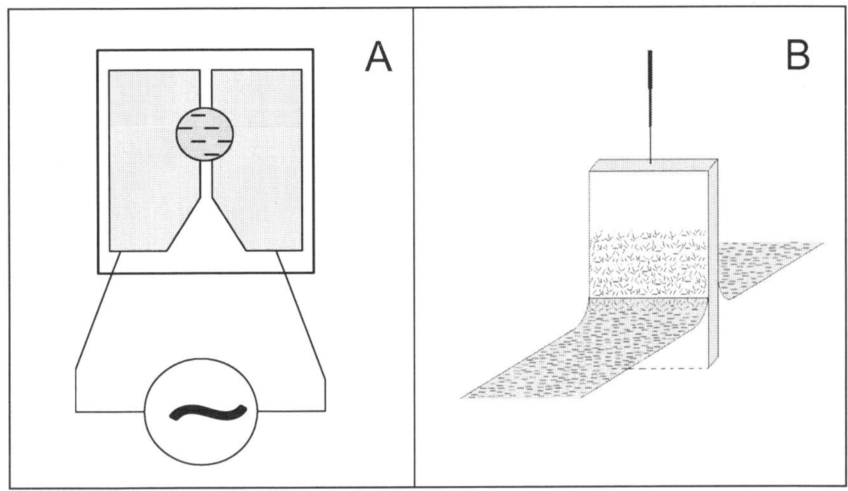

Figure 1. Diagram showing the experimental set up for whisker deposition onto a silicon wafer in an electric field (A) or by the Langmuir–Blodgett method (B).

Characterization of Cellulose Films

Optical Characterization

A Uvisel (Horiba-Jobin Yvon, Longjumeau, France) spectroscopic phase modulated ellipsometer was used to measure the thickness of the transferred layer using an angle of incidence of 70°. The diameter of the light beam was set to 1 mm. The two experimental ellipsometric angles Ψ and Δ are linked to the two reflectivity coefficients r_p and r_s in the directions parallel and perpendicular to the incident plane, respectively, by (*33*):

$$\frac{r_p}{r_s} = \tan(\Psi)\exp(i\Delta) \qquad [1]$$

The ellipticity coefficient, $\bar{\rho}_B$, measured in Brewster conditions, is defined by:

$$\bar{\rho}_B = \tan(\Psi)\sin(\Delta) \qquad [2]$$

The Brewster angle, θ_B, is given by $\tan(\theta_B) = n_2/n_0$, where n_0 and n_2 are the refractive indexes of the upper phase (air) and lower phase (water), respectively. When an adsorption layer begins to form, according to the Drude equation, the ellipticity coefficient is proportional to the layer thickness, h, and increases with the difference (n_1-n_2), where n_1 is the refractive index of this layer (*34*):

$$\bar{\rho}_B \propto h\frac{\left(n_1^2 - 1\right)\left(n_1^2 - n_2^2\right)}{n_1^2} \qquad [3]$$

The experimental Ψ and Δ spectra were collected in a wavelength range of 1.5–5 eV (240–820 nm) every 0.02 eV. The indexes and thicknesses of the cellulose films were calculated by fitting the experimental spectra to theoretical models using a Cauchy dispersion model for the experiments at the air–water interface and the Lorentz oscillator dispersion model (*35*) for the experiments on solid substrates (by the least-square fitting procedure). Brewster angle microscopy (BAM) was performed on the aqueous substrate ($\theta_B \approx 53°12$) to visualize the morphology of the adsorption layer in situ at the micrometer scale without optical probe and before the dipping step of the Langmuir–Blodgett procedure. The ellipticity coefficient at the Brewster angle, $\overline{\rho}_B$ (eq 2), is linked to the reflectivity coefficients according to the relation:

$$r_p = i r_s \overline{\rho}_B \qquad [4]$$

Since the reflectance, $R\ (= I_r/I_0)$, of the interface is proportional to $\overline{\rho}_B{}^2$ with I_r and I_0 being the reflected and incident intensities, respectively, the uncovered substrate appears dark ($r_p \approx 0$) whereas all areas covered by adsorption layers appear more or less bright ($r_p \neq 0$).

Tensiometry Measurements

Rheological properties of DODA–cellulose mixed layers at the liquid–air interface were determined with a dynamic drop tensiometer (Teclis-IT Concept, Longessaigne, France). The surface tension was calculated through shape analysis of an air bubble formed at the tip of a stainless steel needle dipped into water. The needle was attached to a syringe whose plunger was precisely controlled by a micrometer screw driven by an electric motor. The bubble was illuminated by a beam of parallel light and the image was recorded by a CCD camera and digitized. The interfacial tension, γ, of the air–liquid interface was determined by analyzing the profile of the bubble according to the Laplace equation (*36*). The surface pressure, π, is as usual the difference between the surface tension of the pure solvent, $\gamma_0 = 72.6$ mN/m, and that of the interface with surface active molecules, γ. The surface modulus, ε, is defined as the ratio between the variation of the surface pressure, $d\pi$, and the relative change of the surface area, $dA/A = d\ln(A)$ (eq 5) (*37*).

$$\varepsilon = \frac{d\gamma}{d\ln(A)} = -\frac{d\pi}{d\ln(A)} \qquad [5]$$

ε was determined during sinusoïdal fluctuations of the area of the bubble at a chosen amplitude (less than 15% of the mean area (*38*)) and a frequency of 0.1 Hz. A Fourier transform of the data was performed and only the first harmonic was retained. The DODA layer at the surface of the air bubble was formed before injection of the whisker suspension into the aqueous substrate. The amount of DODA and the bulk concentration of whiskers were selected to give surface pressures close to those used in the Langmuir–Blodgett technique. All measurements were done in an air-conditioned room at $20 \pm 1°C$.

Atomic Force Microscopy

AFM imaging was performed using a Nanoscope IIIa atomic force microscope from Digital Instruments (Santa Barbara, CA). The AFM was placed on an active vibration isolation table so that eventual external vibrations did not affect the imaging process. A scanner, calibrated following the standard procedures provided by Digital Instruments, with a maximum scan area of $120~\mu m^2$ was used. Experiments were made at constant room temperature around 20 °C. The samples were mounted onto stainless steel disks using sticky tape. Imaging was done in tapping mode. Commercial 225 μm long cantilevers from Veeco Instruments (France) with a resonant frequency around 190 kHz were used. Scanning rates of 1 Hz or 0.5 Hz depending on the image size and a resolution of 512 × 512 data points were used. During the scans, proportional and integral gains were increased to values just below those at which the feedback started to oscillate. Images were processed only by flattening to remove background slopes. A free software was used to analyze the image.

Results and Discussion

This section discusses the effects of (1) the AC electric field and (2) the Langmuir–Blodgett procedure on the formation of ordered cellulose nanocrystal films. Nanocrystals with high and low aspect ratio (tunicin and ramie whiskers, respectively) were studied to test whether the aspect ratio influences the degree of alignment of these nanoparticles in the films.

Whisker Orientation by Electric Field

Orientation of the tunicin and ramie nanocrystals in the electric field was achieved by depositing drops of dilute whisker suspensions (0.05% (w/v)) into a gap of 10–100 μm in width between two electrodes. The amplitude and frequency of the AC electric field were varied from 250 V/cm to 20 kV/cm and from 1 kHz to 2 MHz, respectively. The dry tunicin film prepared at 20 kV/cm and 1 MHz was analyzed by AFM (Figure 2a). Whiskers deposited directly onto the electrodes were randomly oriented. Between the two electrodes, two patterns were visible. On the film surface, several randomly oriented whiskers could be seen. Below the film surface, however, a high degree of orientation of the nanocrystals along the direction of the field was visible when a gap of 10 μm was used. An analysis of the film topography in the direction perpendicular to the electric field showed surface elevations between 10 and 30 nm (Figure 2b). The same analysis performed in the direction of the electric field exhibited a smoother profile with most surface elevations between 10 and 20 nm. Considering the width of the tunicin whiskers, which is close to 10 nm, these height profiles are consistent with the occurrence of a single layer of whiskers on the silicon substrate and with some whiskers forming a second discontinuous layer. Some whiskers may also lie on top of whiskers of the second layer, generating a third layer, and so on. The difference between the height profiles perpendicular and parallel to the electric field is strongly consistent with the occurrence of an ordered whisker film on the silicon substrate. When the electric field was much lower (1 kV/cm), whisker alignment was not evident in AFM images or from a comparison of height profiles parallel and perpendicular to the electric field (Figure 2c). Parallel organization of whiskers in the film takes place during the few seconds after the drop has been deposited into the gap and before the excess liquid is removed and the film dried in a stream of air. The number of randomly oriented nanocrystals on the surface of the aligned film seemed to depend on the time it took to remove the excess suspension after the drop had been deposited. For ramie whiskers, alignment was observed at a field strength of 5 kV/cm (Figure 2d). The difference in the patterns formed by tunicin and ramie whiskers (Figures 2a and 2d) is probably related to the difference in aspect ratio (~ 70 for tunicin and ~ 35 for ramie whiskers).

For large gaps, when the field was lowered, the thickness of the films did not change. However, preferred orientation of the whiskers in the direction of the electric field was only observed in the vicinity of the electrodes but not in the center of the gap (Figure 3). For fields of approximately 500 V/cm, preferred orientation of the whiskers was no longer noticed. A lack of whisker alignment was also caused by a shift in frequency of the electric field. Whisker alignment was incomplete for frequencies smaller than 1 kHz or larger than 2 MHz. The degree of orientation of tunicin whiskers was calculated from AFM analysis over the entire width of the gap by visual determination of the boundary between the area of aligned whiskers and that of randomly oriented ones. Alignment of the tunicin cellulose nanocrystals with the electric field direction was strong and maximal above 2 kV/cm in the frequency range 10^5–10^6 Hz.

In conclusion, alignment of cellulose nanocrystals in an electric field is an efficient and quick method. Under certain conditions (voltage, frequency), an alternating electric field induces perfect orientation of the rods parallel to the external field. This process is linked to the permittivity (or dielectric constant) of cellulose nanocrystals (*39*), and consequently to their polarizability in aqueous solution. The alignment process is governed by the induced dipole moment of the suspension and the applied electric field. Nevertheless, our results seem to indicate that obtaining extensive, highly ordered nanocrystal monolayers (> 100 μm^2) would require the use of high electric fields. Moreover, the thickness of the whisker film in the gap is not uniform. As a consequence, it is presently difficult to obtain an authentic whisker "monolayer". In addition, the area between the electrodes is presently limited to a strip of a few tens of micrometers in width. The small width of the gap is an important limitation for the analysis of the surface properties of these cellulose films.

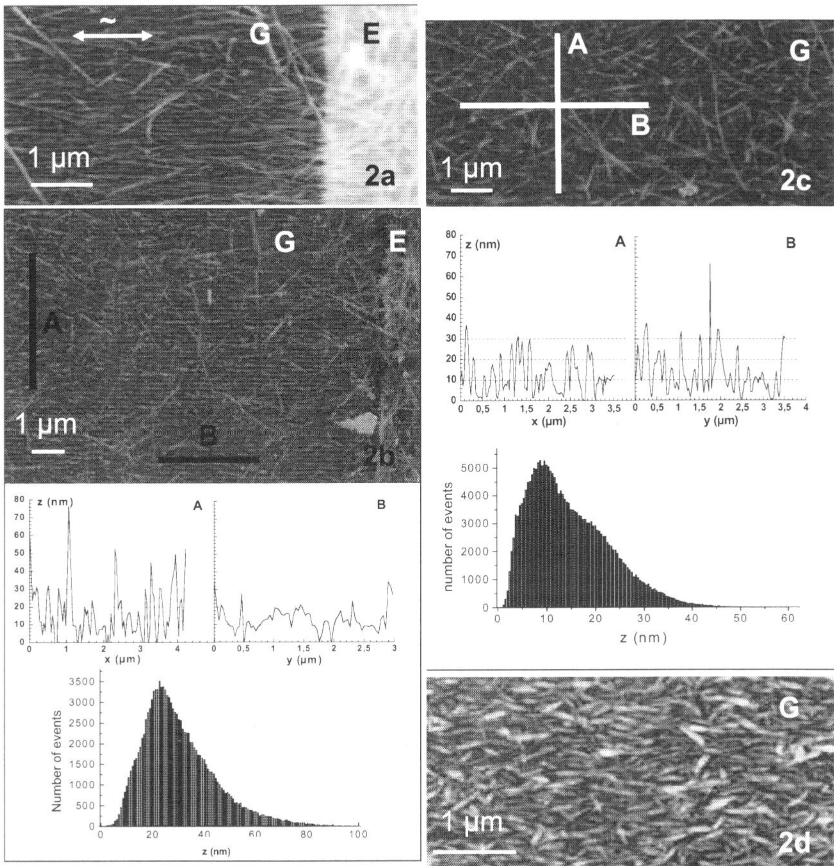

Figure 2. AFM images of electric field-ordered cellulose nanocrystal films of tunicin (2a, 2b, and 2c) and ramie (2d). The electric field strengths were respectively 20, 10, and 1 kV/cm for tunicin and 5 kV/cm for ramie. The frequencies were 1 MHz for (2a), 2 MHz for (2b) and (2c), and 100 kHz for (2d) with a gap width of 10, 20, and 100 μm, respectively, for tunicin, and 20 μm for ramie, at a concentration of 0.05% (w/v). The arrow indicates the direction of the electric field in the gap (G) between the electrodes (E). The topography (area of 340 μm^2) and the profile analysis were made from image (2b and 2c).

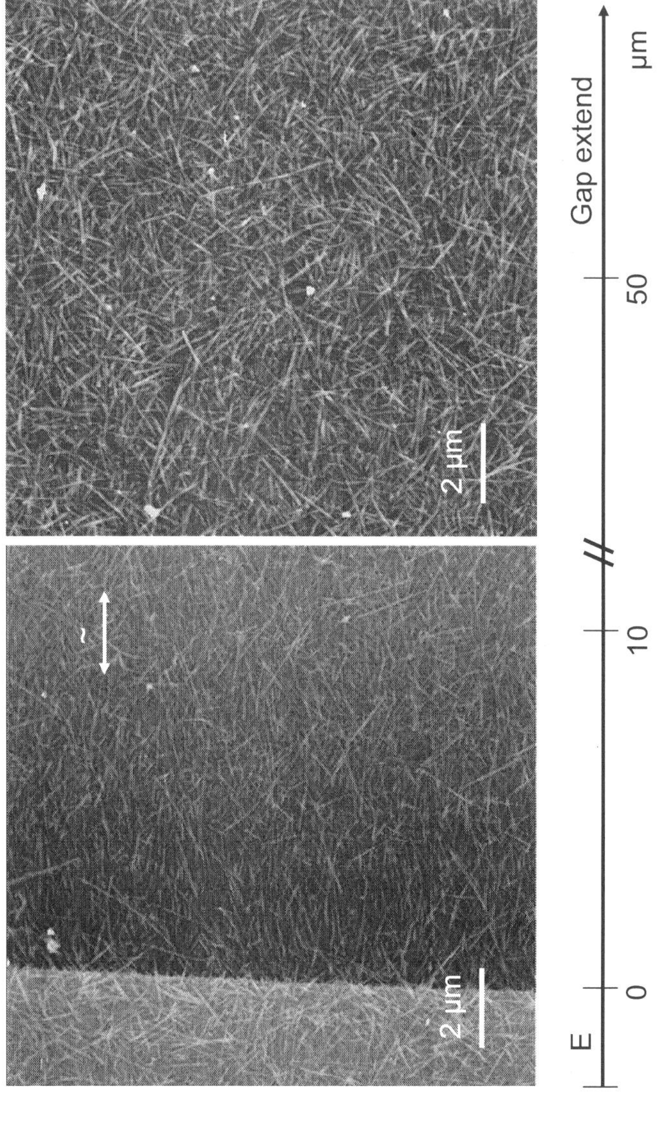

Figure 3. Example of incomplete alignment of tunicin cellulose nanocrystals in a large gap at 2 kV/cm and 1 MHz. Experimental details as in Figure 2.

Whisker Orientation by the Langmuir–Blodgett Technique

Cellulose Whisker Films at the Air–Liquid Interface

A previous and preliminary study showed that in the presence of DODA cellulose whiskers form stable layers at the air–liquid interface (*40*). The compression isotherm of DODA on water exhibits three regions: a liquid expanded (LE) phase followed by a liquid condensed (LC) one, and a collapse at 50 mN/m (Figure 4a), as described elsewhere (*41*). When cellulose whiskers from tunicin or ramie were present in the subphase before DODA deposition, at a nanocrystals/DODA weight ratio of 250, the surface pressure isotherm was significantly modified (Figure 4a). At an area below 80 $Å^2$/molecule and a surface pressure above 25 mN/m, the isotherm of the DODA–whisker mixture diverged from that of pure DODA. The most significant difference between pure DODA and the DODA–whisker mixed layer was the evolution of the surface pressure in the region of the LC phase. The collapse of the layer at the end of the LC region was not clearly apparent by surface pressure measurements in the case of the mixture. These differences clearly point to the formation of a mixed DODA–whisker layer.

In order to get information on the structure of the layers, the formation of the film was followed by ellipsometry (Figure 4b). In the case of pure DODA, just after deposition at a surface concentration of about 0.8 mg/m^2, the ellipticity was -1×10^{-3}. After compression to a molecular area of 40 $Å^2$/molecule, the ellipticity of -6×10^{-3} corresponded to a DODA surface concentration of ~ 2.7 mg/m^2. The refractive index and the thickness of the layer at a surface pressure of 50 mN/m were estimated to 1.473 and 19 ± 1 Å, respectively (Table I). In the case of a mixed layer, the absolute ellipticity values were much higher than those for pure DODA and increased from 1×10^{-2} to 3×10^{-2} during compression of the film (Figure 4b).

The three regions, observed on the surface pressure isotherm, could be tentatively distinguished in the ellipticity curve in the same range of molecular area. Moreover, in the LC phase, between 40 and 80 $Å^2$/molecule, the ellipticity signal fluctuated strongly most likely due to the formation of two-dimensional domains in the mixed layer. Using BAM, the surface morphology was found to change drastically when DODA was spread on the aqueous solution containing cellulose nanocrystals (Figure 4c). The mobility of the dark dot domains in the mixed layer was significantly reduced when the surface pressure was larger than 30 mN/m and consequently when the molecular area was lower than 80 $Å^2$/molecule. In the LC phase region, at a surface pressure of 50 mN/m, the refractive index of the mixed layer was estimated by ellipsometry to 1.512 ± 0.003 for tunicin and 1.506 ± 0.004 for ramie, assuming a thickness layer of 110 Å and 90 Å, respectively (Table I). These values for the refractive index are consistent with a layer structure including a DODA layer and a single layer of nanocrystals (*19*). The film remained very thin and transparent during compression. Interference peaks due to the birefringent nature of the crystals were not detected by ellipsometry at the air–liquid interface with a beam size of 1 mm^2.

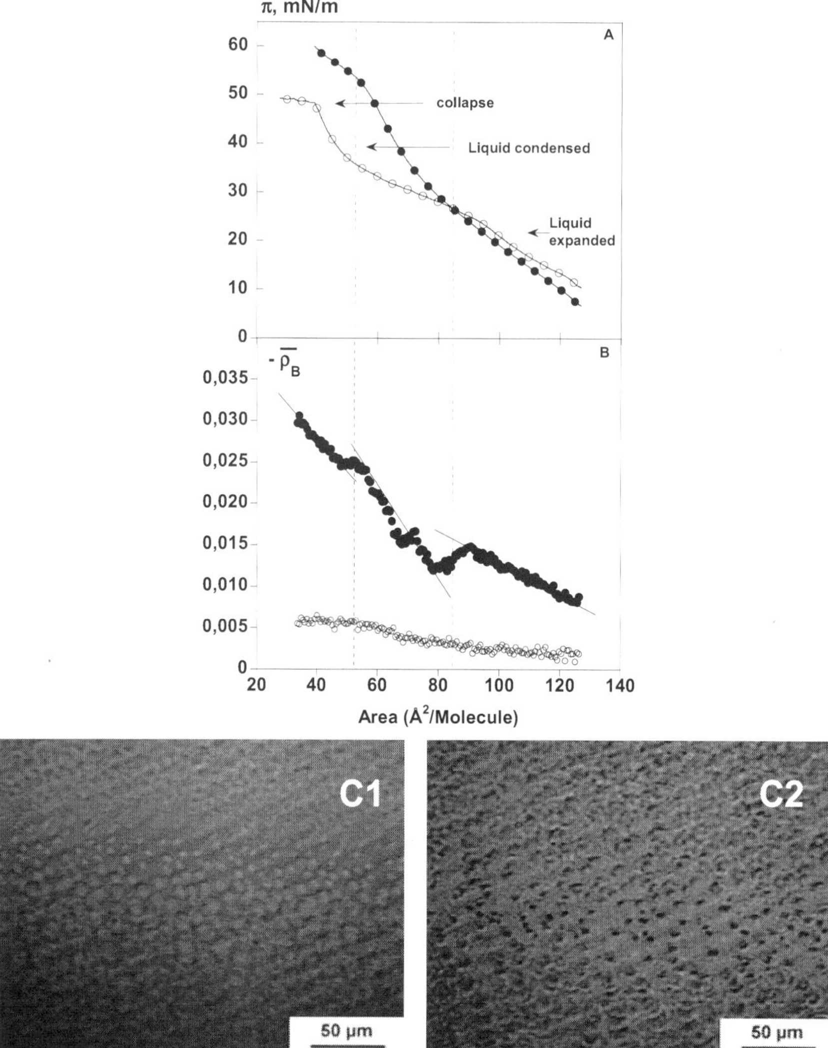

Figure 4. Isotherms of surface pressure, π (A), and ellipticity, ρ̄_B (B), during the compression of pure DODA on water (○) and on tunicin cellulose nanocrystal suspension (●). The weight ratio of cellulose whiskers to DODA is 250. Brewster angle microscopy images of the surfaces at the end of each isotherm: pure DODA (C1) and mixed (C2) layers.

Table I : Refractive indexes (at 589 nm) deduced from ellipsometry spectra of DODA–cellulose nanocrystal mixed layers before and after transfer onto silicon wafers[a]

Sample	Air-liquid interface			Air-silicon interface		
	n_1	n_2	h	n_1	n_2	h
DODA	1.473	1.333	19.5±1	1.469	3.967	12 ± 1
Tunicin	1.512 ±0.003	1.341 ±0.002	110[b]	1.502 ± 0.013 (1.500 ± 0.002)	3.967	108 ± 2 (95 ± 2)
Ramie	1.506 ±0.004	1.341 ±0.002	90[b]	1.513 ± 0.005 (1.513 ± 0.005)	3.967	65 ± 1 (59 ± 2)

[a] The weight ratio of nanocrystals to DODA was 250. Data in brackets were obtained after chloroform and NaOH washing. Data represent mean values of five transfer repetitions. n_1 and h are the refractive index and the thickness of the layer, respectively. n_2 is the refractive index of the substrate.

[b] Refractive indexes were deduced from ellipsometry spectra at 50 mN/m of DODA and DODA–cellulose nanocrystals. The fixed layer thicknesses in the model were chosen to be equal to those of a pure DODA and cellulose whisker layer (90 and 70 Å for tunicin and ramie whiskers, respectively).

Surface dilational properties of the mixed layer were measured by dynamic bubble tensiometry during adsorption of the nanocrystals on the DODA layer at the air–liquid interface (Figure 5) in the surface pressure range 7 to 50 mN/m. The variations of the dilational modulus, ε, seemed to occur in two regimes in the surface pressure ranges of the LE and LC phases apparent in the Langmuir trough experiments. Upon compression of the film, ε increased from 40 mN/m to 60 mN/m during the LE phase (7 mN/m $< \pi <$ 25 mN/m) and then from 60 to 150 mN/m during the LC phase (25 mN/m $< \pi <$ 37 mN/m). In these two regimes, the imaginary part, ε'', corresponding to the viscosity of the layer, remained close to zero, indicating that the mixed layer was mostly elastic. Beyond 37 mN/m, the bubble of the tensiometer became rigid and deformed, and was no longer suitable for surface tension and surface modulus calculations. The data fluctuation above 37 mN/m corresponds to a gel-like collapsed layer, to which the tensiometric technique is not applicable. These dilational rheology data obtained at experimental conditions different from those of the Langmuir trough are consistent with the occurrence of several phases in the layer and point to the fact that at large enough surface pressure, the layer is probably gel-like instead of liquid.

In conclusion, it can be stated that cellulose nanocrystals can adsorb from the bulk beneath the DODA layer formed at the air–liquid interface. The elastic fluid layer obtained at moderate surface pressure becomes more compact and gel-like at 50 mN/m on compression.

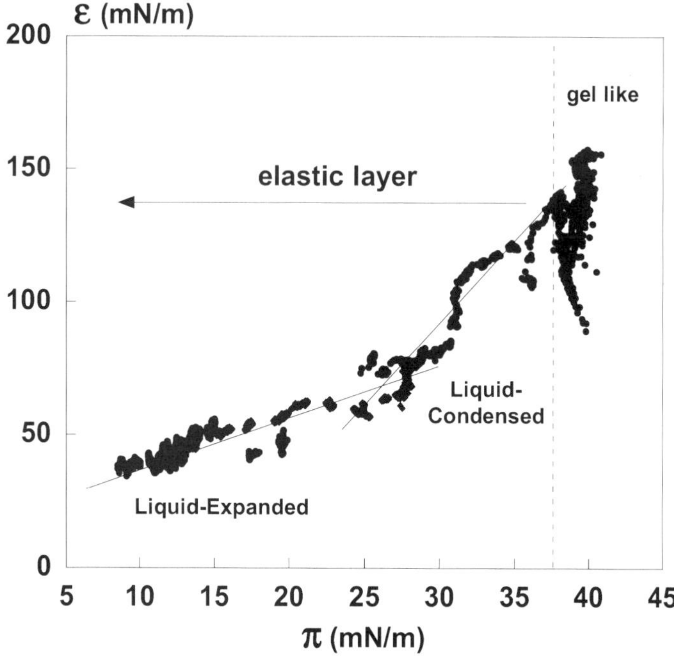

Figure 5. Relationship between the dilational modulus, ε, and the surface pressure, π of the mixed DODA–whisker layer. The dilational modulus and the surface pressure were recorded simultaneously during the adsorption of the nanocrystals on the DODA layer at the air–liquid interface. Several concentrations of DODA and nanocrystals were used to measure ε across the entire surface pressure range.

Cellulose Whisker Langmuir–Blodgett Films on Silicon Substrates

DODA–cellulose films, formed at the air–liquid interface, were transferred onto silicon wafers by a vertical transfer method. The conditions maintained were a weight ratio of 250 both for ramie and tunicin, a surface pressure of 50 mN/m, and a dipping speed of 2 mm/min. After transfer, the films were washed successively with chloroform and a 1% NaOH solution. The thickness of the transferred films was evaluated by spectroscopic ellipsometry using a one layer model for the transferred film (Table I). By fitting the theoretical laws for ellipsometry spectra to the experimental data using the Lorentz oscillator dispersion model, it was found that the thickness of the pure DODA layer on the silicon wafer surface, which was covered by a 20 Å SiO_2 layer, was 12 Å. The thickness of the DODA–cellulose nanocrystal layer was found to be 65 ± 1 Å for ramie and 108 ± 2 Å for tunicin, and was comparable to their average section size, respectively. After washing with chloroform, the thickness was unchanged. After washing with chloroform and NaOH, the thickness of the nanocrystal layer was 59 ± 2 Å for ramie and 95 ± 2 Å for tunicin, respectively. Thus, the DODA molecules covered the nanocrystals with a layer of 9 ± 4 Å thickness and could only be removed from the nanocrystal layer by the combined treatments: alkali and chloroform. This thickness seems in reasonably good agreement with the 12 ± 1 Å measured on silicon wafer.

Morphology of Langmuir–Blodgett Whisker Films

AFM images of the films transferred onto freshly cleaned silicon wafers (total surface coverage: 1 cm^2) were obtained for DODA alone as well as for the DODA–cellulose nanocrystal layers of tunicin and ramie. The pure DODA layer (Figure 6a) had a smooth wavy morphology. This wavy pattern corresponds to the BAM image showing brighter domains (Figure 4c-1). It may be characteristic for DODA molecules that are not in an organized state even at high surface pressure (*42*). The wavy morphology of the pure DODA layer was very different from that of the DODA–cellulose nanocrystal layers (Figures 6b and c). Moreover, after washing with chloroform and NaOH, no difference was apparent between the unwashed and the washed nanocrystal layers (compare Figure 6 with Figures 7 and 8) although a significant difference of ellipsometry thickness was observed (Table I). Height scans of the AFM images gave mean values of the thickness which were close to the values obtained by ellipsometry. These scans also showed that some holes occurred and that these holes were more numerous in the case of tunicin and when the transfer surface pressure was smaller than 50 mN/m or at the beginning of the transfer (Figure 7a). The depth of these holes was practically equal to the layer thickness determined by ellipsometry (10 nm for tunicin and 7 nm for ramie) (Figures 7 and 8). Further away from the wafer surface, several peaks were noticed in the height scans with a maximal height very close to double the layer thickness, indicating that in these regions two whiskers were stacked. Higher peaks were not observed. On average, the surface roughness calculated from AFM images was estimated at 5.8 and 2.9 nm for tunicin and ramie, respectively. These results indicate that the cellulose nanocrystal films were relatively smooth but not completely flat. It can be concluded from these data, combined with the thickness evaluation by ellipsometry, that the films, after washing, consisted primarily of a monolayer of nanocrystals.

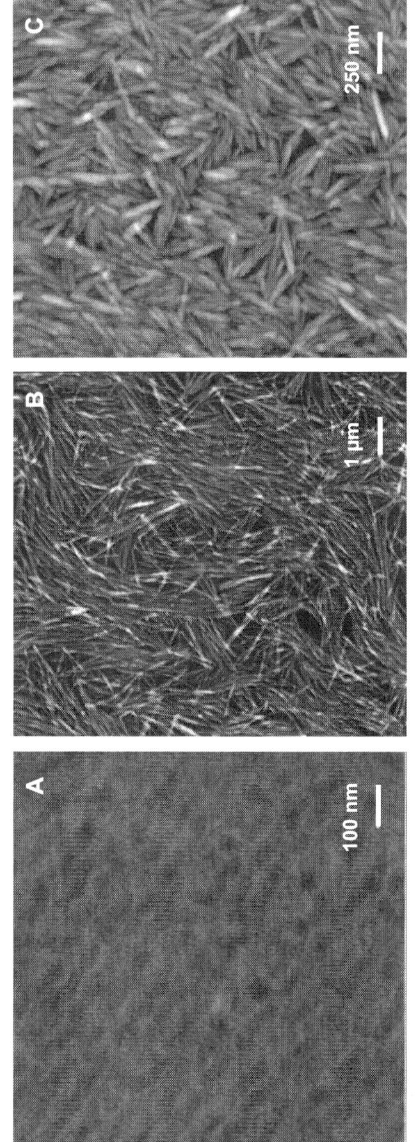

Figure 6. AFM images of DODA (A), tunicin whisker (B), and ramie cellulose nanocrystal (C) films, transferred onto silicon substrates at a surface pressure of 50 mN/m and 2 mm/min, before chloroform and dilute NaOH washing.

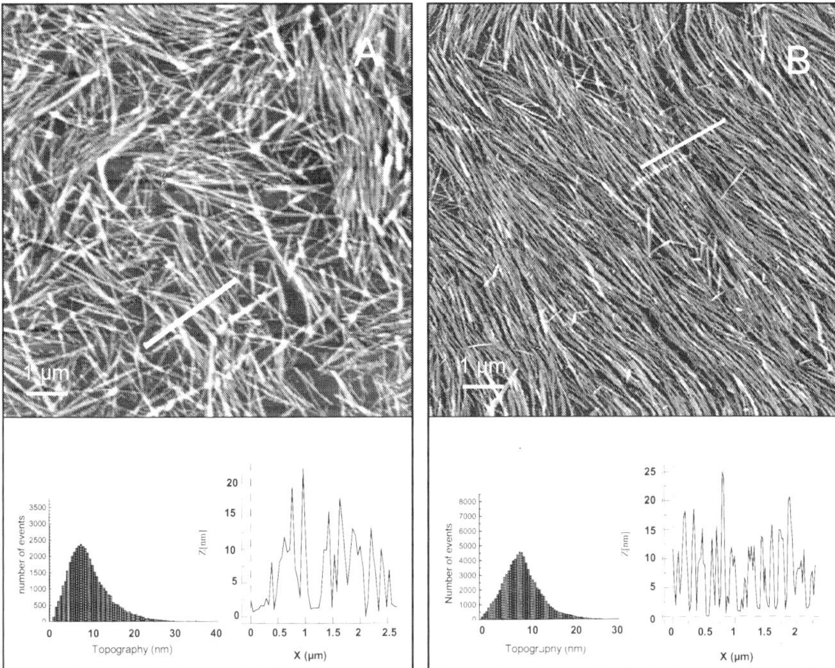

Figure 7. AFM images of the upper (A) and lower (B) end of a tunicin whisker film on a silicon substrate after washing.

Whisker Orientation in Langmuir–Blodgett Films

AFM analysis of the transferred films showed that the cellulose nanocrystals were more or less oriented. In the case of tunicin, bundles of parallel whiskers were observed at the upper end of the transferred film. Their orientation was random with respect to the transfer direction. At the lower end of the transferred film, a higher degree of orientation was noticed with parallel whiskers covering areas larger than 1 mm^2 (Figure 7). The ramie layers exhibited uniform behavior over the whole wafer with domains of mostly parallel whiskers spanning areas of about 1 μm^2 (Figure 8). With ramie and tunicin, such domains were observed in a quite reproducible way. However, a depolarization factor due to the birefringent nature of the crystals in the layer was not detected by ellipsometry on silicon wafers with a beam size of 1 mm^2. Nevertheless, this pattern of oriented nanocrystals may result from 1) micro-areas covered by parallel whiskers at the air–liquid interface before transfer to the wafer or 2) from an orientation mechanism linked to the transfer process. The first case should reflect a quasi-crystalline phase behavior of the whiskers at the fluid interface occurring in the LC state of the layer. In that case, the crystalline domains should have a random orientation and a random distribution throughout the interface plane. The pattern of these domains should be tunable by thermodynamic parameters. In the second case, the orientation should be tuned by the transfer parameters. Since some ordered micro-domains were observed randomly, the phase explanation cannot be ruled out completely. However, in the case of tunicin, the frequent occurrence of ordered domains at the lower end of the transferred film suggests that flow of the mixed layer or the sub phase may influence whisker orientation during the transfer process. Moreover, the axis of the whiskers in the domains was frequently observed to be 45° with respect to the dipping direction, a fact which may point to an effect of the transfer process on the orientation. Whatever the case, the same kind of organization has been observed in Langmuir–Blodgett films of rod-like particles or polymers such as DNA (*43, 44*), polyglutamate (*45–47*), polysiloxane (*48, 49*), alkylated cellulose (*50*), and discotic crystals (*51*). In some instances, the organization of the films occurred during the compression process on the subphase (*52*). In other cases, it was attributed to the deposition process of monolayers (*53–56*). As discussed in a previous report, cellulose nanocrystals can display spontaneously ordered liquid crystalline phases (nematic and chiral nematic) in aqueous suspension when the concentration reaches a critical value (*57*). This critical concentration might have been reached in the mixed film at the upper boundary of the LC phase, allowing the formation of ordered two-dimensional domains.

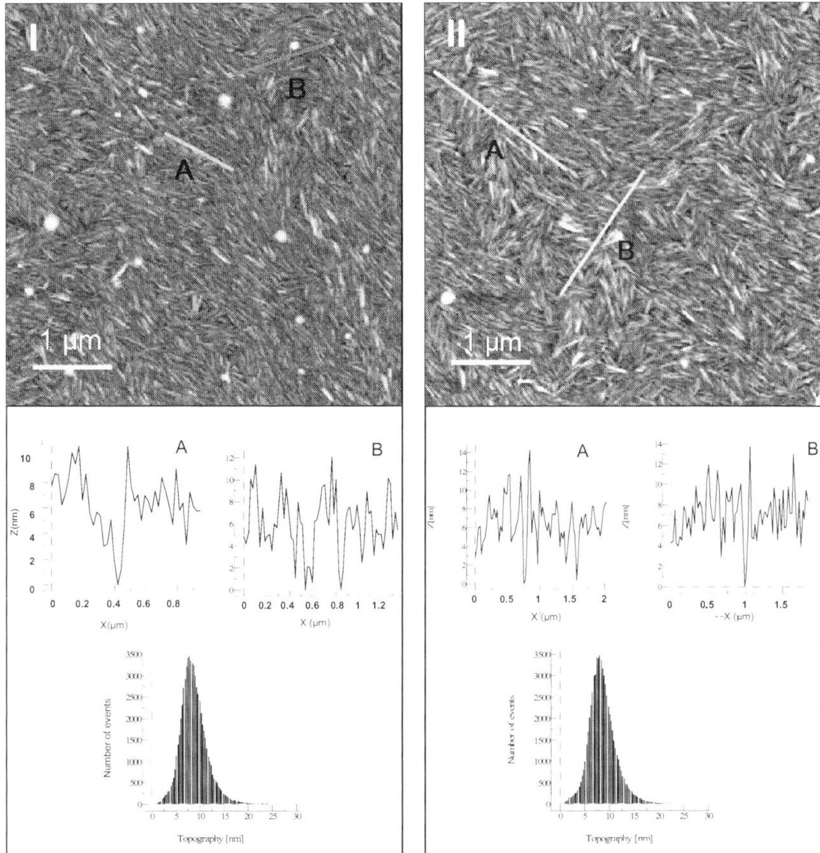

Figure 8. AFM images of the upper (I) and lower (II) end of a ramie cellulose nanocrystal film on a silicon substrate after washing.

Conclusions

From these two experimental ways to prepare model cellulose surfaces, it can be concluded that the alignment of contiguous nanocrystals in homogeneous layers is possible in an electric field. This procedure seems to be an efficient, quick, and inexpensive method compared to the Langmuir–Blodgett technique. Nevertheless, the quality of the alignment decreases when the width of the gap is too large (> 50 µm), and when the field and the frequency are too low (< 2 kV/cm and < 1 kHz, respectively). With the Langmuir–Blodgett method, it is possible to control the compactness of the monolayer before transfer onto silicon wafers and consequently the homogeneity of the monolayer films. However, an overall orientation of whiskers in the layer is not achieved and the ordered micro-domains are oriented randomly. Nevertheless, the partly ordered

134

Langmuir–Blodgett films or the limited ordered areas in the electrode gap are positive experimental starting points for the preparation of structured cellulosic surfaces including other plant cell wall polymers such as hemicelluloses and lignins.

References

1. Vian, B.; Roland, J.-C.; Reis, D. *Int. J. Plant Sci.* **1993,** *154,* 1–9.
2. Vian, B.; Reis, D.; Darzens, D.; Roland, J. C. *Protoplasma* **1994,** *180,* 70–81.
3. Roland, J.-C.; Reis, D.; Vian, B. *Tissue and Cell* **1992,** *24,* 335–345.
4. Carpita, N.; McCann, M. In *Biochemistry & Molecular Biology of Plants;* Buchanan, B. B., Gruissem, W., Jones, R. L., Eds.; The American Society of Plant Physiologists (US), Wiley: West Sussex, United Kingdom, 2000; pp 1367.
5. Holmberg, M.; Berg, J.; Stemme, S.; Ödberg, L.; Rasmusson, J.; Claesson, P. *J. Colloid Interface Sci.* **1997,** *186,* 369–381.
6. Kontturi, E.; Thüne, P. C.; Niemantsverdriet, J. W. *Langmuir* **2003,** *19,* 5735–5741.
7. Rehfeldt, F.; Tanaka, M. *Langmuir* **2003,** *19,* 1467–1473.
8. Gunnars, S.; Wagberg, L.; Cohen Stuart, M. A. *Cellulose* **2002,** *9,* 239–249.
9. Faelt, S.; Wagberg, L.; Vesterlind, E. L.; Larsson, P. T. *Cellulose* **2004,** *11,* 151–162.
10. Sczech, R.; Riegler, H. *J. Colloid Interface Sci.* **2006,** *301,* 376–385.
11. Buchholz, V.; Wegner, G.; Stemme, S.; Oedberg, L. *Adv. Mater.* **1996,** *8,* 399–402.
12. Penfold, J.; Richardson, R. M.; Zarbakhsh, A.; Webster, J. R. P.; Bucknall, D. G.; Rennie, A. R.; Jones R. A. L.; Cosgrove, T.; Thomas, R. K.; Higgins, J. S.; Fletcher, P. D. I.; Dickinson, E.; Roser, S. J.; McLure, I. A.; Hillman, A. R.; Richards, R. W.; Staples, E. J.; Burgess, A. N.; Simister, E. A.; White, J. W. *J. Chem. Soc. Faraday T.* **1997,** *93,* 3899–3917.
13. Tammelin, T.; Saarinen, T.; Oesterberg, M.; Laine, J. *Cellulose* **2006,** *13,* 519–535.
14. Poptoshev, E.; Claesson, P. M. *Langmuir* **2002,** *18,* 1184–1189.
15. Leporatti, S.; Sczech, R.; Riegler, H.; Bruzzano, S.; Storsberg, J.; Loth, F.; Jaeger, W.; Laschewsky, A.; Eichhorn, S.; Donath, E. *J. Colloid Interface Sci.* **2005,** *281,* 101–111.
16. Notley, S. M.; Norgren, M. *Langmuir* **2006,** *22,* 11199–11204.
17. Linder, A.; Bergman, R.; Bodin, A.; Gatenholm, P. *Langmuir* **2003,** *19,* 5072–5077.
18. *Paananen, A.; Oesterberg, M.; Rutland, M.; Tammelin, T.; Saarinen, T.; Tappura, K.; Stenius, P. In Hemicelluloses: Science and Technology;* Gatenholm P., Tenkanen M., Eds.; ACS Symposium Series 864; American Chemical Society: Washington, DC, 2004; pp 269–290.
19. Notley, S. M.; Eriksson, M.; Wagberg, L.; Beck, S.; Gray, D. G. *Langmuir* **2006,** *22,* 3154–3160.

20. Eriksson, J.; Malmsten, M.; Tiberg, F.; Callisen, T. H.; Damhus, T.; Johansen, K. S. *J. Colloid Interface Sci.* **2005,** *285,* 94–99.
21. Eriksson, J.; Malmsten, M.; Tiberg, F.; Callisen, T. H.; Damhus, T.; Johansen, K. S. *J. Colloid Interface Sci.* **2005,** *284,* 99–106.
22. Edgar, C. D.; Gray, D. G. *Cellulose* **2003,** *10,* 299–306.
23. Kontturi, E.; Johansson, L.-S.; Kontturi, K. S.; Ahonen, P.; Thuene, P. C.; Laine, J. *Langmuir* **2007,** *23,* 9674–9680.
24. Cranston, E. D.; Gray, D. G. *Biomacromolecules* **2006,** *7,* 2522–2530.
25. Orts, W. J.; Godbout, L.; Marchessault, R. H.; Revol, J. F. *Macromolecules* **1998,** *31,* 5717–5725.
26. Kimura, F.; Kimura, T.; Tamura, M.; Hirai, A.; Ikuno, M.; Horii, F. *Langmuir* **2005,** *21,* 2034–2037.
27. Dong, X. M.; Gray, D. G. *Langmuir* **1997,** *13,* 3029–3034.
28. Sugiyama, J.; Chanzy, H.; Maret, G. *Macromolecules* **1992,** *25,* 4232–4234.
29. Yoshiharu, N.; Shigenori, K.; Masahisa, W.; Takeshi, O. *Macromolecules* **1997,** *30,* 6395–6397.
30. Ebeling, T.; Paillet, M.; Borsali, R.; Diat, O.; Dufresne, A.; Cavaille, J. Y.; Chanzy, H. *Langmuir* **1999,** *15,* 6123–6126.
31. Bordel, D.; Putaux, J.-L.; Heux, L. *Langmuir* **2006,** *22,* 4899–4901.
32. Wise, L. E.; Murphy, M.; D'Addieco, A. A. *Paper Trade Journal* **1946,** *122,* 35–43.
33. Azzam, R. M. A.; Bashara, N. M. *Ellipsometry and Polarized Light;* Elsevier: Amsterdam, 1987; pp 1–539.
34. Drude, P. *Ann. Phys. Chem.* **1891,** *43,* 126–157.
35. Tompkins, H. G.; McGahan, W. A. *Spectroscopic ellipsometry and reflectometry;* Wiley: New York, 1999; pp 1–228.
36. Labourdenne, S.; Gaudry-Rolland, N.; Letellier, S.; Lin, M.; Cagna, A.; Esposito, G.; Verger, R.; Rivière, C. *Chem. Phys. Lipids* **1994,** *71,* 163–173.
37. Lucassen, J. In *Anionic surfactants: physical chemistry of surfactant action*; Lucassen-Reynders E. H., Ed.; Surfactant Science Series 11; Marcel Dekker: New York, 1981; pp 217–265.
38. Hambardzumyan, A.; Aguié-Béghin, V.; Panaïotov, I.; Douillard, R. *Langmuir* **2003,** *19,* 72–78.
39. Klemm, D.; Philip, B.; Heinze, T.; Heinze, U.; Wagenknecht, W. In *Comprehensive Cellulose Chemistry, Volume 1: General Principles & Analytical Methods*; Wiley-VCH: Weinheim, Germany, 1998; p. 300.
40. Habibi, Y.; Foulon, L.; Aguié-Béghin, V.; Molinari, M.; Douillard, R. *J. Colloid Interface Sci.* **2007,** *316,* 388–397.
41. Cuvillier, N.; Bernon, R.; Doux, J. C.; Merzeau, P.; Mingotaud, C.; Delhaès, P. *Langmuir* **1998,** *14,* 5573–5580.
42. Goubard, F.; Fichet, O.; Teyssie, D.; Fontaine, P.; Goldmann, M. *J. Colloid Interface Sci.* **2007,** *306,* 82–88.
43. Tanaka, K.; Okahata, Y. *J. Am. Chem. Soc.* **1996,** *118,* 10679–10683.
44. Okahata, Y.; Kobayashi, T.; Tanaka, K. *Langmuir* **1996,** *12,* 1326–1330.
45. Duda, G.; Schouten, A. J.; Arndt, T.; Lieser, G.; Schmidt, G. F.; Bubeck, C.; Wegner, G. *Thin Solid Films* **1988,** *159,* 221–230.
46. Hickel, W.; Duda, G.; Jurich, M.; Kroehl, T.; Rochford, K.; Stegeman, G. I.; Swalen, J. D.; Wegner, G.; Knoll, W. *Langmuir* **1990,** *6,* 1403–1407.

47. Auduc, N.; Ringenbach, A.; Stevenson, I.; Jugnet, Y.; Tran Minh, D. *Langmuir* **1993,** *9,* 3567–3573.
48. Erbach, R.; Hoffmann, B.; Schaub, M.; Wegner, G. *Sens. Actuators, B* **1992,** *B6,* 211–216.
49. Yase, K.; Schwiegk, S.; Lieser, G.; Wegner, G. *Thin Solid Films* **1992,** *210–211,* 22–25.
50. Gaines, G. L., Jr. *Langmuir* **1991,** *7,* 834–839.
51. Karthaus, O.; Ringsdorf, H.; Tsukruk, V. V.; Wendorff, J. H. *Langmuir* **1992,** *8,* 2279–2283.
52. Jones, R.; Tredgold, R. H. *J. Phys. D: Appl. Phys.* **1988,** *21,* 449–453.
53. Schwiegk, S.; Vahlenkamp, T.; Xu, Y.; Wegner, G. *Macromolecules* **1992,** *25,* 2513–2525.
54. Minari, N.; Ikegami, K.; Kuroda, S.; Saito, K.; Saito, M.; Sugi, M. *J. Phys. Soc. Jpn.* **1989,** *58,* 222–231.
55. Maruyama, T.; Friedenberg, M.; Fuller, G. G.; Frank, C. W.; Robertson, C. R.; Ferencz, A.; Wegner, G. *Thin Solid Films* **1996,** *273,* 76–83.
56. Sugi, M.; Minari, N.; Ikegami, K.; Kuroda, S.; Saito, K.; Saito, M. *Thin Solid Films* **1989,** *178,* 157–164.
57. De Souza Lima, M. M.; Borsali, R. *Macromol. Rapid Commun.* **2004,** *25,* 771–787.

Chapter 6

Optical Characterization of Cellulose Films via Multiple Incident Media Ellipsometry

Ufuk Karabiyik[1,3], Min Mao[1,3], Maren Roman[2,3], Thomas Jaworek[4], Gerhard Wegner[4], and Alan R. Esker[1,3,*]

[1]Department of Chemistry, [2]Department of Wood Science and Forest Products, and the [3]Macromolecules and Interfaces Institute, Virginia Polytechnic Institute and State University, Blacksburg VA 24061
[4]Max-Planck-Institute for Polymer Research, Ackermannweg 10, 55128 Mainz, Germany

Ellipsometry measures the relative intensity and the phase shift between the parallel and perpendicular components of polarized light reflecting from a surface. Single wavelength ellipsometry measurements at Brewster's angle provide a powerful technique for characterizing ultrathin polymeric films. In this study multiple incident media ellipsometry is utilized to simultaneously obtain the refractive indices and thicknesses of thin films of trimethylsilylcellulose (TMSC), regenerated cellulose, and cellulose nanocrystals. Experiments were conducted in air and water for TMSC, and in air and hexane for regenerated cellulose and cellulose nanocrystals. The refractive indices of TMSC, regenerated cellulose, and cellulose nanocrystals are found to be 1.46 ± 0.01, 1.51 ± 0.01, and 1.51 ± 0.01, respectively.

Introduction

Model cellulose surfaces are important for elucidating how cellulose, hemicellulose, and lignin self-assemble to form the hierarchical structure of cell walls (*1–4*). Likewise, model surfaces provide model substrates for studying the enzymatic degradation of lignocellulosic materials (*5, 6*). On the other hand, the swelling behavior of cellulose surfaces in aqueous media attracts great attention

138

from both fundamental and applied sciences in terms of performance and potential applications of cellulose based materials in the paper and textile industries (*7, 8, 9*). It is clear that in such applications where the cellulose is in contact with a liquid medium, techniques that are applicable for in situ characterization are desirable. In addition, prior to further surface treatment and subsequent surface analysis, it is important to characterize and explore initial surface characteristics, e.g. film thickness and refractive index. Many techniques have been developed that can be used to measure the thicknesses and refractive indices of thin films, such as refractometry (*10*), waveguide prism couplers (*11*), surface plasmon resonance spectroscopy (*12, 13*), polarizing interference microscopy (*14*), variable-angle single wavelength ellipsometry (*15, 16*), and spectroscopic ellipsometry (*17*). Of these, ellipsometry as a rapid, non-contact, and non-destructive method is ideal for measuring thickness and refractive index in nanoscale coatings through changes in polarization upon the reflection of light from the surface. In addition, the simultaneous determination of film thickness and refractive index is possible trough multiple incident media (MIM) ellipsometry (*18, 19*).

Thickness determinations via ellipsometry are complicated by the need to know the film's optical properties. Refractive index and thickness are correlated parameters in ellipsometry, hence, it is not possible to uniquely obtain both parameters through a single measurement at a constant wavelength for thin films (*20*). Spectroscopic ellipsometers overcome this problem by conducting measurements at multiple wavelengths. However, the refractive index of the film needs to be optically modeled as a function of wavelength. As a consequence, some prior knowledge of the refractive index of the film at some point in the sampled wavelength window is usually desired. Another complication is that the bulk refractive indices may not be applicable for thin films with thicknesses <5 nm (*21*). In order to avoid these problems for single wavelength instruments MIM ellipsometry can be utilized. This technique has previously been applied to silicon surfaces with an oxide layer (*22*), self-assembled monolayers on silicon substrates (*19, 23, 24*), and water adsorbed on chromium slides (*25*). MIM ellipsometry requires two ambient media whose refractive indices are different from each other. Additionally, the ambient media should be chemically and physically inert to the surface. Moreover, the liquid sample cell should be compatible with a variable angle ellipsometry setup. The most common cell design reported in the literature has a trapezoidal shape that allows the incident and reflected light to enter and leave the sample cell at normal incidence thereby avoiding changes in the polarization state of the light. Another cell design is a hollow prism (*25*). In this study, a cylindrical quartz sample cell, schematically shown in Figure 1, has been used with a phase modulated ellipsometer to conduct MIM ellipsometry measurements on cellulose based films. This cylindrical cell design does not require a fixed incident angle, therefore, it is easy to scan Brewster's angle.

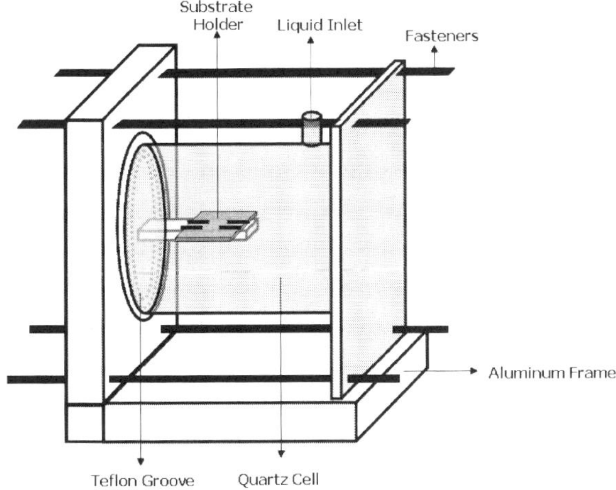

Figure 1. Schematic representation of the multiple incident media (MIM) ellipsometry sample cell.

The focus of this chapter is the use of MIM ellipsometry to probe the optical properties of model cellulose surfaces. Cellulose is a naturally abundant polymer in the cell walls of plants and is widely used in the wood, paper, and textile industries. One of the greatest complications associated with obtaining model cellulose surfaces is that cellulose is insoluble in most common organic solvents. Recently, there have been efforts to directly prepare model cellulose surfaces (*26–29*). However, issues with uniformity and surface roughness for these approaches can complicate detailed optical characterization of the resulting films. Therefore, instead of using cellulose directly, more readily soluble cellulose derivatives have been used for the preparation of some model surfaces in this study. Trimethylsilylcellulose (TMSC) has been deposited onto substrates by the Langmuir–Blodgett (LB) technique (*30*) and spin coating (*31*). The resulting TMSC films are easily converted back into cellulose via vapor phase HCl hydrolysis (*31*). As a basis for comparison, model cellulose surfaces are also obtained by spin coating aqueous cellulose nanocrystal suspensions (*32*). TMSC LB-films are ideal for testing the MIM ellipsometry technique because quantitative LB-transfer by Y-type deposition yields films whose thicknesses linearly increase as a function of the number of layers (*30, 33–38*). Hence, results for TMSC LB-films help to validate subsequent MIM ellipsometry studies on spin-coated TMSC, regenerated cellulose, and cellulose nanocrystal films.

Experimental

Materials

The synthesis and the preparation of TMSC and cellulose nanocrystals are provided elsewhere (*20, 39*). The TMSC used in this study had a degree of substitution of DS = 2.01, and a polystyrene equivalent number average molar mass of M_n = 44,000 g·mol^{-1}. Chloroform (HPLC grade, EMD Chemicals) was used to prepare ~0.5 mg·g^{-1} TMSC solutions for LB-deposition. In order to obtain complete dissolution, the samples were prepared and stored for 24 h at room temperature in sealed vials to avoid the evaporation of chloroform. Spin-coated films of TMSC and cellulose nanocrystals were prepared from different weight percent concentration solutions in toluene (HPLC grade, EMD Chemicals) or dispersions in water (18.2 MΩ·cm, Millipore, MilliQ Gradient A-10, <10 ppb organic impurities), respectively. The water used in all steps of the experiments was ultrapure water. In addition, hexane (HPLC grade, EMD Chemicals) was used as the second ambient medium for regenerated cellulose and cellulose nanocrystal films. The chemicals used for substrate cleaning: H_2O_2 (30% by volume), H_2SO_4 (conc.), and NH_4OH (28% by volume), were purchased from EM Science, VWR International, and Fisher Scientific, respectively. 4" silicon wafers (100) were purchased from Waferworld, Inc.

Film Preparation

Silicon wafers were used for both LB and spin-coated films. Substrates for spin coating were cut into ~15 x 15 mm^2 pieces. LB-films were prepared from 40 x 40 mm^2 substrates. All substrates were cleaned in a 5:1:1 (by volume) boiling mixture of H_2O:H_2O_2:NH_4OH for 1 h. After the substrates were rinsed with Millipore water and dried with nitrogen, the substrates were placed in a 7:3 (by volume) mixture of H_2SO_4:H_2O_2 for 3 h. The substrates were then rinsed with copious amounts of water and dried with nitrogen. At this point, the surface is a hydrophilic silica surface that can be used to create spin-coated films of cellulose nanocrystals. In order to obtain a hydrophobic silicon surface, substrates were dipped into buffered HF solutions (J. T. Baker) for 5 min followed by a short dip into a buffered NH_4F (J. T. Baker) solution. Both LB and spin-coated films of TMSC were prepared on hydrophobic silicon surfaces. LB-films were prepared on a standard LB-trough (KSV 2000, KSV Instruments, Inc.) resting on a floating table inside a Plexiglas box. The temperature of the subphase (Millipore water) was maintained at 22.5 °C by a water circulation bath. Surface pressure, Π, was monitored via the Wilhelmy plate technique. The trough was filled with Millipore water and the TMSC spreading solution was spread to Π = 8 to 12 mN·m^{-1}; i.e. below the transfer and collapse pressures to avoid multilayer formation. After allowing ~20 min for the spreading solvent to evaporate, the TMSC was compressed to a constant transfer Π below the collapse Π. The transfer Π = 25 mN·m^{-1} was approached with a compression rate of 10 mm·min^{-1} and the maximum forward and reverse barrier speeds were 10 mm·min^{-1}. The dipping rates of the substrate for both up- and down-strokes were set to 10 mm·min^{-1}. Transfer proceeded by Y-type deposition to prepare multilayer films of TMSC. Following deposition, LB-films were cut into ~15 x 15 mm^2 pieces for ellipsometry measurements. Spin-coated films of TMSC were prepared from solutions of varying wt % TMSC in toluene and were spun onto hydrophobic silicon wafers at 2000 rpm for 60 s. In contrast, spin-coated films of cellulose nanocrystals were prepared from dispersions of varying wt %

cellulose nanocrystals in water and were spun onto hydrophilic silicon wafers at 2000 rpm for 60 s.

Multiple Incident Media (MIM) Ellipsometry

MIM ellipsometry measurements were carried out with a phase modulated ellipsometer (Beaglehole Instruments, Wellington, New Zealand) at a wavelength of 632.8 nm (HeNe laser) at Brewster's angle. The sample cell is depicted in Figure 1. Measurements of the ellipticity, ρ, in air were performed at several different positions in order to confirm the uniformity and the quality of the films through the quartz cell ($\Delta\rho < 1\%$). Measurements with water and hexane as the ambient media were performed in the same quartz cell. The design of the quartz cell allows us to fill the sample cell with liquid after completing the air measurements without removing the substrate, thereby allowing measurements on the same position of the wafer. The fundamental equation for the reflection coefficient in ellipsometry is ($23, 40$)

$$r = \frac{r_p}{r_s} = \mathrm{Re}(r) + i\,\mathrm{Im}(r) = \tan\Psi\exp(i\Delta) \tag{1}$$

where r_p and r_s are the reflection coefficients for p and s polarized light, respectively, and Ψ and Δ are the ellipsometric parameters. At Brewster's angle, $\mathrm{Re}(r) = 0$, which is equivalent to $\Delta = 90°$. Under these conditions, eq 1 simplifies to an equation for the ellipticity (41)

$$\rho = r = \frac{r_p}{r_s} = i\,\mathrm{Im}(r) = \tan\Psi \tag{2}$$

For the case where the film thickness, D, is much smaller than the wavelength of light, λ, the ellipticity can be expressed as a power series in terms of (D/λ). The first term of this power series provides Drude's equation ($42, 43$):

$$\rho = \frac{\pi}{\lambda}\frac{(\varepsilon_1 + \varepsilon_2)^{1/2}}{(\varepsilon_1 - \varepsilon_2)}\int_0^D \frac{(\varepsilon - \varepsilon_1)(\varepsilon - \varepsilon_2)}{\varepsilon}\,dz \tag{3}$$

where ε is the dielectric constant at position z in the film, ε_1 is the dielectric constant of the ambient medium, and ε_2 is the dielectric constant of the substrate. For the case of homogenous ultrathin films with negligible surface roughnesses, ε is constant and eq 3 becomes

$$\rho = \frac{\pi}{\lambda}\frac{(\varepsilon_1 + \varepsilon_2)^{1/2}}{(\varepsilon_1 - \varepsilon_2)}\frac{(\varepsilon - \varepsilon_1)(\varepsilon - \varepsilon_2)}{\varepsilon}D \tag{4}$$

As Mao et al. (24) noted, measurements with two different ambient media, A and B (denoted as superscripts in eqs 5 and 6), allow one to eliminate D by taking the ratio:

$$\frac{\rho^A}{\rho^B} = \frac{(\varepsilon_1^A + \varepsilon_2)^{1/2}(\varepsilon_1^B - \varepsilon_2)}{(\varepsilon_1^B - \varepsilon_2)^{1/2}(\varepsilon_1^A - \varepsilon_2)}\frac{(\varepsilon - \varepsilon_1^A)}{(\varepsilon - \varepsilon_1^B)} \tag{5}$$

which leads to an analytical expression for ε:

$$\varepsilon = \frac{\rho^A \sqrt{\varepsilon_1^B + \varepsilon_2}\,(\varepsilon_1^A - \varepsilon_2)(\varepsilon_1^B) - \rho^B \sqrt{\varepsilon_1^A + \varepsilon_2}\,(\varepsilon_1^B - \varepsilon_2)(\varepsilon_1^A)}{\rho^A \sqrt{\varepsilon_1^B + \varepsilon_2}\,(\varepsilon_1^A - \varepsilon_2) - \rho^B \sqrt{\varepsilon_1^A + \varepsilon_2}\,(\varepsilon_1^B - \varepsilon_2)} \tag{6}$$

Once ε is known, D can be obtained from eq 4 for either ambient medium. Error estimates on D and the refractive index, $n = \varepsilon^{1/2}$, for MIM ellipsometry measurements are obtained as $\pm$ one standard deviation via a propagation of error calculation starting from eqs 4 and 6.

Multiple Angle of Incidence (MAOI) Ellipsometry Measurements

MAOI ellipsometry measurements were carried out with a phase modulated ellipsometer (Beaglehole Instruments, Wellington, New Zealand) at a wavelength of 632.8 nm (HeNe laser).

Spectroscopic Ellipsometry (SE) Measurements

SE measurements were carried out with a phase modulated ellipsometer (Beaglehole Instruments, Wellington, New Zealand) at a 60° angle of incidence and a wavelength range of 230 nm to 800 nm (halogen and deuterium lamp). X and Y data measured with a phase modulated system can be converted to the traditional parameters Ψ and Δ using eqs 7 through 10:

$$\mathrm{Re}(r) = \frac{1 \pm \sqrt{1 - X^2 - Y^2}}{X^2 + Y^2} X \tag{7}$$

$$\mathrm{Im}(r) = \frac{1 \pm \sqrt{1 - X^2 - Y^2}}{X^2 + Y^2} Y \tag{8}$$

$$\tan \Psi = \sqrt{\mathrm{Re}(r)^2 + \mathrm{Im}(r)^2} \tag{9}$$

$$\tan \Delta = \mathrm{Im}(r) / \mathrm{Re}(r) \tag{10}$$

In most practical cases the values of $\mathrm{Re}(r)$ and $\mathrm{Im}(r)$ are small enough to use the negative root in the conversion equations (eqs 7 and 8). Nonetheless, TFCompanion[TM] software enables direct analysis of X and Y data without conversion to obtain thickness and refractive index values. The most commonly used electronic model to analyze ellipsometry data is the Harmonic Oscillator Approximation (HOA) (*44*). Here, the results utilize the Critical Point Exciton (CPE) material approximation (*45*), an extension to the HOA model commonly used for polymeric systems. The generalized expression for the line shape in the CPE model is:

$$\varepsilon(E) = \mathrm{UV}_{\mathrm{term}} - \sum_{1}^{N} A_j e^{i\phi_j} (E_{c_j} - E + \Gamma_j)^{-1} \tag{11}$$

where, $\varepsilon(E)$ is the dielectric constant, and the $\mathrm{UV}_{\mathrm{term}}$ is a constant that represents the UV peaks, A_j is the amplitude, E_c is the critical point energy, Γ_j is the line broadening, and ϕ_j is the phase of the jth transition. In eq 11, N is the number of critical points and determines the number of parameters. For our studies $N = 1$

was sufficient to obtain realistic optical constant dispersions for the cellulose derivatives.

Results and Discussion

MIM Ellipsometry for TMSC LB-Films

Figure 2 shows ellipticity, ρ, as a function of the number of LB-layers for data obtained in air ($n_2 = 1$, $\varepsilon_2 = n_2^2 = 1$) and water ($n_2 = 1.333$, $\varepsilon_2 = n_2^2 = 1.777$). A linear relationship between ρ and the number of TMSC LB-layers is observed for both data sets. Here it should be noted that LB-deposition allows precise control over the film thickness by varying the number of deposited layers. Furthermore, Figure 2 indicates that film thicknesses are in a regime where eq 4 should be valid for analyzing the data. Analysis of the MIM ellipsometry data in Figure 2 can proceed in two ways: Approach 1 – The refractive index and thickness of each film can be determined according to eqs 6 and 4; and Approach 2 – The slope of each curve in Figure 2 can be used to obtain ρ^{air}/layer and ρ^{water}/layer, thereby allowing one to deduce the refractive index and thickness per layer, d, through eqs 6 and 4, respectively. The total thickness of the film, D, is then calculated from eq 4 using n derived from the slopes and ρ data measured in air or water. In this study ρ data measured in air was used to compute D.

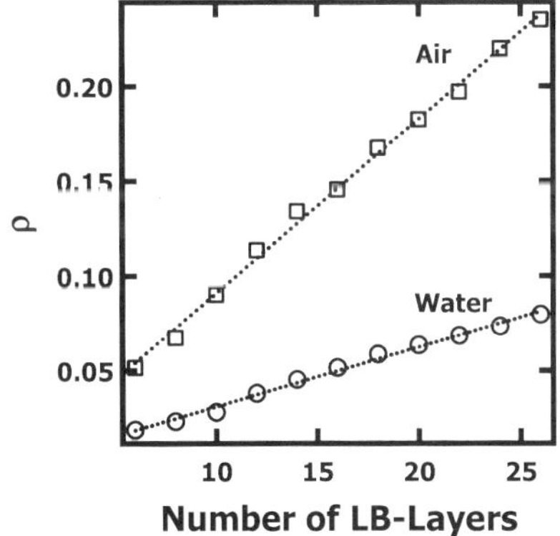

Figure 2. Ellipticity vs. the number of layers in TMSC LB-films measured in air (□) and water (○) at a wavelength of 632 nm.

Approach 1

Table 1 contains thickness and refractive index values for each film obtained from measurements in air and water utilizing eqs 4 and 6 by Approach 1. Table 1 shows that n values are independent of the number of layers transferred, with an average value of $n = 1.46 \pm 0.01$. The refractive index value obtained by Approach 1 is in agreement with a previous study of TMSC ($n = 1.44 \pm 0.01$) by Holmberg et al. (*33*).

Approach 2

The slopes of ρ^{air}/layer = $(9.20 \pm 0.04) \times 10^{-3}$ and ρ^{water}/layer = $(3.1 \pm 0.1) \times 10^{-3}$ in Figure 2 yield $d = 0.95 \pm 0.01$ nm and $n = 1.46 \pm 0.01$ utilizing eqs 4 and 6, respectively. The monolayer thickness obtained via Approach 2 is in agreement with the published values for TMSC LB-films (*35*). Utilizing the n value obtained from Approach 2 and ρ values obtained from measurements in air, it is possible to calculate D for each film. These values are also summarized in Table 1. The conclusion is clear, so long as Drude's equation is valid, the MIM ellipsometry results provide unambiguous values of refractive index and film thickness that agree well with the literature when a nonswelling nonsolvent is used.

Table 1. MIM ellipsometry analysis of TMSC LB-films[a]

#of Layers	D/nm		n[b]
	Approach 1	*Approach 2*[c]	
4	5.2 ± 0.2	5.3 ± 0.2	1.47 ± 0.01
6	6.9 ± 0.1	6.8 ± 0.2	1.46 ± 0.01
8	9.4 ± 0.3	9.1 ± 0.1	1.45 ± 0.02
10	11.7 ± 0.1	11.2 ± 0.4	1.46 ± 0.01
12	13.8 ± 0.8	13.4 ± 0.3	1.46 ± 0.02
14	14.9 ± 0.3	14.8 ± 0.2	1.47 ± 0.01
16	17.2 ± 0.2	17.4 ± 0.1	1.47 ± 0.01
18	18.7 ± 0.2	19.1 ± 0.1	1.46 ± 0.01
20	20.2 ± 0.3	20.6 ± 0.2	1.46 ± 0.02
22	22.7 ± 0.5	23.0 ± 0.3	1.46 ± 0.01
24	24.2 ± 0.4	24.8 ± 0.3	1.46 ± 0.01
		Average	1.46 ± 0.01
		Approach 2	1.46 ± 0.01

[a]One standard deviation error bars

[b]Utilizing Approach 1

[c]ρ data measured in air

MIM Ellipsometry Studies for Spin-coated TMSC Films

After obtaining a value of $n = 1.46 \pm 0.01$ from MIM ellipsometry measurements on LB-films of TMSC via the slope-based Approach 2, the data analysis procedure now could be applied to spin-coated systems of TMSC. Figure 3a has ρ plotted as a function of the spin-coating solution concentration (wt % TMSC in toluene) for measurements in air and water. As expected for a spin-coated film, the plot of ρ vs. the wt % concentration of the TMSC spin-coating solution is essentially linear, but more scattered than the LB-films. This result is not surprising since the preparation of the spin-coated films does not allow one to control film thickness as well as the LB-technique. However, Approach 1 can be used to deduce the refractive index and thickness of each film through eqs 6 and 4, respectively. The results for D and n are summarized in Table 2, and the ellipticity values from Figure 3a are now plotted against D in Figure 3b.

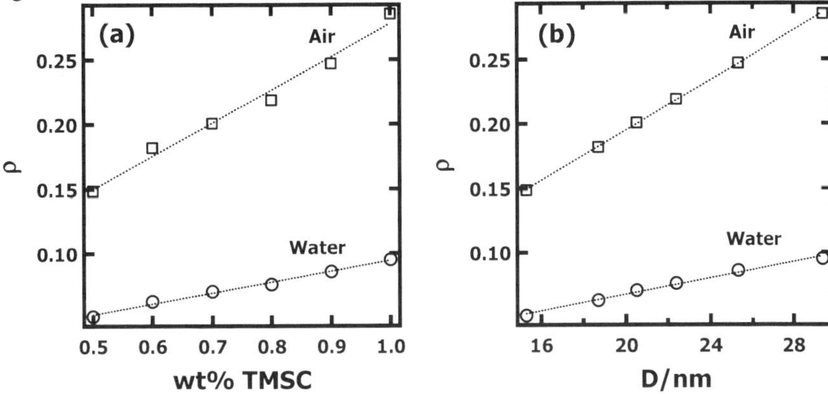

Figure 3. (a) Ellipticity vs. wt % concentration of the TMSC spin-coating solution. (b) Ellipticity vs. film thickness obtained from MIM ellipsometry data utilizing Approach 1 for spin-coated TMSC films in (a). Symbols correspond to measurements in air (□) and water (○) at a wavelength of 632 nm.

Table 2. Thickness and refractive index values for spin-coated TMSC films[a]

Concentration of spin-coating solution/wt %	D/nm[b]	n[b]
0.5	15.3 ± 0.2	1.46 ± 0.01
0.6	18.7 ± 0.1	1.46 ± 0.01
0.7	20.5 ± 0.3	1.47 ± 0.01
0.8	22.4 ± 0.2	1.46 ± 0.01
0.9	25.3 ± 0.2	1.46 ± 0.01
1.0	29.4 ± 0.2	1.46 ± 0.01
	Average	1.46 ± 0.01
	Approach 2	1.45 ± 0.01

[a]One standard deviation error bars

[b]Utilizing Approach 1

Figure 3b shows a nearly linear relationship between ρ and D as expected from eq 3. Next the data in Figure 3b are fit with a linear relationship to obtain $\rho^{air}/nm = (9.64 \pm 0.08) \times 10^{-3} \, nm^{-1}$ and $\rho^{water}/nm = (3.1 \pm 0.3) \times 10^{-3} \, nm^{-1}$. From the values of ρ^{air}/nm and ρ^{water}/nm it is now possible to deduce $n = 1.45 \pm 0.01$ from eq 6 for spin-coated TMSC films via Approach 2. This value is in excellent agreement with the value of $n = 1.46 \pm 0.01$ obtained from TMSC LB-films by Approach 2 and the literature value of $n = 1.44 \pm 0.01$ (*33*). Therefore, we can conclude that MIM ellipsometry can also be applied to spin-coated films.

MIM Ellipsometry Studies for Cellulose Films Regenerated from TMSC Films

Next, the MIM ellipsometry method is applied to regenerated cellulose films derived from the LB and spin-coated TMSC films of the previous sections. Here, hexane ($n_2 = 1.375$, $\varepsilon_2 = n_2^2 = 1.890$) serves as a nonswelling nonsolvent. Analogous plots to Figure 3a and 3b in air and hexane are provided as Figure 4a through 4d for regenerated cellulose films. The refractive index and thickness results are summarized in Tables 3 and 4 for LB and spin-coated films, respectively. Here it should be noted that upon acid catalyzed hydrolysis of TMSC the thickness values decrease by ~56%. The refractive index value for regenerated cellulose is found to be $n = 1.51 \pm 0.01$. In addition, the "monolayer thickness" reduces from 0.95 ± 0.01 nm for TMSC to 0.38 ± 0.01 nm for the regenerated cellulose. These results agree well with previously reported values for monolayer thicknesses for TMSC and regenerated cellulose (*35*). The refractive index of the thin film changes from 1.46 ± 0.01 for TMSC to 1.51 ± 0.01 for the regenerated cellulose films. The measurements reveal no significant difference between the refractive indices of the cellulose obtained from the LB or spin-coated TMSC films. The refractive index value reported for cotton cellulose ranges from 1.53 to 1.58 and is anisotropic (values perpendicular vs. parallel to the cellulose backbone) (*46*). However the lower refractive index value ($n = 1.51 \pm 0.01$) obtained via MIM ellipsometry for regenerated cellulose is consistent with the previously reported value ($n = 1.49 \pm 0.02$) from ellipsometry studies (*33*).

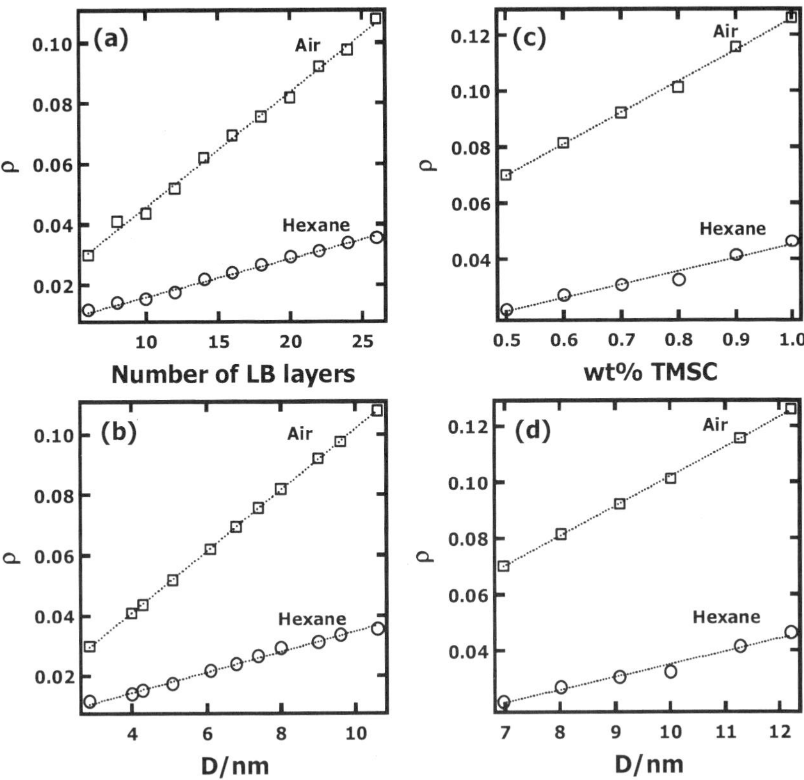

Figure 4. (a) Ellipticity vs. number of LB-layers in the precursor TMSC film and (b) ellipticity vs. film thickness obtained from MIM ellipsometry data utilizing Approach 1 for cellulose films regenerated from TMSC LB-films. (c) Ellipticity vs. wt % concentration of TMSC in the spin-coating solution and (d) ellipticity vs. film thickness for MIM ellipsometry data utilizing Approach 1 for cellulose films regenerated from spin-coated TMSC films. Symbols correspond to measurements in air (□) and hexane (○) at a wavelength of 632 nm.

Table 3. Thickness and refractive index values for cellulose films regenerated from TMSC LB-films[a]

#of Layers	D/nm Approach 1	D/nm Approach 2[c]	n[b]
4	2.9 ± 0.1	3.1 ± 0.2	1.54 ± 0.01
6	3.8 ± 0.3	3.6 ± 0.5	1.52 ± 0.01
8	4.3 ± 0.4	4.1 ± 0.2	1.52 ± 0.01
10	5.1 ± 0.2	5.2 ± 0.3	1.51 ± 0.01
12	6.1 ± 0.2	6.4 ± 0.2	1.52 ± 0.01
14	6.8 ± 0.2	6.2 ± 0.1	1.52 ± 0.01
16	7.4 ± 0.2	7.1 ± 0.4	1.52 ± 0.01
18	8.0 ± 0.1	8.1 ± 0.1	1.52 ± 0.01
20	9.0 ± 0.1	9.2 ± 0.2	1.51 ± 0.01
22	9.6 ± 0.2	9.8 ± 0.2	1.52 ± 0.01
24	10.6 ± 0.2	10.8 ± 0.3	1.51 ± 0.01
		Average	1.52 ± 0.01
		Approach 2	1.51 ± 0.01

[a]One standard deviation error bars

[b]Utilizing Approach 1

[c]ρ data measured in air

Table 4. Thickness and refractive index values for cellulose films regenerated from spin-coated TMSC films[a]

Concentration of spin-coating solution/wt %	D/nm[b]	n[b]
0.5	7.0 ± 0.1	1.50 ± 0.01
0.6	8.0 ± 0.8	1.51 ± 0.01
0.7	9.1 ± 0.3	1.51 ± 0.01
0.8	10.0 ± 0.1	1.50 ± 0.01
0.9	11.3 ± 0.4	1.52 ± 0.01
1.0	12.2 ± 0.3	1.53 ± 0.02
	Average	1.51 ± 0.01
	Approach 2	1.51 ± 0.01

[a]One standard deviation error bars

[b]Utilizing Approach 1

MIM Ellipsometry Studies of Cellulose Nanocrystal Films

The MIM ellipsometry method is also applicable to cellulose nanocrystal thin films by making measurements in air and hexane. Plots analogous to Figure 3a and b in air and hexane are provided as Figure 5a and b for thin films of cellulose nanocrystals. The refractive index and thickness results for cellulose nanocrystals are summarized in Table 5. MIM ellipsometry yields refractive index values for the cellulose nanocrystals that are the same as cellulose films regenerated from TMSC within experimental error. These results suggest that the regenerated cellulose and cellulose nanocrystal films have similar degrees of crystallinity.

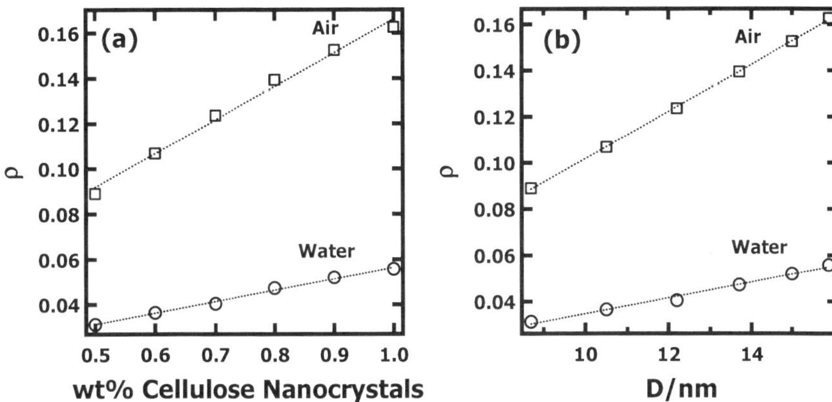

Figure 5. (a) Ellipticity vs. wt % concentration of cellulose nanocrystals in spin-coating dispersions and (b) ellipticity vs. film thickness obtained from MIM ellipsometry data utilizing Approach 1 for spin-coated cellulose nanocrystal films. Symbols correspond to measurements in air (□) and hexane (○) at a wavelength of 632 nm.

Cellulose crystals show birefringence with refractive index values along the direction parallel to the cellulose backbone ($n_{\parallel}$) being higher than those perpendicular to the cellulose backbone ($n_{\perp}$) (47) and similar behavior is expected for cellulose nanocrystals (48). However, due to the nature of MIM ellipsometry and the TFCompanion[TM] software used in this study, estimates of refractive index anisotropy are not possible. Nonetheless, MIM ellipsometry results for cellulose nanocrystals are close to the smaller refractive index reported for $n_{\perp}$ (46–49).

150

Table 5. Thickness and refractive index values for spin-coated films of cellulose nanocrystals[a]

Concentration of spin-coating dispersion/wt %	D/nm^{b}	n^{b}
0.5	8.7 ± 0.1	1.52 ± 0.01
0.6	10.5 ± 0.1	1.51 ± 0.01
0.7	12.2 ± 0.3	1.51 ± 0.01
0.8	13.7 ± 0.1	1.51 ± 0.01
0.9	15.0 ± 0.2	1.51 ± 0.01
1.0	15.9 ± 0.2	1.52 ± 0.01
	Average	1.51 ± 0.01
	Approach 2	1.51 ± 0.01

[a]One standard deviation error bars

[b]Utilizing Approach 1

MIM Ellipsometry versus SE and MAOI Ellipsometry Measurements

Direct attempts to simultaneously determine the thickness and refractive index values for the TMSC thin films used in this study via SE or MAOI ellipsometry measurements yield estimates with extremely large errors (representative data are provided in Table 6). The data in Table 6 represent the use of the TFCompanion^{TM} algorithm without any attempts to constrain D or n. A better approach is to prepare thicker TMSC and regenerated cellulose films and use SE and MAOI ellipsometry to determine the film thickness and refractive index. Once n is known, it serves as a fixed parameter for determining D for the thinner films. Therefore, a thicker spin-coated film of TMSC was prepared and measured, and the same film was used to form a regenerated cellulose film. For SE, it is necessary to model $\varepsilon(\lambda)$. In this study, the CPE material approximation is used (eq 11). CPE parameters for TMSC and cellulose are provided in Table 7. The thicknesses and refractive index values for the thick films are summarized in Table 8. As seen in Table 8, there is an ~62% reduction in the thickness after cellulose is regenerated from TMSC. The refractive index values as a function of wavelength are also provided through Figure 6 and its legend. Using these fixed refractive index values, the thicknesses of the thin films are obtained from SE and MAOI ellipsometry measurements and summarized in Table 9. It is important to note that there are no significant differences for n between the three ellipsometric methods (MIM, MAOI, and SE). Furthermore, D values for thin films deduced from SE and MAOI methods after n is fixed, yield D values that agree with the MIM ellipsometry measurements within experimental error.

Table 6. Attempts to simultaneously determine the thickness and refractive index values for representative thin TMSC LB-films without any constraints on their values[a]

# of Layers	D/nm		n^b
	SE	MAOI	
4	16.0 ± 55.3	14.8 ± 89.1	1.05 ± 2.01
24	47.3 ± 25.1	25.8 ± 11.3	1.22 ± 1.01

[a]One standard deviation error bars

[b]MAOI ellipsometry at $\lambda = 632.8$ nm

Table 7. CPE parameters for TMSC and regenerated cellulose

	UV_{term}	A_j	E_{cj}	Γ_j	ϕ_j
TMSC	1.89	-0.74	7.1	-0.038	9.3
Regenerated Cellulose	2.12	-0.86	7.4	0.97	3.0

Table 8. Thickness and refractive index values for a thick spin-coated film of TMSC and the corresponding regenerated cellulose film[a]

	D/nm		n^b
	SE	MAOI	
TMSC	228.8 ± 4.1	226.6 ± 1.1	1.45 ± 0.01
Cellulose	86.3 ± 3.6	85.4 ± 0.7	1.52 ± 0.01

[a]One standard deviation error bars

[b]MAOI ellipsometry at $\lambda = 632.8$ nm

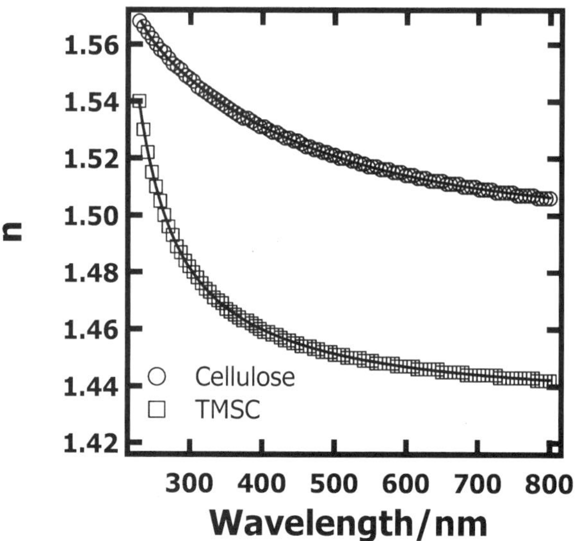

Figure 6. Refractive index of regenerated cellulose and TMSC films as a function of wavelength obtained via SE ellipsometry. CPE fitting parameters are summarized in Table 7. Emprical relationships for n as a function of wavelength via the Cauchy equations (solid lines) are provided for regenerated cellulose and TMSC as $n_{Cellulose}(\lambda) = 1.4953 + 7628.8/\lambda^2 - 3.5445 \cdot 10^8/\lambda^4 + 8.3012 \cdot 10^{12}/\lambda^6$ and $n_{TMSC}(\lambda) = 1.436 + 4155.3/\lambda^2 - 1.1466 \cdot 10^8/\lambda^4 + 9.712 \cdot 10^{12}/\lambda^6$, respectively. Deviations between the emprical Cauchy equations and the n values obtained from SE ellipsometry are <0.001 for 230 nm < λ < 800 nm.

Table 9. Thicknesses of TMSC LB-films obtained from SE and MAOI ellipsometry measurements utilizing the optical constants in Table 7 compared to MIM ellipsometry results[a]

# of Layers	D/nm		
	MIM[b]	SE	MAOI
4	5.2 ± 0.2	5.5 ± 1.3	5.1 ± 1.1
6	6.9 ± 0.1	6.4 ± 2.4	6.0 ± 0.3
8	9.4 ± 0.3	9.0 ± 1.8	9.8 ± 0.5
10	11.7 ± 0.1	10.9 ± 1.6	11.4 ± 1.2
12	13.8 ± 0.8	14.6 ± 2.8	13.6 ± 0.4
14	14.9 ± 0.3	15.6 ± 2.3	14.0 ± 0.8
16	17.2 ± 0.2	17.9 ± 1.7	18.3 ± 0.6
18	18.7 ± 0.2	19.3 ± 3.1	19.4 ± 0.4
20	20.2 ± 0.3	21.1 ± 1.3	20.1 ± 0.9
22	22.7 ± 0.5	23.2 ± 3.2	22.9 ± 1.2
24	24.2 ± 0.4	25.8 ± 2.3	25.8 ± 0.3

[a]One standard deviation error bars

[b]Utilizing Approach 1

Conclusion

Multiple incident media ellipsometry provides a rapid (<5 min for a single film) and unambiguous method for obtaining both film thicknesses and refractive indices of ultrathin films of TMSC, regenerated cellulose, and cellulose nanocrystals. Thickness and refractive index values obtained via the MIM ellipsometry method are in excellent agreement with literature values. Moreover, the MIM ellipsometry results are in quantitative agreement with more traditional ellipsometric techniques (SE and MAOI ellipsometry) within experimental error. Furthermore, it is observed that the value of $n = 1.51 \pm 0.01$ for regenerated cellulose and cellulose nanocrystals via MIM ellipsometry is lower than the parallel component and is consistent with the perpendicular component of the anisotropic refractive index reported for cellulose systems. The fact that there is no difference in n between the regenerated TMSC and cellulose nanocrytal films may indicate that they have similar degrees of crystallinity.

Acknowledgements

The authors would like to thank the National Research Initiative of the USDA Cooperative State Research, Education and Extension Service, Grant Number 2005-35504-16088, the National Science Foundation (CHE-0724126), and the Virginia Tech Aspires Program for financial support.

References

1. Richards, G. N.; Blake, J. D. *Carbohydr. Res.* **1971,** *18,* 11–21.
2. Saake, B.; Kruse, T.; Puls, J. *Bioresour. Technol.* **2001,** *80,* 195–204.
3. Esker, A.; Becker, U.; Jamin, S. Beppu, S.; Renneckar, S.; Glasser, W. In *Hemicelluloses: Science and Technology;* Gatenholm, P.; Tenkanen, M., Eds.; ACS Symposium Series 864; American Chemical Society: Washington, DC, 2004; pp 198–219.
4. Gradwell, S. E.; Renneckar, S. Esker A. R.; Heinze, T.; Gatenholm, P.; Vaca-Garcia, C.; Glasser, W. *C. R. Biol.* **2004,** *327,* 945–953.
5. Eriksson, J.; Malmsten, M.; Tiberg, F.; Callisen, T. H.; Damhus, T.; Johansen, K. S. *J. Colloid Interface Sci.* **2005,** *285,* 94–99.
6. Eriksson, J.; Malmsten, M.; Tiberg, F.; Callisen, T. H.; Damhus, T.; Johansen, K. S. *J. Colloid Interface Sci.* **2005,** *284,* 99–106.
7. Freudenberg, U.; Zimmermann, R.; Schmidt, K.; Behrens S. H.; Werner C. *J. Colloid Interface Sci.* **2007,** *309,* 360–365.
8. Karlsson, J. O.; Andersson, N.; Berntsson, P.; Chihani, T.; Gatenholm P. *Polymer* **1998,** *39,* 3589–3595.
9. Kowalczuk J.; Tritt-Goc, J.; Pislewski N. *Solid State Nucl. Magn. Reson.* **2004,** *25,* 35–41.
10. Hardaker, S. S.; Moghazy, S.; Cha, C. Y.; Samuels, R. J. *J. Polym. Sci., Part B: Polym. Phys.* **1993,** *31,* 1951–1963.
11. Chiang, K.S.; Cheng S.Y.; Liu, Q.; *J. Lightwave Technol.* **2007,** *25,* 1206–1212.
12. Salamon, Z.; Macleod, H. A.; Tollin, G. *Biochim. Biophys. Acta, Biomembr.* **1997,** *1331,* 117–129.
13. Salamon, Z.; Macleod, H. A.; Tollin, G. *Biophys. J.* **1997,** *73,* 2791–2797.
14. Sadik, A. M.; Ramadan, W. A.; Litwin, D. *Meas. Sci. Technol.* **2003,** *14,* 1753–1759.
15. Krzyzanowska, H.; Kulik, M.; Zuk, J. *J. Lumin.* **1998,** *80,* 183–186.
16. Matsuhashi, N.; Okumoto, Y.; Kimura, M.; Akahane, T. *Jpn. J. Appl. Phys., Part 1* **2002,** *41,* 4615–4619.
17. Woollam, J. A.; Bungay, C.; Hilfiker, J.; Tiwald, T. *Nucl. Instrum. Methods Phys. Res., Sect. B* **2003,** *208,* 35–39.
18. Azzam, R. M. A.; Bashara, N. M. *Ellipsometry and Polarized Light;* Elsevier: Amsterdam, 1987.
19. Kattner, J.; Hoffmann, H. *J. Phys. Chem. B* **2002,** *106,* 9723–9729.
20. Arwin, H.; Aspnes, D. E. *Thin Solid Films* **1986,** *138,* 195–207.
21. Irene, E. A. *Solid-State Electron.* **2001,** *45,* 1207–1217.
22. Landgren, M.; Jonsson, B. *J. Phys. Chem.* **1993,** *97,* 1656–1664.
23. Ayupov, B. M.; Sysoeva, N. P. *Cryst. Res. Technol.* **1981,** *16,* 503–512.
24. Mao, M.; Zhang, J. H.; Yoon, R. H.; Ducker, W. A. *Langmuir* **2004,** *20,* 1843–1849.
25. McCrackin, F. L.; Passaglia, E.; Stromberg, R. R.; Steinberg, H. L. *J. Res. Natl. Bur. Std.* **1963,** *67,* 363–367.
26. Gunnars, S.; Wagberg, L.; Stuart, M. A. C. *Cellulose* **2002,** *9,* 239–249.
27. Rosenau, T.; Potthast, A.; Sixta, H.; Kosma, P. *Prog. Polym. Sci.* **2001,** *26,* 1763–1837.

28. Dawsey, T. R.; McCormick, C. L. *JMS-Rev. Macromol. Chem. Phys.* **1990,** *30,* 405–440.

29. Nehl, I. Wagenknecht, W.; Philipp, B. *Cellul. Chem. Technol.* **1995,** *29,* 243.

30. Schaub, M.; Wenz, G.; Wegner, G.; Stein, A.; Klemm, D. *Adv. Mater.* **1993,** *5,* 919–922.

31. Kontturi, E.; Thune, P. C.; Niemantsverdriet, J. W. *Polymer* **2003,** *44,* 3621–3625.

32. Beck-Candanedo, S.; Roman, M.; Gray, D. G. *Biomacromolecules* **2005,** *6,* 1048–1054.

33. Holmberg, M.; Berg, J.; Stemme, S.; Odberg, L.; Rasmusson, J.; Claesson, P. *J. Colloid Interface Sci.* **1997,** *186,* 369–381.

34. Loscher, F.; Ruckstuhl, T.; Jaworek, T.; Wegner, G.; Seeger, S. *Langmuir* **1998,** *14,* 2786–2789.

35. Buchholz, V.; Adler, P., Backer, M.; Holle, W.; Simon, A.; Wegner, G. *Langmuir* **1997,** *13,* 3206–3209.

36. Rehfeldt, F.; Tanaka, M. *Langmuir* **2003,** *19,* 1467–1473.

37. Buchholz, V.; Wegner, G.; Stemme, S.; Odberg, L. *Adv. Mater.* **1996,** *8,* 399–402.

38. Blodgett, K. B. *J. Am. Chem. Soc.* **1934,** *56,* 495–495.

39. Muller, F.; Beck, U. *Das Papier* **1978,** *32,* 25–31.

40. Keddie, J. L. *Curr. Opin. Colloid Interface Sci.* **2001,** *6,* 102–110.

41. Beaglehole, D.; Christenson, H. K. *J. Phys. Chem.* **1992,** *96,* 3395–3403.

42. Drude, P. *Anal. Phys.* **1889,** *36,* 532–560.

43. Lekner, J. Theory of Reflection; Nijhoff: Amsterdam, 1988.

44. Adachi, S. *Optical Properties of Crystalline and Amorphous Semiconductors;* Kluwer Academic: London, 1999.

45. Arwin, H.; Jansson, R. *Electrochim. Acta* **1994,** *39,* 211–215.

46. Brandrup, J.; Immergut, E. H.; Grulke, E. A. *Polymer Handbook, 4th ed.;* Wiley: New York, 1999.

47. Bordel, D.; Putaux, J. L.; Heux, L. *Langmuir* **2006,** *22,* 1899–1901.

48. Roman, M.; Gray, D. G. *Langmuir* **2005,** *21,* 5555–5561

49. Cranston, E. D.; Gray, D. G. *Biomacromolecules* **2006,** *7,* 2522–2530.

Deposition of Cellulose Nanocrystals by Inkjet Printing

Maren Roman and Fernando Navarro

Macromolecules and Interfaces Institute and Department of Wood Science and Forest Products, Virginia Polytechnic Institute and State University, Blacksburg, VA 24061

The present study investigates the use of inkjet technology for the deposition of cellulose nanocrystals onto flat substrates. Aqueous suspensions of cellulose nanocrystals were printed onto glass substrates using a commercial, piezoelectric, drop-on-demand inkjet printer. Poor wetting of the glass substrates impeded the generation of continuous films. However, printing of microdot arrays yielded regular microscale arrays of nanocrystal deposits. Radial, outwards capillary flow in the drying droplets led to ring formation but could be suppressed by altering the surface chemistry of the glass substrate. Co deposition of a cellulose nanocrystal suspension and a chitosan solution produced uniform, two-component deposits.

Introduction

A number of techniques have been studied for the deposition of cellulose nanocrystals from aqueous suspensions onto flat substrates, including spin coating (*1–5*), layer-by-layer deposition (*6–10*), and Langmuir Blodgett deposition (*11*). These methods, generally, yield continuous films with a thickness below 100 nm. Spin coating can also generate regular, discontinuous or submonolayer films of cellulose nanocrystals (*4, 5*), useful for analyzing the average size and size distribution of cellulose nanocrystals, for example.

The present study investigates the use of inkjet technology as a novel, alternative deposition method for cellulose nanocrystals. In recent years, inkjet printing has generated considerable interest as a deposition method for polymer solutions and colloidal suspensions of ceramic, metal, and polymer particles due to its low cost, ease of controllability, versatility, contact free nature, and potential for high throughput (*12*). It has been explored as a tool to fabricate organic and inorganic electronic devices, 3D ceramic structures, and bio-microarrays, among others (*13–17*). With respect to the deposition methods mentioned above, inkjet printing provides additional possibilities, including patterning and simultaneous co-deposition of different materials. Yet, despite its high promise as a technique for depositing and patterning functional materials in the liquid phase onto a substrate, the phenomena associated with inkjet printing of colloidal systems remain incompletely understood.

The objectives for the present study were (1) to test whether inkjet printing could be used to generate cellulose nanocrystal films, micropatterns, and multi-component layers on flat substrates, and (2) to determine which experimental factors govern the deposition process.

Materials and Methods

Cellulose Nanocrystal Preparation

Cellulose nanocrystals were prepared by sulfuric acid hydrolysis of dissolving-grade softwood sulfite pulp. Lapsheets of the pulp (Temalfa 93A-A), kindly provided by Tembec, Inc., were cut into small pieces of 1 cm by 1 cm and milled in a Wiley mill (Thomas Wiley Mini-Mill) to pass a 60 mesh screen. The milled pulp was hydrolyzed under stirring with 60 wt % sulfuric acid (10 ml/g cellulose) at 50 °C for 60 min. The hydrolysis was stopped by diluting the reaction mixture 10-fold with deionized water (Millipore Direct-Q 5, 18.2 MΩ·cm). The nanocrystals were collected and washed once with deionized water by centrifugation for 10 min at 25 °C and 4550 G (Thermo IEC Centra-GP8R) and then dialyzed (Spectra/Por 4 dialysis tubing) against deionized water until the pH of fresh dialysis medium stayed constant over time. The nanocrystal suspension was sonicated (Sonics & Materials Model VC-505) for 10 min at 200 W under ice-bath cooling and filtered through a 0.45 μm polyvinylidene fluoride (PVDF) syringe filter (Millipore) to remove any aggregates present. An AFM image of the obtained cellulose nanocrystals is shown in Figure 1.

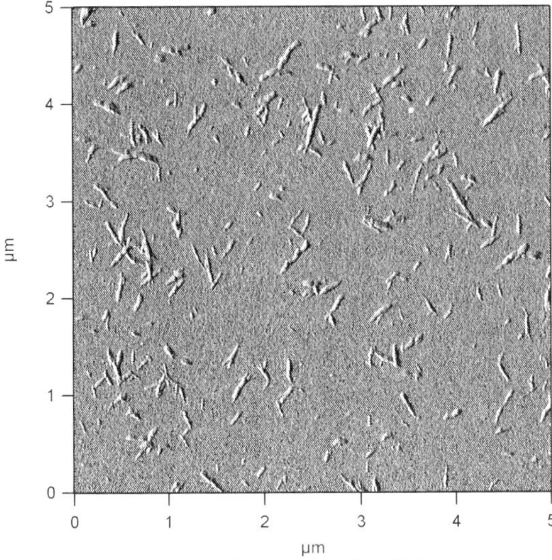

Figure 1. AFM amplitude image of cellulose nanocrystals

Preparation of Chitosan Solution

A 0.05% chitosan solution was prepared by dissolving low molecular weight chitosan (Sigma-Aldrich) over night under stirring in 0.1 N hydrochloric acid (Certified, Fisher Chemical, 0.0995-0.1005 N). The solution was consecutively filtered through 541 Whatman filter paper, a 0.45 µm PVDF syringe filter, and a 0.2 µm PVDF syringe filter (Millipore).

Substrate Cleaning Procedure

Conventional 3 x 1 in. glass microscope slides were used as substrates. The glass slides were used as received, after removing any loose particles with a pressurized gas cleaner (Clean Jet 100 Dust Remover), or cleaned using one of the following three methods:

Soap Cleaning Procedure

A 4 wt % solution of Alconox in deionized water was degassed for 30 min at 45 °C in an ultrasonic bath (Branson 3510 Ultrasonic Cleaner) and then poured into a glass staining dish loaded with microscope slides. The staining dish was placed into the ultrasonic bath and subjected to ultrasound for 10 min at 45 °C. Next, the soap solution in the staining dish was replaced with deionized water and the ultrasound treatment was repeated. The slides were washed twice more with deionized water and an ultrasound treatment of 5 and 2 min, respectively. Finally, the slides were dried in a stream of nitrogen and stored in a slide storage box.

Aqua Regia Cleaning Procedure

A 3:1 mixture by volume of hydrochloric acid (NF/FCC, Fisher Chemical, 36.5-38%) and nitric acid (NF, Fisher Chemical, 69-70%) was poured into a glass staining dish loaded with microscope slides. The staining dish was covered and left undisturbed for 30 min. Then, the acid mixture was replaced with deionized water and the slides were washed three times as described above. Finally, the slides were dried in a stream of nitrogen and stored in a slide storage box.

Organic Solvent Cleaning Procedure

Four microscope slides at a time were carefully placed with minimal contact into the extraction tube of a Soxhlet apparatus. A 1:1 mixture by volume of ethanol (ACS, Acros Organics, 99.5%) and chloroform (Reagent ACS, Acros Organics, 99.8%) was brought to a boil in the still pot. After four reflux cycles, the glass slides were placed into a glass staining dish and rinsed three times with deionized water in an ultrasonic bath as described above. Finally, the slides were dried in a stream of nitrogen and stored in a slide storage box.

Inkjet Printing

Inkjet printing was done with an Epson Stylus Photo R800 piezoelectric drop-on-demand inkjet printer (Figure 2) using the CD tray and refillable, clear, spongeless MIS cartridges (MIS Associates) (Figure 3). To ensure contact free movement of the substrate during printing, several paper feed rollers had been removed from the printer. Commercial inkjet printers generally mix inks from several cartridges to obtain the desired color. Unintentional mixing was avoided by using the CMYK (cyan, magenta, yellow, black) color space with the ICC (International Color Consortium) color profile and color enhancement options turned off in the Corel Draw (version 12.0) and printer software.

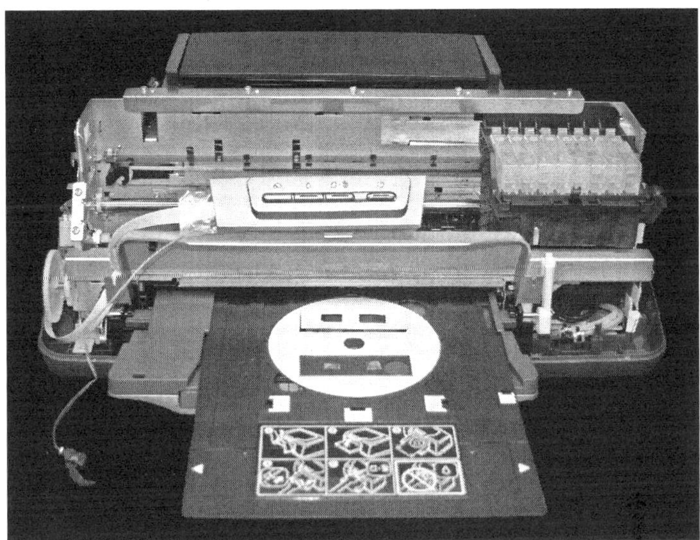

Figure 2. Epson Stylus Photo R800 piezoelectric drop-on-demand inkjet printer with refillable cartridges, CD tray, and cardboard substrate mount

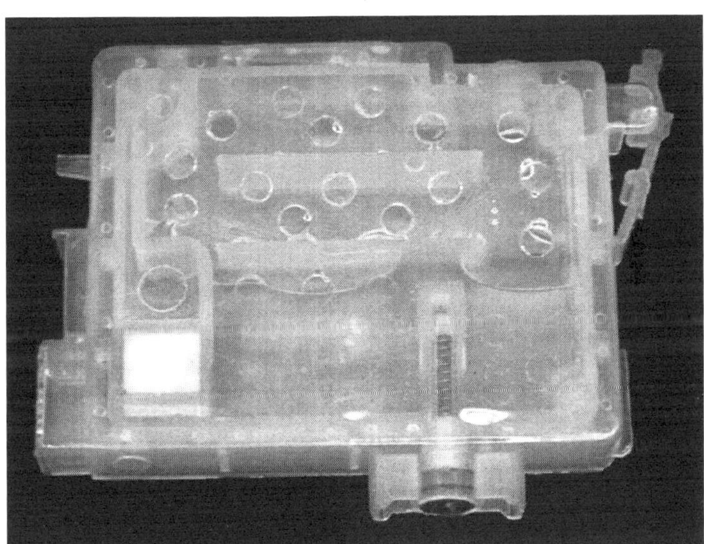

Figure 3. Refillable inkjet cartridge

162

Polarized Light Microscopy

For polarized light microscopy, a Zeiss Axioskop 40 A POL was used. Images were recorded with a Canon EOS 20D digital single-lens reflex camera (8.2 megapixels) mounted onto the microscope.

Atomic Force Microscopy

Atomic force microscopy (AFM) was performed with an Asylum Research MFP-3D mounted onto an Olympus IX 71 inverted fluorescence microscope. Samples were scanned in intermittent contact mode in air with Olympus OMCL-AC160TS tips (nominal tip radius: <10 nm).

Results and Discussion

Inkjet printing of functional materials offers many interesting opportunities. The following sections present our initial results on the fabrication of cellulose nanocrystal micropatterns and the preparation of cellulose nanocrystal multi-component layers by inkjet co-deposition. Further provided is a discussion of the fundamental phenomenon of cellulose nanocrystal transport in drying droplets.

Micropatterning of Cellulose Nanocrystals

Poor wetting of glass substrates by cellulose nanocrystal suspensions was found to prevent inkjet deposition of thin, continuous films of cellulose nanocrystals onto glass substrates. Inkjet printed liquid films, initially covering large areas or continuous lines, retracted rapidly into droplets of various sizes before drying. Thus, inkjet fabrication of cellulose nanocrystal thin films and continuous micropatterns, such as square grid patterns, on glass substrates requires the use of wetting agents or adhesion layers on the substrate. However, the fabrication of microdot patterns should be possible. Figure 4 shows the deposit pattern obtained by printing a 0.76 wt % cellulose nanocrystal suspension into a regular array of dots with diameters of 70 μm and a spacing of 280 μm onto an untreated glass slide. A regular array of ring-like deposits of 68–74 μm in diameter can be observed. The formation of ring-like deposits, as opposed to evenly filled circles, is a result of the so-called "coffee drop effect", a common phenomenon observed when drops containing dispersed solids evaporate on a solid substrate (*18, 19*). The formation of contact line deposits from inkjet printed cellulose nanocrystal droplets is discussed in the next section.

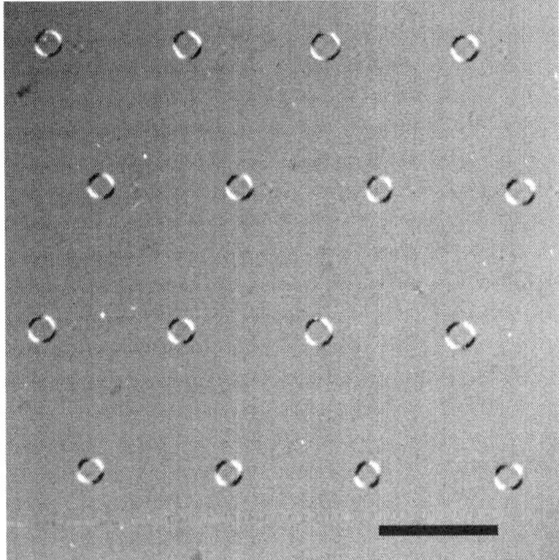

*Figure 4. Cellulose micropattern on glass obtained by printing a 0.76 wt %
cellulose nanocrystal suspension into a square array of dots. Scale bar: 300 μm*

Particle Transport in Drying Cellulose Nanocrystal Droplets

The drying of a droplet on a solid surface can proceed by one of two mechanisms: (i) the droplet can maintain its shape (constant contact angle) and decrease in size or (ii) the droplet can maintain its diameter and decrease in height. The former case is less likely as it requires the contact line, that is the line between wet and dry substrate, to move freely across the surface. More often, the contact line of a drying droplet is pinned to its initial position as a result of surface roughness or chemical heterogeneities. In order for a droplet to maintain its diameter during drying, the edge of the droplet has to be constantly replenished with liquid from the center, that is, a drying droplet with a pinned contact line experiences a type of capillary flow from the center of the droplet to its edge upon evaporation of the liquid. The resulting outward flow carries any dispersed material to the edge of the drop, leading to dense deposits along its perimeter. The phenomenon of ring stain formation from drying droplets is called "coffee drop effect" in reference to the well known appearance of dried coffee drops. "Self pinning" is the reinforcement of contact line pinning through particle deposition at the edge of the droplet upon evaporation of the liquid, making translocation of the contact line progressively less likely.

Contact line deposits were observed with inkjet droplets of cellulose nanocrystal suspensions under most conditions. Figure 5 shows a 3D AFM image of a cellulose nanocrystal ring deposit, similar to the deposits shown in Figure 4, formed by an individual inkjet droplet of a 0.76 wt % suspension.

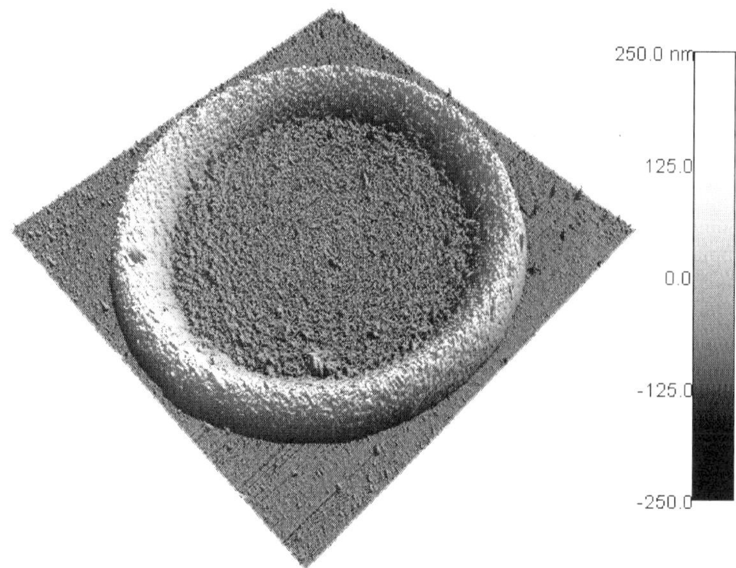

*Figure 5. 3D AFM height image of a dried inkjet droplet from a 0.76 wt %
suspension of cellulose nanocrystals on a glass substrate, scan size: 80 μm*

To determine the factors governing the formation of contact line deposits, glass substrates with different surface energies were prepared using three different cleaning methods, described in the methods section. The surface energies, determined by contact angle measurements using the method by van Oss and Giese (*20*), were 59.6 mJ/m^2 for the soap-cleaned substrate, 52.9 mJ/m^2 for the aqua regia-cleaned one, and 46.1 mJ/m^2 for the organic solvent-cleaned one. Figures 6–8 show AFM images and line scans of deposits from individual inkjet droplets of a 1 wt % cellulose nanocrystal suspension on a soap, aqua regia, and organic solvent-cleaned glass substrate, respectively. The deposits on the soap and organic solvent-cleaned substrates showed distinct contact line deposits, whereas the deposit on the aqua regia-cleaned substrate was nearly uniform in thickness along the diameter of the deposit.

The fact that the magnitude of the coffee drop effect showed no apparent correlation with the surface energy of the substrate suggested that other factors were involved. The observed difference in behavior might be due to different electrostatic interactions between the glass substrate and the nanocrystals. Quartz is known to undergo protonation and deprotonation in acidic and basic media, respectively (*21*). Glass substrates cleaned with aqua regia, therefore, might have positively charged silanol groups (~SiOH$_2^+$) on the surface, whereas those cleaned with soap might have negatively charged groups (~SiO$^-$). Cellulose nanocrystals produced with sulfuric acid are negatively charged in aqueous media due to dissociated sulfate groups on their surface (*22, 23*). Thus, an aqua regia-cleaned glass substrate might exert attractive electrostatic forces on cellulose nanocrystals whereas a soap-cleaned one might exert repulsive forces. An organic solvent-cleaned glass substrate would be expected to have minimal surface charge and thus minimal electrostatic interactions with the nanocrystals.

Multi-Component Films by Co-Deposition

The ability to simultaneously print from different cartridges allows the fabrication of multi-component films by co-deposition. We have explored this possibility by co-depositing a 0.29 wt % cellulose nanocrystal suspension and a 0.05 wt % chitosan solution. Figure 9 shows the interior of a dried droplet of the chitosan solution printed onto an aqua regia-cleaned glass slide. A network of ridges could be observed. When the cellulose nanocrystal suspension and chitosan solution were co-deposited at a ratio of 40:60, instead of individual ridges, uniform coverage with a root mean square surface roughness of 12 nm was obtained. The coffee drop effect, observed in droplets of a 0.29 wt % cellulose nanocrystal suspension on aqua regia-cleaned glass substrates, was entirely suppressed. The absence of ridges and ring formation suggests that the negatively charged cellulose nanocrystals and positively charged chitosan molecules interact rapidly upon mixing on the substrate, possibly forming an extended network that is held together by electrostatic forces and withstands the radial capillary flow in the droplet upon drying.

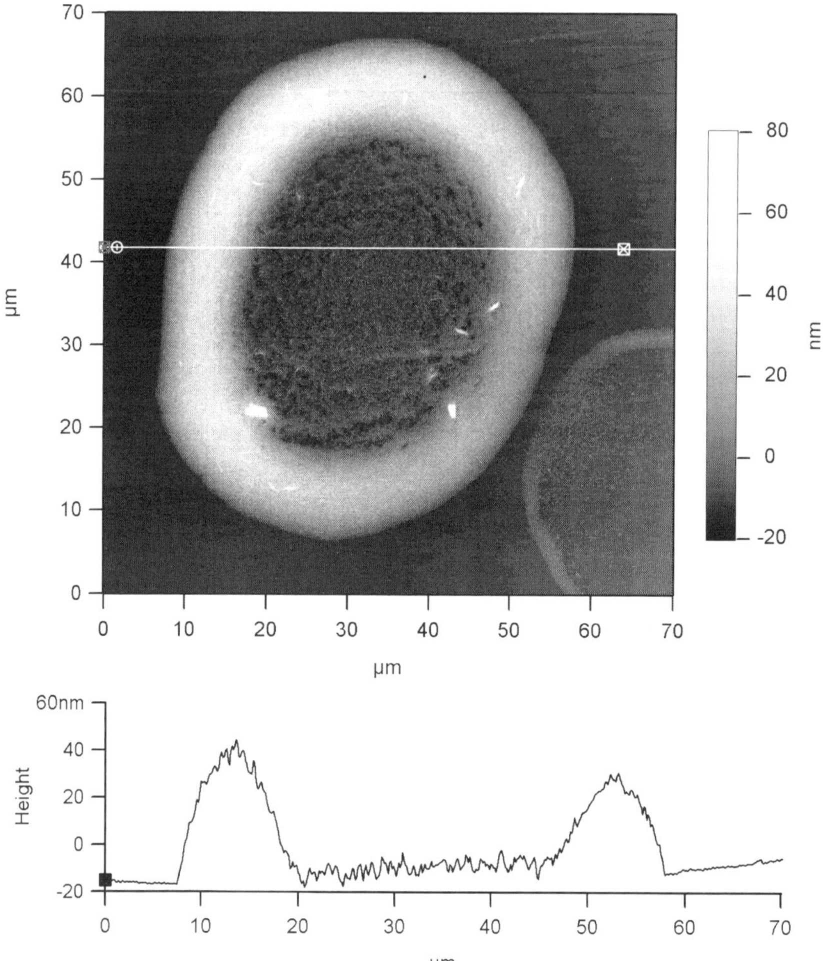

Figure 6. Dried inkjet droplet obtained by printing a 1 wt% cellulose nanocrystal suspension onto a soap-cleaned glass substrate. Top: AFM height image, Bottom: Line scan indicated in the height images

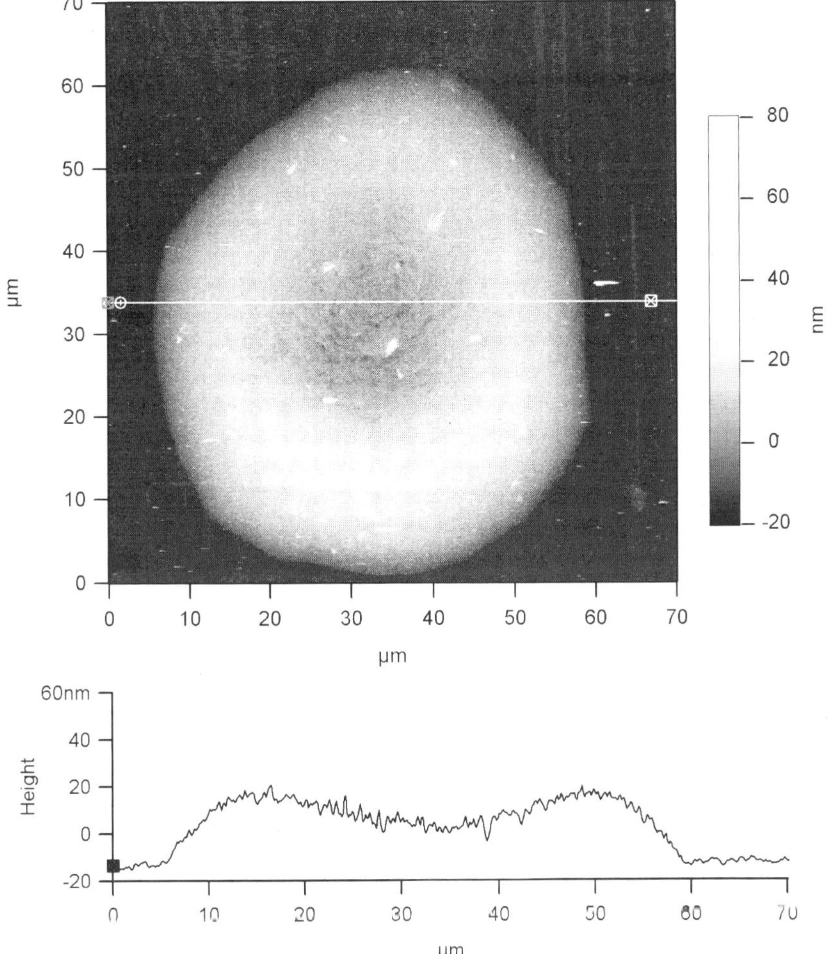

Figure 7. Dried inkjet droplet obtained by printing a 1 wt% cellulose nanocrystal suspension onto an aqua regia-cleaned glass substrate. Top: AFM height image, Bottom: Line scan indicated in the height images

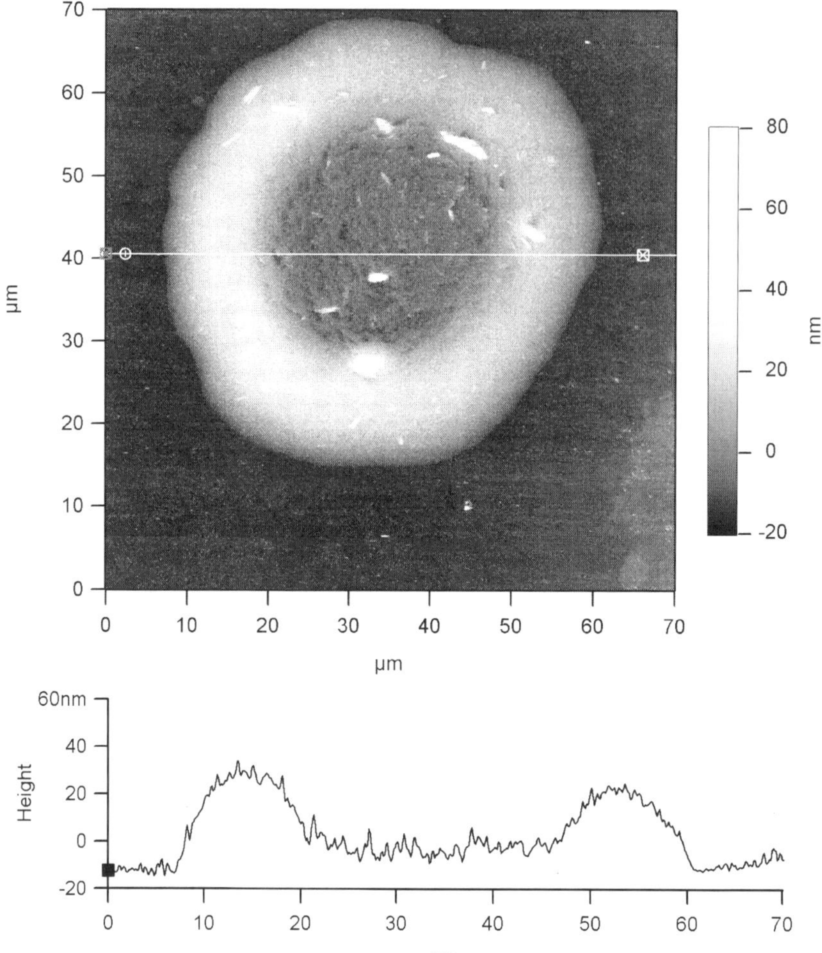

Figure 8. Dried inkjet droplet obtained by printing a 1 wt% cellulose nanocrystal suspension onto a solvent-cleaned glass substrate. Top: AFM height image, Bottom: Line scan indicated in the height images

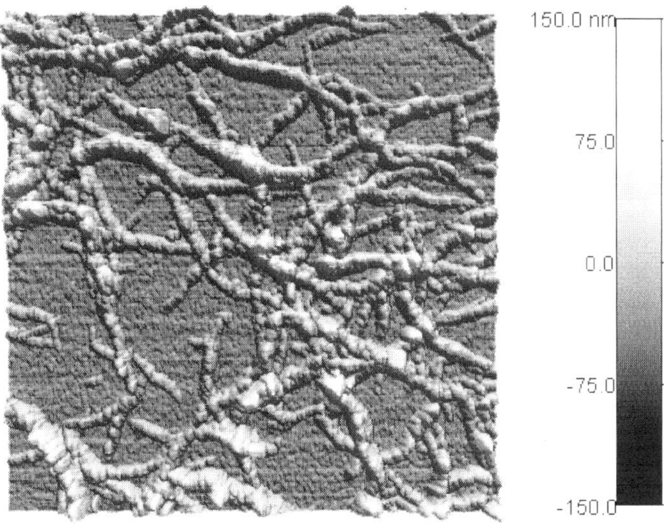

Figure 9. *3D AFM height image of chitosan printed onto an aqua regia-cleaned glass substrate from a 0.05 wt % solution in 0.1 N HCl, scan size: 5 μm*

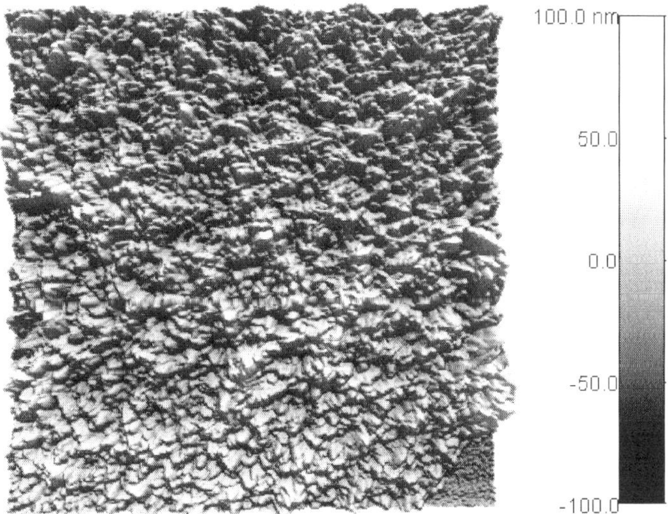

Figure 10. *3D AFM height image of a mixed cellulose chitosan film obtained by printing a 0.29 wt % cellulose nanocrystal suspension and a 0.05 wt % chitosan solution in 0.1 N HCl at a 40:60 ratio, scan size: 5 μm*

Conclusions

Our results confirm that inkjet printing holds great promise for micropatterning and co-deposition of cellulose nanocrystals onto flat substrates. Direct printing of thin films and continuous micropatterns of cellulose nanocrystals onto glass substrates requires the use of wetting agents or adhesion on the substrate to improve wetting. The formation of contact line deposits can be suppressed through attractive electrostatic interactions between the nanocrystals and the substrate.

Acknowledgments

The authors gratefully acknowledge generous scholarship support for F. N. by the Secretariat of Public Education (Secretaría de Educación Pública) of Mexico through its Professional Development for Teachers program (PROMEP). This project was further supported by the National Research Initiative of the USDA Cooperative State Research, Education and Extension Service, grant number 2005-35504-16088. Additional support from OMNOVA, Inc. and Tembec, Inc. are also acknowledged.

References

1. Edgar, C. D.; Gray, D. G. *Cellulose* **2003,** *10,* 299–306.
2. Lefebvre, J.; Gray, D. G. *Cellulose* **2005,** *12,* 127–134.
3. Eriksson, M.; Notley, S. M.; Wagberg, L. *Biomacromolecules* **2007,** *8,* 912-919.
4. Kontturi, E.; Thune, P. C.; Alexeev, A.; Niemantsverdriet, J. W. *Polymer* **2005,** *46,* 3307-3317.
5. Kontturi, E.; Johansson, L. S.; Kontturi, K. S.; Ahonen, P.; Thune, P. C.; Laine, J. *Langmuir* **2007,** *23,* 9674-9680.
6. Podsiadlo, P.; Choi, S. Y.; Shim, B.; Lee, J.; Cuddihy, M.; Kotov, N. A. *Biomacromolecules* **2005,** *6,* 2914–2918.
7. Notley, S. M.; Eriksson, M.; Wagberg, L.; Beck, S.; Gray, D. G. *Langmuir* **2006,** *22,* 3154–3160.
8. Cranston, E. D.; Gray, D. G. *Sci. Technol. Adv. Mater.* **2006,** *7,* 319–321.
9. Cranston, E. D.; Gray, D. G. *Biomacromolecules* **2006,** *7,* 2522–2530.
10. Podsiadlo, P.; Sui, L.; Elkasabi, Y.; Burgardt, P.; Lee, J.; Miryala, A.; Kusumaatmaja, W.; Carman, M. R.; Shtein, M.; Kieffer, J.; Lahann, J.; Kotov, N. A. *Langmuir* **2007,** *23,* 7901-7906.
11. Habibi, Y.; Foulon, L.; Aguie-Beghin, V.; Molinari, M.; Douillard, R. *J. Colloid Interface Sci.* **2007,** *316,* 388-397.
12. Geissler, M.; Xia, Y. N. *Adv. Mater.* **2004,** *16,* 1249-1269.
13. Calvert, P. *Chem. Mater.* **2001,** *13,* 3299-3305.
14. Sirringhaus, H.; Shimoda, T. *MRS Bull.* **2003,** *28,* 802-803.
15. Schubert, U. S. *Macromol. Rapid Commun.* **2005,** *26,* 237-237.

16. Yap, F. L.; Zhang, Y. *Biosens. Bioelectron.* **2007,** *22,* 775-788.

17. Barbulovic-Nad, I.; Lucente, M.; Sun, Y.; Zhang, M. J.; Wheeler, A. R.; Bussmann, M. *Crit. Rev. Biotechnol.* **2006,** *26,* 237-259.

18. Deegan, R. D.; Bakajin, O.; Dupont, T. F.; Huber, G.; Nagel, S. R.; Witten, T. A. *Nature* **1997,** *389,* 827-829.

19. Deegan, R. D.; Bakajin, O.; Dupont, T. F.; Huber, G.; Nagel, S. R.; Witten, T. A. *Phys. Rev. E* **2000,** *62,* 756-765.

20. van Oss, C. J.; Giese, R. F. *J. Dispersion Sci. Technol.* **2004,** *25,* 631-655.

21. Duval, Y.; Mielczarski, J. A.; Pokrovsky, O. S.; Mielczarski, E.; Ehrhardt, J. J. *J. Phys. Chem. B* **2002,** *106,* 2937-2945.

22. Dong, X. M.; Revol, J.-F.; Gray, D. G. *Cellulose* **1998,** *5,* 19–32.

23. Roman, M.; Winter, W. T. *Biomacromolecules* **2004,** *5,* 1671–1677.

Chapter 8

Hydroxypropyl Xylan Self-Assembly at Air/Water and Water/Cellulose Interfaces

Abdulaziz Kaya[1], Daniel A. Drazenovich[1], Wolfgang G. Glasser[2], Thomas Heinze[3], and Alan R. Esker[1,*]

[1]Department of Chemistry and the Macromolecules and Interfaces Institute, and [2]Department of Wood Science and Forest Products, Virginia Polytechnic Institute and State University, Blacksburg, VA 24061
[3]Center of Excellence for Polysaccharide Research, Friedrich Schiller University of Jena, Humboldtstraße 10, Jena, 07743 Germany

Hydroxypropylation of polysaccharides is one strategy for enhancing aqueous solubility. The degree of hydroxypropyl substitution can be controlled through the pH of the hydroxypropylation reaction. Surface tension measurements of aqueous solutions of hydroxypropyl xylan (HPX), synthesized from barley husk xylans, by the Wilhelmy plate technique show that surface tension changes ($\Delta\gamma = \gamma_{water} - \gamma_{HPX(aq)}$) increase and critical aggregation concentrations generally decrease with increasing degree of substitution. Hence, even though hydroxypropyl substitution is necessary to induce aqueous solubility, excessive hydroxypropylation promotes aggregation in water. While surface tension studies reveal HPX affinity for the air/water interface, surface plasmon resonance spectroscopy studies indicate that HPXs do not adsorb significantly onto model regenerated cellulose surfaces (submonolayer coverage). Likewise, the HPXs do not show significant adsorption onto hydroxyl-terminated self-assembled monolayers of 11-mercapto-1-undecanol (SAM-OH). In contrast, HPX does adsorb (~monolayer coverage) onto methyl-terminated self-assembled monolayers of 1-dodecanethiol (SAM-CH$_3$). These results show hydroxypropylation is a sound approach for creating soluble xylan derivatives, suitable for further chemical modification.

174

Introduction

Cellulose is one of the most important natural polymers and is used extensively in the textile and paper industries (*1*). In nature, cellulose is located in the core of plant cell walls (*2*) and is associated with hemicellulose and lignin in a hierarchial (composite) superstructure (*3*). Hemicelluloses, which serve as a matrix for the cellulose superstructure, are lower molar mass polysaccharides containing short side chains (*4*). These polysaccharides consist of various five (D-xylose, L-arabinose) and six carbon (D-glucose, D-galactose, D-mannose etc.) sugars (*5*). Xylans are the most common hemicelluloses and are considered to be the second most abundant biopolymer in land plants (*6*). Structurally, xylans are a class of heteropolysaccharides consisting of poly(anhydroxylose) with varying degrees of 4-O methyl glucuronic acid, acetyl groups, and anhydroarabinose substituents depending on the source and isolation procedures used to obtain the xylan (*7*). During the past several years, the need for effective biomass utilization has renewed interest in the exploitation of xylans as sources of biopolymers. This interest is aided by the fact that xylans are readily available as organic wastes from renewable forest and agricultural residues, such as wood meal and shavings, stems, stalks, hulls, cobs, and husks (*8*). Even though the isolation of xylans from biomass is relatively easy, the potential application of xylans has not yet been completely realized (*8–11*). Possible reasons for the lack of xylan utilization as a material stream include a shortage of high molar mass xylans on an industrial scale (*9*), heterogeneity of xylan structures within even a single plant (*8*), and the partial degradation of hemicelluloses during pulping processes (*12*).

Another complication hindering widespread use of xylans is that they are usually difficult to dissolve in aqueous media and aprotic solvents even when they are isolated by aqueous extraction. Hence, investigations of xylan solution properties and molecular weight determinations are difficult (*13*). The substitution of a xylan's hydroxyl groups by alkoxy or acetoxy groups enhances solubility in water and/or organic solvents (*11*). Therefore, chemical modification of xylans provides one avenue to make soluble xylans for molecular weight determinations and producing materials with interesting physical properties (*11, 14–18*). Glaudemans and Timmel prepared xylan acetate that was completely soluble in chloroform and chloroform–ethanol mixtures. These polymers had a degree of polymerization of ~200 (*14*). In addition to xylan acetates, other esters of xylans, such as benzoate, caprate, laurate, myristate, and palmitate have been synthesized (*15*). In another study, xylans fully substituted with carbamate groups showed thermoplastic behavior at high temperatures (*16*). Likewise, Jain et al. prepared water-soluble hydroxypropyl xylans and acetoxypropyl xylans that showed thermoplastic behavior and solubility in most organic solvents (*11*). Trimethylammonium-2-hydroxypropyl xylan prepared from beechwood and corn cob xylan show promise for papermaking additives by improving the strength of bleached hardwood kraft pulp and unbleached thermomechanical pulp, and by increasing the retention of fiber fines (*17, 18*).

The enhancement of pulp properties by some xylan derivatives provides strong incentive for studying xylan self-assembly onto model cellulose and cellulose fiber surfaces. Mora et al. investigated xylan retention on cellulose fibers and concluded that the driving force for xylan aggregate sorption and retention on cellulose fibers was hydrogen bonding between cellulose fibers and the xylans (*19*). Henrikkson et al. also invoked hydrogen bonding along with changes in colloidal stability to explain the adsorption behavior of autoclaved xylans onto cellulose fibers at elevated temperatures under alkaline conditions (*20*). In another study, it was observed that commercial birch xylan adsorbed slowly and irreversibly onto model cellulose surfaces at pH 10 (*21*). However, it was argued that the driving force for adsorption was a combination of weak van der Waals' attractions and an entropically favorable release of solvent molecules when the polymer chains adsorbed. Recently, Esker et al. have shown that cationic and hydrophobic modification of xylan enhances xylan adsorption onto regenerated cellulose films prepared by the Langmuir–Blodgett technique (*22*). This result demonstrates that the hydrophobic forces and electrostatic interactions also influence xylan self-assembly onto cellulose surfaces.

In this study, the adsorption of hydroxypropyl xylans (HPXs) onto model surfaces is studied as a function of the degree of hydroxypropyl (HP) substitution (DS). The source of the "parent" xylans for the HPX derivatives is barley husks (*Hordeum* spp.) (*11*). HPX self-assembly at the air/water interface is probed through the Wilhelmy plate technique, whereas surface plasmon resonance (SPR) spectroscopy studies allow quantification of HPX adsorption onto regenerated cellulose, and self-assembled monolayers (SAMs) of 11-mercaptoundecanol (SAM-OH) and 1-dodecanethiol (SAM-CH$_3$) on gold substrates. These studies provide insight into molecular factors influencing HPX self-assembly at surfaces and potential use of further-derivatized water-soluble HPX derivatives to modify surfaces and interfaces.

Experimental

Materials

Ultrapure water was used in all experiments (Millipore, Milli-Q Gradient A-10, 18.2 MΩ·cm, <5 ppb organic impurities) to make HPX solutions. Trimethylsilyl cellulose (DS 2.71) was synthesized as described previously (*23*). 11-mercapto-1-undecanol (SAM-OH) and 1-dodecanethiol (SAM-CH$_3$) were purchased from Aldrich. Details of the HPX synthesis are provided elsewhere (*11*). In this study, HPX derivatives are named according to the pH of the aqueous solution used for the hydroxypropylation of xylan. HPX120, HPX125, HPX127, and HPX130 correspond to pH = 12.0, 12.5, 12.7, and 13.0, respectively. In general, the DS increases with pH (*11*).

Acetylation of HPX

In order to determine the DS, HPX was acetylated with acetic anhydride following the method of Carson and Maclay with minor modification (*24*). HPX (0.5 g) was dissolved in 10 mL of dry formamide (Riedel-de Haën) at 45–50 °C with vigorous stirring for 30 min. Next, 5 mL of pyridine (Alfa Aesar) was added at 45–50 °C and the mixture was allowed to stir for 30 min. Afterwards, the reaction mixture was cooled to 30 °C, and 5 mL of acetic anhydride (Fluka) was added. Following overnight stirring, the reaction mixture was precipitated in 200 mL of a cold aqueous 2 wt % HCl solution. Finally, the precipitated solids were filtered and rinsed with copious amounts of cold aqueous 0.5 wt % HCl solution and cold Millipore water. The resulting product was dried overnight at 35 °C under vacuum.

NMR Spectroscopy

For ^{1}H NMR analysis, 3–5 mg of acetylated HPX was dissolved in 5 mL of $CDCl_3$. ^{1}H NMR spectra were obtained on a 400 MHz Varian Inova spectrometer.

Surface Tension Measurements

The surface tension of aqueous HPX solutions was determined by the Wilhemy plate method using a paper plate attached to a Cahn 2000 electrobalance. A fixed amount of 20 mL of water was placed in a specially designed glass jar that consisted of an inner cup containing the solution and an outer jacket containing 20.0 °C water flowing from a thermostated circulating bath. Next, HPX stock solution (~150 mg·L^{-1}) was added to the pure water incrementally with a digital variable volume pipettor to vary the HPX solution concentration. The sample cell was inside a Plexiglass box to prevent water evaporation and maintain a constant relative humidity (~73%).

Refractive Index Increment Measurements

The refractive index increments (dn/dc) of HPX solutions were determined by using a Wyatt Optilab rEX differential refractometer. The experiments were carried out at $\lambda = 690$ nm and at 20.0 °C over the concentration range of 0–150 mg·L^{-1}. The instrument's software was used to calculate dn/dc for each polymer in water.

Model Cellulose Film Preparation

Smooth, uniform films of regenerated cellulose were prepared on 12.0 mm × 12.0 mm × 0.9 mm sensor slides using trimethylsilylcellulose (TMSC). Sensor slides consisted of a glass slide covered with 20 Å of chromium and 480 Å of gold. Sensor slides were obtained by first depositing the chromium and then gold onto precleaned soda lime float glass (Specialty Glass Products, Inc.) at 3×10^{-6} torr using an electron beam evaporator system (Thermionic Vacuum Products). Spin coating was used to prepare TMSC surfaces. Each sensor slide was cleaned by immersion in a 7:3 by volume solution of sulfuric acid:hydrogen peroxide (piranha solution) for 1 h and rinsed exhaustively with Millipore water prior to spin coating. TMSC was spin coated onto cleaned sensor slides with a spinning speed of 4000 rpm from 10 $g \cdot L^{-1}$ TMSC solutions in toluene (*25*). Trimethylsilyl groups of TMSC were cleaved by exposing the gold slide to the vapor of an aqueous 10 wt % HCl solution for 2 min. This process yields a regenerated cellulose surface (*26*).

Self-Assembled Monolayer (SAM) Preparation

The sensor slides underwent the same cleaning procedure described for model cellulose surface film preparation. Once dry, a sensor slide was placed in a 1 mM solution of the appropriate SAM-forming molecule (1-mercapto-1-undecanol or 1-dodecanethiol) in absolute ethanol for at least 24 h (*27*). Once the SAM sensor was needed, the slide was removed from the 1 mM ethanolic solution, rinsed with absolute ethanol to remove excess SAM-forming molecules, and dried with nitrogen. Finally, the SAM sensor slide was washed with Millipore water and dried with nitrogen.

Surface Plasmon Resonance (SPR) Spectroscopy

HPX adsorption onto regenerated cellulose and SAM surfaces was investigated by SPR spectroscopy. After preparing the desired film (cellulose or SAM) on the sensor slide, the slide was refractive index-matched to the prism of a Reichert SR 7000 surface plasmon resonance refractometer using immersion oil (n_D = 1.5150). This system used a laser diode with an emission wavelength of 780 nm. The flow cell body was equipped with a Viton gasket (Dupont Dow Elastomers, LLC) and was mounted on top of the sensor slide. Solutions were pumped into the flow cell at a flow rate of 0.25 $mL \cdot min^{-1}$ via Teflon tubing connected to a cartridge pump (Masterflex) at 20.0 °C. The pump was linked to a switch valve that made it possible to switch between the HPX solutions and buffer without introducing bubbles into the system. For SPR experiments, polymer stock solutions were prepared by dissolving the polymer in Millipore water to ~150 $mg \cdot L^{-1}$. From stock solutions, HPX solutions were prepared by dilution with Millipore water and were degassed before SPR experiments. Prior to data acquisition, the cellulose surface was allowed to reach equilibrium swelling by flowing only Millipore water through the system. Once a stable baseline was established, HPX solutions were pumped into the flow cell. Each solution was allowed to flow until a new baseline was achieved before switching to water via a solvent selection valve. The process of acquiring data systematically is depicted in Figure 1. Once a new baseline was achieved by flowing water through the flow cell, a solution with the next higher concentration was allowed to flow over the sensor. This process was repeated in succession from the lowest to the highest concentration. Each SPR experiment was run three times for a given HPX and surface.

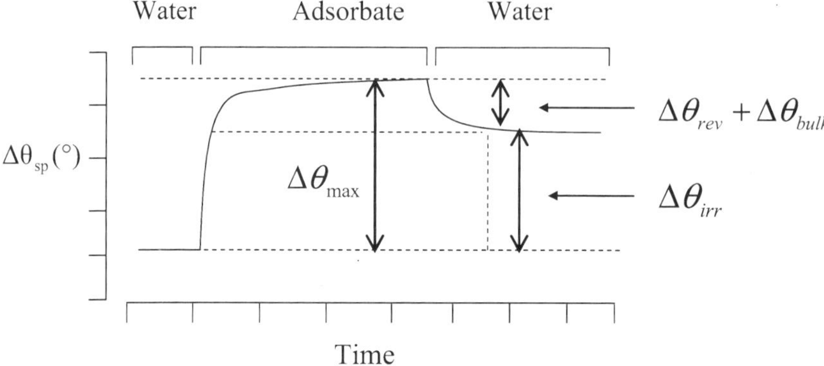

Figure 1. Schematic representation of raw SPR data for the case in which water is used to establish a baseline value of $\Delta\theta_{sp}$. A solution containing adsorbate produces a maximum change in $\Delta\theta_{sp}$ ($\Delta\theta_{max}$). Water eliminates the bulk contribution ($\Delta\theta_{bulk}$) and removes reversibly bound adsorbate ($\Delta\theta_{rev}$) to yield irreversibly bound adsorbate ($\Delta\theta_{irr}$).

Analysis of SPR Data

SPR is commonly used in the fields of chemistry and biochemistry to characterize biological surfaces and monitor binding events (*28, 29*). Although there are many types of SPR sensing systems, most use gold-coated sensors in a Kretchman prism configuration. SPR sensing systems detect refractive index changes in the vicinity of the gold surface by monitoring changes in the resonance angle (θ_{sp}), the incident angle at which the reflected light intensity is a minimum (*30*). Because the SPR system can monitor the change in the refractive index of the medium within ~200 nm of the surface, it is sensitive to adsorbed molecules as well as the surrounding medium. This latter effect (bulk effect) produces a displacement in θ_{sp} with respect to increasing analyte concentration. Figure 1 shows a schematic depiction for the adsorption of analyte to the sensing surface and the desorption of the analyte from the sensing surface by washing with water. Here, the maximum increase in resonance angle ($\Delta\theta_{max}$) corresponds to the observed $\Delta\theta_{sp}$ caused by reversible adsorption ($\Delta\theta_{rev}$), irreversible adsorption ($\Delta\theta_{irr}$), and the bulk effect ($\Delta\theta_{bulk}$). The decrease in $\Delta\theta_{sp}$ after subsequently flowing water over the surface corresponds to the desorption of some analyte molecules and the elimination of the bulk effect (*31*).

$\Delta\theta_{max}$ and $\Delta\theta_{irr}$ values are retrieved for each solution concentration as depicted in Figure 1. Next, eqs 1 and 2 are used to obtain the change in the resonant angle associated with adsorption (θ_a) by subtracting the contribution of bulk refractive index changes in the dielectric medium:

$$\Delta\theta_a = \Delta\theta_{max} - c \cdot \frac{d\theta_{sp}}{dc} \tag{1}$$

$$\frac{d\theta_{sp}}{dc} = \frac{d\theta_{sp}}{dn} \cdot \frac{dn}{dc} \tag{2}$$

where c is the concentration, dn/dc is the refractive index increment of the adsorbate solution, and $d\theta_{sp}/dn = 61.5°$ is an instrument specific parameter obtained by calibrating the instrument with ethylene glycol standards. $\Delta\theta_a$ is used to calculate the surface excess (Γ) for each concentration by using the de Feijter equation (*32*):

$$\Gamma = \frac{L\,(n_f - n)}{dn/dc} = \frac{\Delta\theta_a}{d\theta/dL}\frac{(n_f - n)}{dn/dc} \tag{3}$$

where, n_f is the refractive index of the film, which is assumed to be 1.45, and $n \approx$ 1.32823 is the refractive index of the solvent (water) (*33*). The other constant in eq 3 is $d\theta/dL$ which can be determined by Fresnel calculations. Theoretical Fresnel calculations are carried out by using a computer simulation program written in Mathlab. The values of n and the thickness of the six layers used in the Fresnel calculations are summarized in Table 1. For model cellulose and SAM surfaces, $d\theta/dL$ is equal to $(4.2 \pm 0.2) \times 10^{-3}$ deg·Å^{-1} and $(3.9 \pm 0.2) \times 10^{-3}$ deg·Å^{-1}, respectively. The SAM value is in excellent agreement with the value of 4×10^{-3} deg·Å^{-1} previously reported by Tulpar et al. (*34*).

180

Table 1. Thicknesses and refractive indices of different layers used to determine dθ/dL[a]

	Layer	Thickness $\mathring{A}$	Refractive Index n	Absorption Coefficient κ
L1	Sapphire prism	5×10^6	1.76074 (*35*)	0
L2	Chromium	20	4.1106 (*36*)	4.3492 (*36*)
L3	Gold	480	0.174 (*36*)	4.86 (*36*)
L4	Cellulose or SAM	195 (*25*) 16 (*37*)	1.44 (*38*) 1.45 (*37*)	0 0
L5	HPX	Variable	1.45[b]	0
L6	Water	500	1.32813 (*33*)	0

[a]Numbers in parentheses correspond to references

[b]Assumed to be 1.45 for an organic medium

Results and Discussion

HPX Characterization

In addition to the properties of the HPX samples reported elsewhere (*11*), the DSs of the specific HPX samples used in this study were deduced from ^{1}H NMR studies by preparing chloroform-soluble acetoxypropyl xylan (APX) derivatives. Figure 2 shows ^{1}H NMR spectra of the four APX derivatives. The DSs for the HPX derivatives were determined by assuming that there were only two available hydroxyl groups for derivatization in each xylose unit. Furthermore, assuming complete acetylation and the addition of only a single HP unit at any given xylose hydroxyl group yields

$$DS = \frac{\left(\text{Integral of the HP} - CH_3 \text{ peak at } \delta \approx 1.2\,\text{ppm}\right)}{\left(\text{Integral of the acetyl} - CH_3 \text{ peak at } \delta \approx 2.0\,\text{ppm}\right)} \times 2$$

DS values are summarized in Table 2. The data are consistent with the general trend of DS increasing with the pH of the hydroxypropylation reaction.

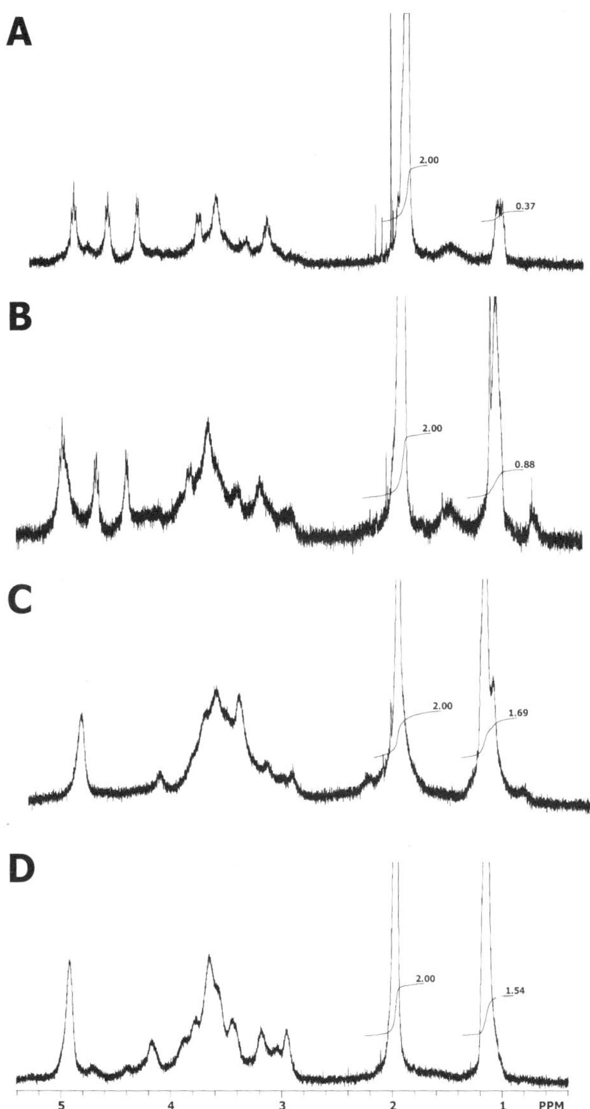

Figure 2. ^{1}H NMR spectra of acetylated HPX derivatives. Spectra A–D correspond to acetylated versions of HPX120, HPX125, HPX127, and HPX130, respectively.

In addition to ^{1}H NMR, one additional bulk solution characterization technique was carried out on the HPX derivatives. Differential refractometer studies were performed in water to determine the refractive index increment (dn/dc) at a wavelength of 690 nm. These values, necessary for analyzing SPR results, are also summarized in Table 2. As seen in the table, dn/dc in water is independent of DS. Additionally, the values are assumed to be relatively independent of wavelength between 690 nm and 780 nm, the wavelength for SPR experiments.

Table 2. Selected properties of HPX derivatives

HPX Acronym	$HP\ DS^a$	dn/dc^b $mL{\cdot}g^{-1}$	CAC $mg{\cdot}L^{-1}$	$\Delta\gamma_{max}$ $mN{\cdot}m^{-1}$	$\Gamma_{max}{}^c$ $\mu mol{\cdot}m^{-2}$
HPX120	0.37	0.128 ± 0.001	~ 90	~ 3	~ 0.8
HPX125	0.88	0.128 ± 0.001	~ 8	~ 9	~ 1.5
HPX127	1.69	0.128 ± 0.001	~ 20	~ 19	~ 11
HPX130	1.54	0.125 ± 0.002	~ 4	~ 18	~ 6

[a]DS from ^{1}H NMR of acetylated HPXs

[b]Water at 20.0 °C and a wavelength of 690 nm

[c]From eq 4

Surface Tension Studies of Aqueous HPX Solutions

Gibbs derived an expression for the surface excess of a solute at constant temperature and pressure at an air/liquid interface (*39*):

$$\Gamma = -\frac{1}{RT}\left(\frac{\partial \gamma}{\partial \ln a}\right)_{T,p} \approx -\frac{1}{RT}\left(\frac{\partial \gamma}{\partial \ln c}\right)_{T,p} \tag{4}$$

where R is the gas constant, γ is the surface tension, and a and c are the activity and concentration of the solute in solution, respectively. Hence, Γ at the air/water interface can be estimated from measurements of γ as a function of c. Additionally, γ–c plots which show drops in γ and a concentration invariant γ at high c can be used to estimate critical micelle concentrations (CMCs) of surfactants. Estimates of the CMC are generally made by extrapolating the sharply decreasing γ region and plateau regions to a common intercept as done in Figures 3 and 4. The sharpness of the transition increases with aggregation number. For random coil polymers, aggregation numbers are small (as small as 1–2 polymer chains) and aggregates are irregularly shaped leading to broad transitions. As a consequence, the concentration of the transition is more appropriately termed a critical aggregation concentration (CAC).

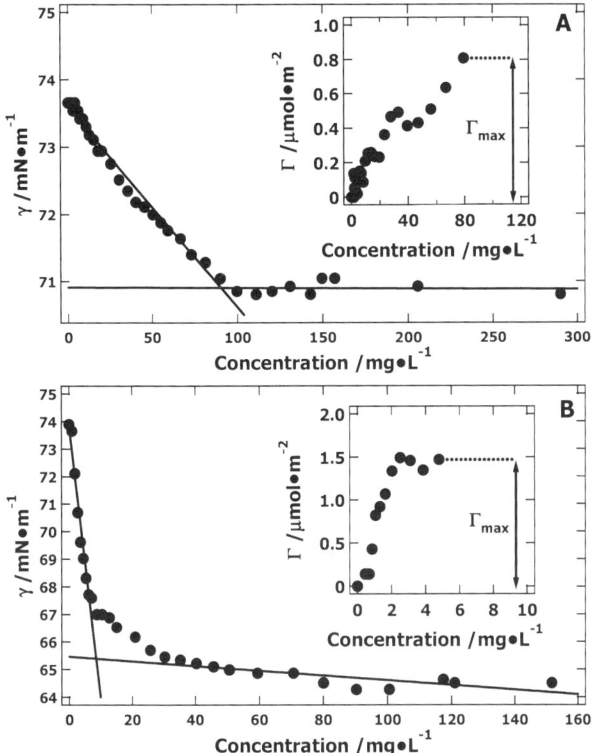

Figure 3. γ–c curves for HPX derivatives at the air/water interface at T = 20.0 °C. Graphs A and B correspond to HPX120 and HPX125, respectively. The solid lines provide estimates of the CAC. The insets provide estimates of Γ–c deduced from eq 4.

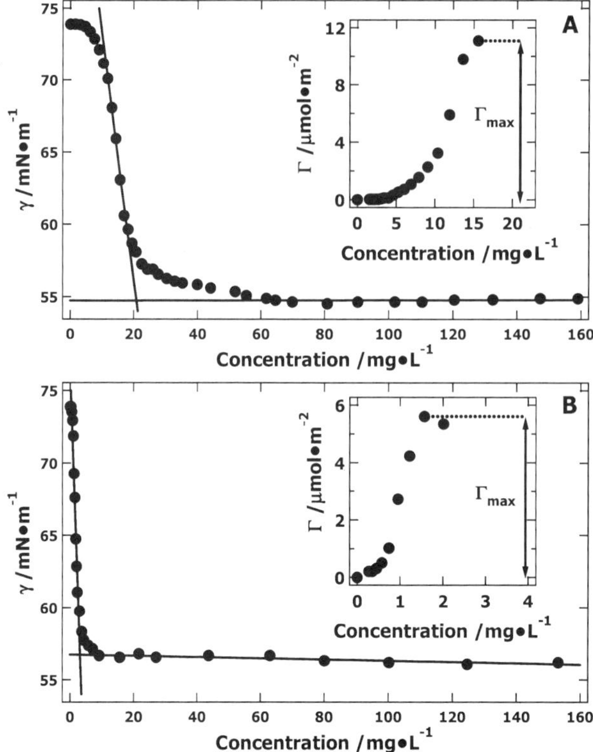

Figure 4. γ–c curves for HPX derivatives at the air/water interface at T = 20.0 °C. Graphs A and B correspond to HPX127 and HPX130, respectively. The solid lines provide estimates of the CAC. The insets provide estimates of Γ–c deduced from eq 4.

Several points in Figures 3 and 4 are worth noting. First, HPX with the lowest DS (HPX120) has the highest CAC, ~90 mg·L^{-1}, and the smallest overall change in surface tension, $\Delta\gamma \approx 2.9$ mN·m^{-1}. As DS increases, there is a shift to smaller CAC values, and $\Delta\gamma$ increases with DS. These values are summarized in Table 2. Similar estimates of the CAC are obtained for plots of γ vs. ln c. Second, Γ calculated from eq 4 (insets of Figures 3 and 4) increases with DS. Maximum values (Γ_{max}) are also provided in Table 2.

On the basis of the data in Figures 3 and 4, it can be concluded that even though hydroxypropylation enhances polysaccharide aqueous solubility and dispersion by breaking up hydrogen bonding, the HP groups enhance the amphiphilic character of the xylan. Similar behavior is seen in hydroxypropyl cellulose (HPC) where $\Delta\gamma \approx 28$ mN·m^{-1} (40). The principle differences between xylan and cellulose are two vs. three hydroxyl groups available for hydroxypropylation, respectively, and a more rigid backbone structure for cellulose. It is also interesting to consider the observed behavior for the HPX derivatives relative to ethylene oxide–propylene oxide copolymers. Block copolymers (Pluronics) show that propylene oxide units enhance the hydrophobic character of the copolymer. These surfactants show increasing $\Delta\gamma$

and decreasing CACs with increasing propylene oxide content (*41–43*), ultimately leading to water insoluble surfactants (*41*). Similar effects are believed to be the cause for the enhancement of HPX surface activity at the air/water interface with increasing DS. As we will show shortly, HPX derivatives have stronger affinity for model hydrophobic surfaces than for cellulose and other hydrophilic model surfaces.

Adsorption onto Regenerated Cellulose and SAM-OH Surfaces

Figure 5 shows representative SPR raw data for HPX adsorption onto model cellulose surfaces. The maximum change in $\Delta\theta_{sp}$ from all sources (reversible adsorption, irreversible adsorption, and the bulk effect) of $\sim0.003°-0.005°$ and the contribution arising solely from irreversible adsorption of $\sim0.002°-0.003°$ is small. A better way to verify this interpretation is to use eqs 1–3 to produce adsorption isotherms (Figure 6 for reversible and irreversible adsorption). The y-axis for Figure 6 is chosen to be 0 to 0.9 mg·m^{-2} for the sake of subsequent comparisons of HPX adsorption onto cellulose and SAM-OH surfaces with HPX adsorption onto SAM-CH$_3$ surfaces. Additionally, two dotted horizontal lines have been added as an estimate of Γ for a flat monolayer (Γ_{mono}) of the HPX derivatives with the lowest and highest DS. These estimates assume that the cross-sectional area of the xylose unit is comparable to the cross-sectional area for a cellulose unit, ~60 Å^2 (*44*). The lower dotted line represents Γ_{mono} for DS ~0.37 and the upper dotted line represents Γ_{mono} for DS ~1.69. In order to better see the trends with respect to DS, insets are provided. Figure 6 clearly indicates that HPX has the same affinity for the cellulose and SAM-OH surfaces, the affinity is insensitive to DS, and HPX adsorbs at submonolayer coverage. The relatively weak adsorption to these hydrophilic surfaces suggests hydrogen bonding interactions are insufficient to promote HPX adsorption from aqueous solution. Moreover, the amphiphilicity observed at the air/water interface for HPX derivatives, where air can be regarded as hydrophobic, clearly did not provide predictive insight into HPX adsorption onto hydrophilic surfaces. In the next section, HPX adsorption onto hydrophobic surfaces is considered.

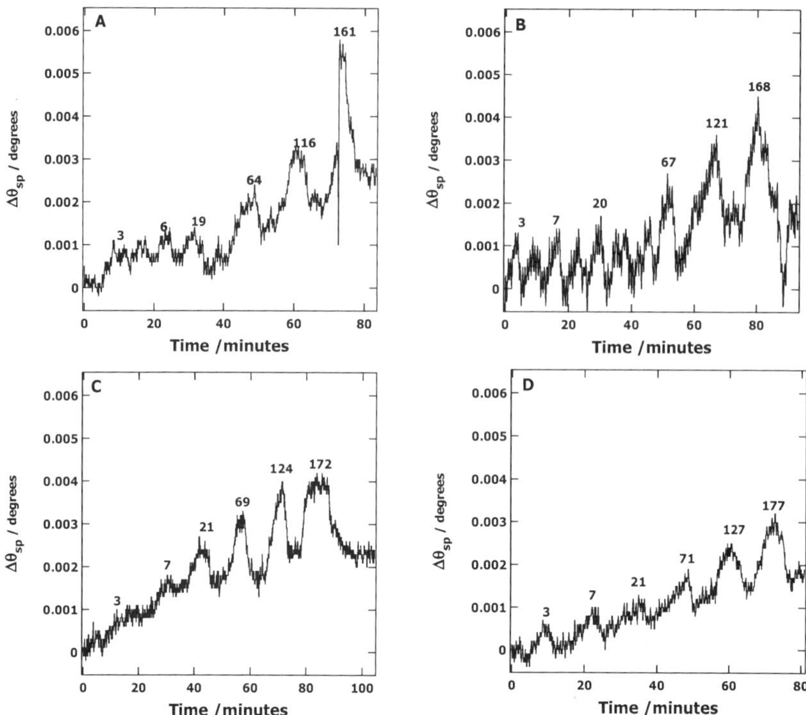

Figure 5. Representative SPR data for HPX adsorption onto cellulose regenerated from spin-coated TMSC films at 20.0 °C. Graphs A–D correspond to HPX120, HPX125, HPX127, and HPX130, respectively. Solution concentrations in $mg \cdot L^{-1}$ correspond to the numbers on A–D. Water is flowed through the SPR instrument before and after each new adsorbate solution leading to the observed saw-tooth pattern.

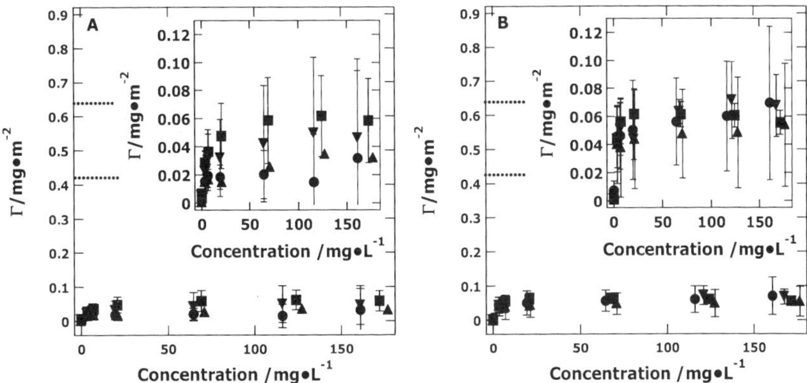

Figure 6. HPX adsorption isotherms (reversible and irreversible) on (A) regenerated cellulose, and (B) SAM-OH at 20.0 °C. Symbols correspond to HPX120 (circles), HPX125 (inverted triangles), HPX127 (squares), and HPX130 (triangles) with one standard deviation error bars. Dotted lines on the main graph correspond to estimates of monolayer coverage for the lowest (bottom line) and highest (top line) DS HPX samples.

Adsorption onto SAM-CH$_3$ Surfaces

Figure 7 shows representative SPR data for HPX adsorption onto SAM-CH$_3$ surfaces. Comparing these curves with Figure 5, there is at least a 10-fold increase in $\Delta\theta_{sp}$ for both total angle change (reversible adsorption, irreversible adsorption, and the bulk effect) and irreversible HPX adsorption onto cellulose and SAM-OH surfaces. Following the procedure outlined in eqs 1–3, the data in Figure 7 is converted into adsorption isotherms (Figure 8) for total (Figure 8A, reversible and irreversible) and irreversible adsorption (Figure 8B). More than 90% of the total adsorbed amount comes from irreversible adsorption. Figure 8 also contains dotted horizontal lines corresponding to Γ values for a flat HPX monolayer of DS ~0.37 (bottom dotted line) and DS ~1.69 (top dotted line). As one can see, HPX adsorption is consistent with an adsorbed monolayer on the SAM-CH$_3$ surface. It is also worth noting that there may be a slight (though statistically insignificant for this study) tendency for HPX adsorption on SAM-CH$_3$ surfaces to increase with decreasing DS. If this trend is real, it would be consistent with poorer HPX solubility for low DS HPX promoting deposition onto hydrophobic surfaces.

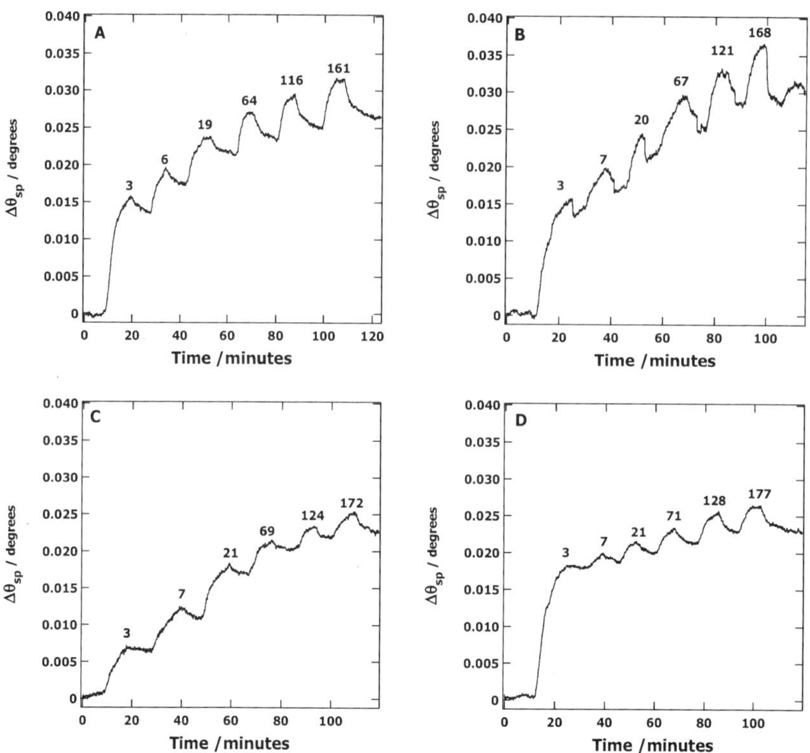

Figure 7. Representative SPR data for HPX adsorption onto SAM-CH₃ surfaces at 20.0 °C. Graphs A–D correspond to HPX120, HPX125, HPX127, and HPX130, respectively. Solution concentrations in mg·L⁻¹ correspond to the numbers on A–D. Water is flowed through the SPR instrument before and after each new adsorbate solution leading to the observed saw-tooth pattern.

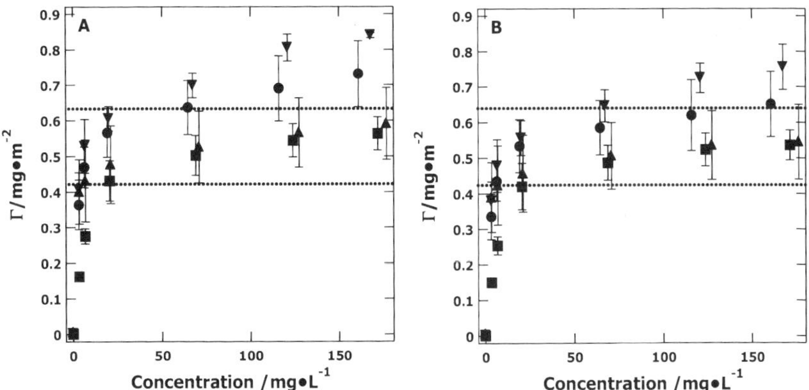

Figure 8. HPX adsorption isotherms on SAM-CH₃ surfaces at 20.0 °C. Graph A corresponds to total adsorption (reversible and irreversible) whereas B corresponds to irreversible adsorption. The symbols correspond to HPX120 (circles), HPX125 (inverted triangles), HPX127 (squares), and HPX130 (triangles) with one standard deviation error bars. The dotted horizontal lines correspond to estimates of monolayer coverage for the lowest (bottom line) and highest (top line) DS HPX derivatives.

Conclusions

Water-soluble HPX derivatives are amphiphilic and exhibit increasing affinity for the air/water interface with increasing DS. This behavior is in stark contrast to hydrophilic cellulose and SAM-OH surfaces, where HPX adsorbs at submonolayer coverage. Hence, even though hydroxypropylation yields water-soluble polysaccharides, the derivatives are not necessarily strongly hydrophilic. This interpretation is supported by the observation of HPX adsorption onto hydrophobic SAM-CH₃ surfaces at essentially monolayer coverage. Collectively, these studies show that hydroxypropylation is a good strategy for making soluble hemicellulose derivatives that could be further derivatized to produce surface modifying agents.

Acknowledgments

We thank Mr. Thomas Ryan from Reichert, Inc. for SPR calibration, Dr. Raymond Dessy at Virginia Tech for help with Fresnel reflectivity calculations, and Russell Mammei at Virginia Tech for helping with electron beam evaporation. The project was supported by the National Research Initiative of the USDA Cooperative State Research, Education and Extension Service, grant number 2005-35504-16088 and the National Science Foundation (CHE-0724126 and DMR-0552661).

References

1. Rinaudo, M. In *Viscoelasticity of Biomaterials;* Glasser, W. G., Hatakeyama, H., Eds.; ACS Symposium Series 489; American Chemical Society: Washington, DC, 1992; pp 24–37.
2. Lerouxel, O.; Cavalier, D. M.; Liepman, A. H.; Keegstra, K. *Curr. Opin. Plant Biol.* **2006,** *9,* 621–630.
3. Reiter, W.-D. *Curr. Opin. Plant Biol.* **2002,** *5,* 536–542.
4. Parham, R. A.; Gray, R. L. In *The Chemistry of Solid Wood;* Rowell, R., Ed.; Advances in Chemistry Series 207; American Chemical Society: Washington, DC, 1984; pp 3–56.
5. Sun, R.; Sun, X. F.; Tomkinson, J. In *Hemicelluloses: Science and Technology;* Gatenholm, P., Tenkanen, M., Eds.; ACS Symposium Series 864; American Chemical Society: Washington, DC, 2004; pp 2–22.
6. Ebringerová, A.; Heinze, T. *Macromol. Rapid Commun.* **2000,** *21,* 542–556.
7. Glasser, W. G.; Ravindran, G.; Jain, R. K.; Samaranayake, G.; Todd, J. *Biotechnol. Progr.* **1995,** *11,* 552–557.
8. Ebringerová, A.; Hromádková, Z. *Biotechnol. Genet. Eng. Rev.* **1999,** *16,* 325–346.
9. Gröndahl, M.; Teleman, A.; Gatenholm, P. *Carbohydr. Polym.* **2003,** *52,* 359–366.
10. Gröndahl, M.; Eriksson, L.; Gatenholm, P. *Biomacromolecules* **2004,** *5,* 1528–1535.
11. Jain, R. K.; Sjöstedt, M.; Glasser, W. G. *Cellulose* **2001,** *7,* 319–336.
12. Gabrielii, I.; Gatenholm, P.; Glasser, W. G.; Jain, R. K.; Kenne, L. *Carbohydr. Polym.* **2000,** *43,* 367–374.
13. Ebringerová, A.; Hromadkova, Z.; Burchard, W.; Dolega, R.; Vorwerg, W. *Carbohydr. Polym.* **1994,** *24,* 161–169.
14. Timell, T. E.; Glaudemans, C. P. J.; Gillham, J. K. *Tappi* **1959,** *42,* 623–634.
15. Carson, J. F.; Maclay, W. D. *J. Am. Chem. Soc.* **1948,** *70,* 293–295.
16. Vincendon, M. *Makromol. Chem.* **1993,** *194,* 321–328.
17. Antal, M.; Ebringerová, A.; Hromádková, Z.; Pikulik, I. I.; Laleg, M.; Micko, M. M. *Papier* **1997,** *51,* 223–226.
18. Ebringerová, A.; Hromádková, Z.; Kacuráková, M.; Antal, M. *Carbohydr. Polym.* **1994,** *24,* 301–307.
19. Mora, F.; Ruel, K.; Comtat, J.; Joseleau, J.-P. *Holzforschung* **1986,** *40,* 85–91.
20. Henriksson, Å.; Gatenholm, P. *Holzforschung* **2001,** *55,* 494–502.
21. Paananen, A.; Österberg, M.; Rutland, M.; Tammelin, T.; Saarinen, T.; Tappura, K.; Stenius, P. In *Hemicelluloses: Science and Technology;* Gatenholm, P., Tenkanen, M., Eds.; ACS Symposium Series 864; American Chemical Society: Washington, DC, 2004; pp 269–290.
22. Esker, A.; Becker, U.; Jamin, S.; Beppu, S.; Renneckar, S.; Glasser, W. In *Hemicelluloses: Science and Technology;* Gatenholm, P., Tenkanen, M., Eds.; ACS Symposium Series 864; American Chemical Society: Washington, DC, 2004; pp 198–219.
23. Rivera-Armentaa, J. L.; Heinze, T.; Mendoza-Martínez, A. M. *Eur. Polym. J.* **2004,** *40,* 2803–2812.

24. Carson, J. F.; Maclay, W. D. *J. Am. Chem. Soc.* **1946**, *68*, 1015–1017.
25. Kontturi, E.; Thune, P. C.; Niemantsverdriert, J. W. H. *Langmuir* **2003**, *19*, 5735–5741.
26. Schaub, M.; Wenz, G.; Wegner, G.; Stein, A.; Klemm, D. *Adv. Mater.* **1993**, *5*, 919–922.
27. Bain, C. D.; Troughton, E. B.; Tao, Y.-T.; Evall, J.; Whitesides, G. M.; Nuzzo, R. G. *J. Am. Chem. Soc.* **1989**, *111*, 321–335.
28. Brockman, J. M.; Nelson, B. P.; Corn, R. M. *Annu. Rev. Phys. Chem.* **2000**, *51*, 41–63.
29. Gradwell, S. E.; Renneckar, S.; Esker, A. R.; Heinze, T.; Glasser, W. G. *C. R. Biol.* **2004**, *327*, 945–953.
30. Toyama, S.; Aoki, K.; Kato, S. *Sens. Actuators, B* **2005**, *108*, 903–909.
31. Sigal, G. B.; Mrksich, M.; Whitesides, G. M. *Langmuir* **1997**, *13*, 2749–2755.
32. Feijter, J. A. d.; Benjamins, J.; Veer, F. A. *Biopolymers* **1978**, *17*, 1759–1772.
33. Harvey, A. H.; Gallagher, J. S.; Sengers, J. M. H. L. *J. Phys. Chem. Ref. Data* **1998**, *27*, 761–774.
34. Tulpar, A.; Ducker, W. A. *J. Phys. Chem. B* **2004**, *108*, 1667–1676.
35. Malitson, I. H. *J. Opt. Soc. Am.* **1962**, *52*, 1377–1379.
36. Palik, E. D. *Handbook of Optical Constants of Solids;* Academic Press: Orlando, FL, 1985.
37. Peterlinz, K. A.; Georgiadis, R. *Langmuir* **1996**, *12*, 4731–4740.
38. Holmberg, M.; Berg, J.; Stemme, S.; Odberg, L.; Rasmusson, J.; Claesson, P. *J. Colloid Interface Sci.* **1997**, *186*, 369–381.
39. Evans, D. F.; Wennerström, H. *The Colloidal Domain: Where Physics, Chemistry, Biology, and Technology Meet,* 2nd ed.; Advances in Interfacial Engineering Series; Wiley-VCH: New York, 1999.
40. McNally, E.; Zografi, G. *J. Colloid Interface Sci.* **1990**, *138*, 61–68.
41. Alexandridis, P.; Hatton, T. A. *Colloids Surf., A* **1995**, *96*, 1–46.
42. Paterson, I. F.; Chowdhry, B. Z.; Leharne, S. A. *Langmuir* **1999**, *15*, 6187–6194.
43. Dong, J.; Chowdhry, B. Z.; Leharne, S. A. *Colloids Surf., A* **2006**, *277*, 249–254.
44. Kawaguchi, T.; Nakahara, H.; Fukuda, K. *Thin Solid Films* **1985**, *133*, 29–38.

Cellulosic Surfaces

Polysaccharide Derivatives for the Modification of Surfaces by Self-Assembly

Thomas Heinze, Stephanie Hornig, Nico Michaelis, and Katrin Schwikal

Centre of Excellence for Polysaccharide Research, Friedrich Schiller University of Jena, Humboldtstraße 10, D-07743 Jena, Germany

Typical reaction pathways for the synthesis of self-assembling polysaccharides (dextran, cellulose, and xylan) are discussed. Dextran- and cellulose derivatives with different sulfur-containing functional groups (thiol-, thioester- and thioether moieties) possess a remarkable ability to self-assemble on gold surfaces. Aminocellulose derivatives, obtained by nucleophilic substitution of cellulose-p-toluenesulfonic acid ester with di- and oligoamines, are able to form monolayers on various substrates. The NH_2 moieties of these biocompatible aminocellulose surfaces can be used to immobilize enzymes while maintaining enzyme activity. A novel approach to aminocellulose through introduction of polyamidoamine based dendrons by the copper-catalyzed Huisgen reaction of azido cellulose with dendrons possessing ethynyl moieties is described. Furthermore, the preparation of cationic xylan derivatives and their adsorption behavior on cellulose and two types of self-assembled monolayers is reviewed. Xylan layers on various pulp fibers were found to enhance paper strength.

Introduction

Polysaccharides are the most important renewable terrestrial resource and, after chemical modification by etherification or esterification, are used in many different application fields (*1*). Cellulose esters and ethers are commercially produced in large quantities. The chemical modification of polysaccharides is one of the most important processing methods available and enables us to take full advantage of this fascinating biopolymer resource. Polysaccharides that are synthesized by bacteria, such as dextran, offer additional opportunities because their structure and possibly even shape may be controlled during biosynthesis (*2*). Structures and processes that rely upon self-assembly are ubiquitous in polysaccharide systems. In fact, they are the basis for the amazing functions and properties of this class of biopolymers in nature. In addition to their inherent ability to form supramolecular structures, the possibility to introduce smart functional groups into their molecular structure that allow controlling and directing their self-assembly behavior opens new avenues to readily-obtainable advanced materials. Controlled nano- and micro structuring of polysaccharide derivatives by self-assembly can be achieved by combining the naturally occurring molecular structures with functional groups that impart the desired interactions between the semi-synthetic polysaccharide derivatives and synthetic polymers or inorganic materials (including noble metals, glass etc.).

This chapter aims to detail several synthetic approaches, including a few of our own, to synthesizing self-assembling polysaccharide derivatives.

Polysaccharide Derivatives for the Modification of Gold Surfaces

Polymer coatings can be used to protect materials against environmental influences or to enhance the biocompatibility of materials for use in, e.g., tissue engineering or bioanalytical devices. Advantages of surface modification with polysaccharides include the accessibility of numerous reactive groups (in particular hydroxyl moieties), which can be functionalized to fit the desired application, and the compatibility of many polysaccharides with living systems. Molecules containing thiol- or disulfide groups are known to spontaneously adsorb from solution onto the surfaces of certain metals, such as gold, and to self-assemble into densely packed monolayers via formation of an adsorbate–substrate sulfur–metal bond (*3*). Mainly two methods have been used to coat gold surfaces with polysaccharides, in particular with cellulose and dextran:

1. Functionalization of polysaccharides with sulfur-containing groups

2. Use of low-molecular-weight self-assembled monolayers (SAMs) as a linker between the substrate and the polysaccharide

In the case of cellulose, the introduction of sulfur-containing groups is the preferred method for modifying gold substrates whereas for dextran both methods have been used. The first method has the advantage that it requires a less extensive preparation procedure and that chemisorption occurs at multiple positions throughout the same molecule yielding more stable monolayers compared to those of low-molecular-weight substances (*4*).

Cellulose Derivatives for the Modification of Gold Surfaces

Sulfur-containing functional groups, capable of chemisorption onto gold, can be introduced into polysaccharide molecules by functionalization with thiol-, thioester-, or thioether moieties. One possible approach is nucleophilic substitution (S_N) of chlorine in chlorine-containing cellulose esters, carbonates, and isocyanates with sodium thiosulfate (Figure 1) (5). The reaction products are soluble in water. The displacement of chlorine with thiosulfate is almost quantitative yielding degrees of substitution (DS) between 0.8 and 1.8.

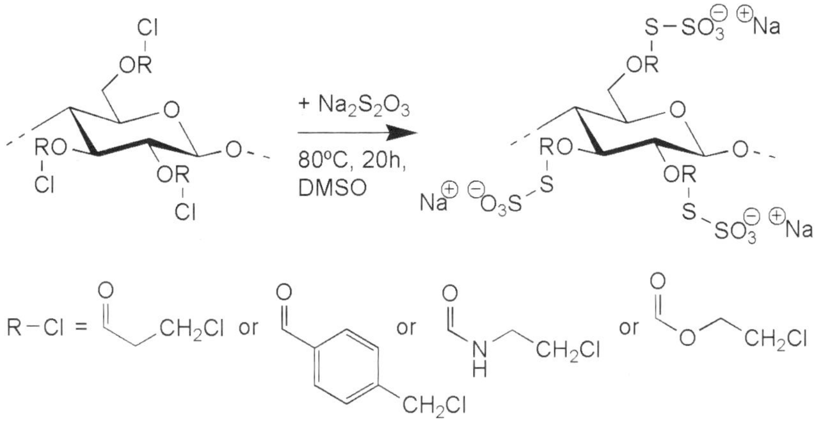

Figure 1. Synthesis of cellulose thiosulfate derivatives

Another option is the addition of thiosulfate to C–C double bonds. 6-*O*-(2",3"-bis(thiosulfato)propyloxy-2'-hydroxypropyl) cellulose was synthesized by addition of potassium tetrathionate to 6-*O*-(3'-allyloxy-2'-hydroxypropyl) cellulose, which had been prepared before by reacting cellulose with allyl glycidyl ether in aqueous NaOH (Figure 2, route A) (4). The thiosulfate derivatives adsorbed from aqueous solutions onto gold surfaces, forming dense monolayers with a mean thickness of about 4 nm. The homolytic cleavage of the substituents' S–S bond proceeds via direct oxidative addition to gold. Adsorption experiments with fibrinogen and bovine serum albumin provided evidence for non-specific surface–protein interactions, after coating of the gold surface. However, in water, the thiosulfate derivatives tended to precipitate after some time. To improve the water solubility, carboxymethyl cellulose was reacted with allyl glycidyl ether and tetrathionate in the same way as described above, yielding a DS of 0.1 with respect to the bisthiosulfate groups (6). The DS could be increased to 0.17 by using a two step procedure with bromine addition to the allyl moiety and subsequent reaction with thiosulfate. The same reaction sequence in water gave 6-*O*-(2"-hydroxy-3"-thiosulfatopropyloxy-2'-hydroxypropyl)-carboxymethyl cellulose with a DS of 0.15 as shown in Figure 2, route B. Surface plasmon resonance (SPR) spectroscopy experiments revealed a decrease in monolayer thickness with increasing thiosulfate content, attributable to the increase in the number of binding sides per molecule available for chemisorption onto gold. The thickness of the layers was in the range of 1.3–2.4 nm.

Introduction of sulfur atoms into the polymer backbone was also achieved by S_N reaction of 6-*O*-tosyl cellulose with sodium methyl mercaptide, leading to a 6-thiomethyl-6-deoxy cellulose derivative with a DS of 0.65 (Figure 3, route B). Subsequent reaction with sodium chloroacetate gave a water soluble thiomethyl carboxymethyl cellulose suitable for SAM formation on gold surfaces. Despite the comparatively high DS, the derivatives showed a lower affinity toward the gold surface than the cellulose thiosulfates, as determined by SPR spectroscopy. The resulting film exhibited a thickness of 1.2 nm. Tosyl cellulose also reacts with sodium thiosulfate yielding 6-thiosulfate-6-deoxy cellulose (Figure 3, route A) (*5*).

The presence of carboxyl groups is not only advantageous for the aqueous solubility of a molecule but also for the nature of its chemical reactions. Biotin was immobilized via a linker using *N,N'*-dicyclohexyl carbodiimide as coupling agent (*7*). 6-Thiomethyl-6-deoxy carboxymethyl cellulose gave a biotin conjugate bearing 0.3 biotin molecules per anhydroglucose unit whereas the conversion with the bisthiosulfate derivative gave a DS of 0.08. The biotin conjugates formed SAMs that had the ability to irreversibly bind the receptor protein streptavidin.

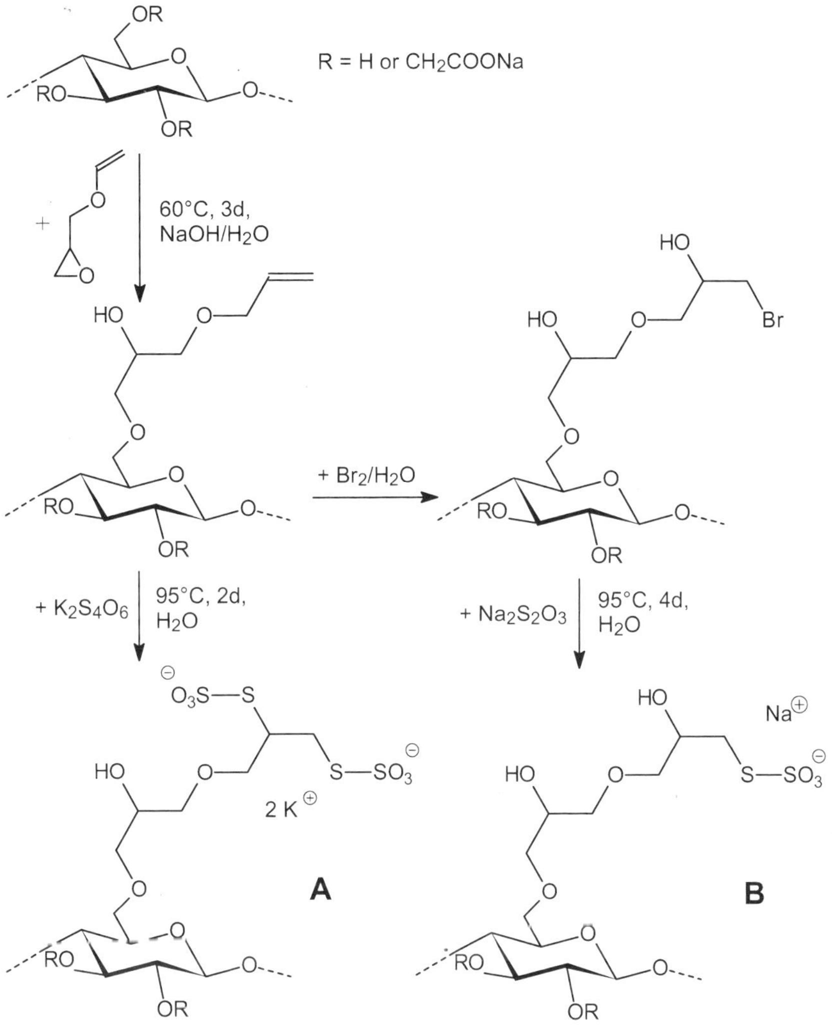

Figure 2. Synthesis of 6-O-(2",3"-bis(thiosulfato)propyloxy-2'-hydroxypropyl) cellulose (route A) and 6-O-(2"-hydroxy-3"-thiosulfatopropyloxy-2'-hydroxypropyl)-carboxymethyl cellulose (route B)

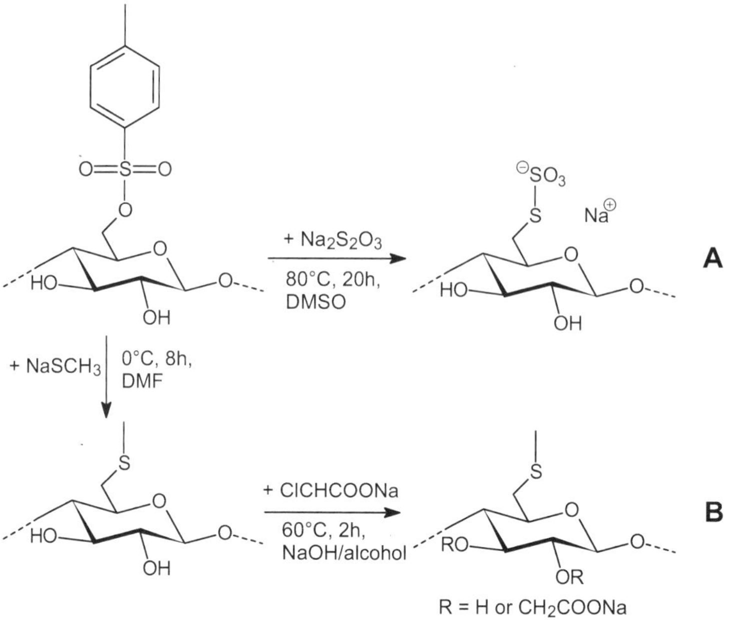

Figure 3. Synthesis of 6-thiosulfate-6-deoxy cellulose (route A) and 6-thiomethyl-6-deoxy carboxymethyl cellulose (route B) from 6-O-tosyl cellulose

A simpler and more efficient method for introducing sulfur-containing groups into polysaccharides is esterification with sulfur-containing carboxylic acids. By using sulfur-containing carboxylic acids and in situ activating agents, highly substituted cellulose esters can be obtained in a one-step synthesis. The antioxidant α-lipoic acid, a biologically active compound, can be covalently bound to cellulose using different methods of in situ activation of the carboxylic acid (*8*). The reaction proceeds homogeneously in *N,N*-dimethylacetamide (DMA)/LiCl with *N,N'*-carbonyldiimidazole (CDI) or *p*-toluenesulfonyl chloride (TosCl) as activating agents (Figure 4). Using this reaction, DS values in the range of 0.11–1.45 are accessible, but only the less substituted derivatives (up to DS 0.18) are soluble in organic solvents. The disulfide function stays intact under the mild reaction conditions as confirmed by IR and NMR spectroscopy. Adsorption studies from dimethyl sulfoxide (DMSO) solutions show the formation of monolayers with a thickness of 2.9–4.9 nm, as determined by SPR spectroscopy.

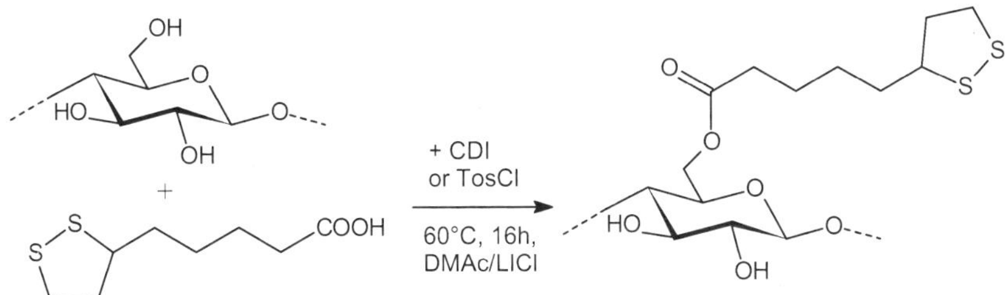

Figure 4. Reaction of cellulose with α-lipoic acid, in situ activated by N,N'-carbonyldiimidazole (CDI) or p-toluenesulfonyl chloride (TosCl)

Methyl cellulose is of considerable technological interest because of its reversible gelation behavior in aqueous solution around a specific lower critical solution temperature (LCST). Physical immobilization of methyl cellulose on a surface gives thermoreversible films, which exhibit thermally responsive wetting characteristics (*9*). The introduction of sulfur-containing groups into methyl cellulose for the immobilization on gold has been achieved by conversion of the reducing end groups with thiosemicarbazole (Figure 5).

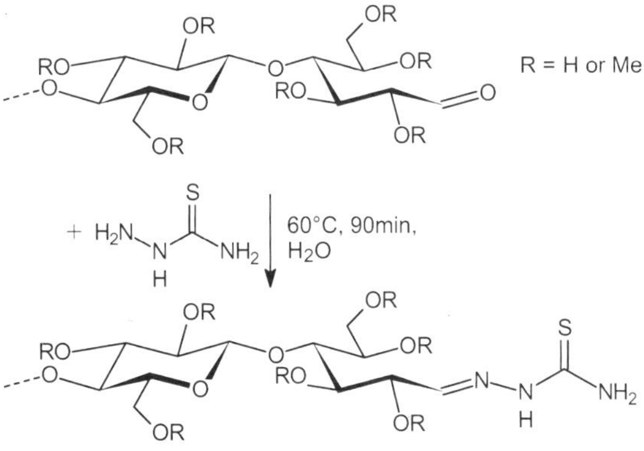

Figure 5. Selective modification of the reducing end group of methyl cellulose with thiosemicarbazide

The sulfur-modified methyl cellulose chains self-assembled into layers with a thickness of 15 nm. The water contact angle was found to increase above a temperature of 65 °C and decrease again below this LCST. The addition of sodium chloride resulted in depression of the LCST down to 30 °C, depending on salt concentration. The modified surfaces might be of interest for artificial scaffolds in tissue engineering.

Dextran Derivatives for the Modification of Gold Surfaces

One of the first commercial applications of dextran-functionalized gold surfaces has been in SPR sensors that interrogate biomolecular interactions. The immobilization of one interacting partner on the sensor surface and the controlled introduction of the other partner via a flow system allows the kinetic analysis of antigen–antibody, receptor–ligand, and DNA–protein interactions (*10*). The interactions are manifested in changes in the SPR angle, which is sensitive to the refractive index near the metal surface. Coating of the surface with the hydrophilic biopolymer dextran reduces non-specific protein adsorption (*11*, *12*). Furthermore, introduction of functional groups, like carboxyl moieties, can enhance the antibody immobilization capacity. Several techniques are used to bind dextran covalently onto gold surfaces either via low-molecular-weight SAMs or through sulfur-containing substituents.

The first commercialized bioanalytical system using dextran-functionalized gold surfaces is based on the SAM of 16-mercapto-hexadecan-1-ol (*13*). The SAM's hydroxyl groups are reacted with epichlorohydrin to introduce epoxy moieties that are subsequently reacted with dextran (Figure 6). Further treatment with bromoacetic acid yields carboxymethyl dextran. Covalent attachment of ligands is achieved by subsequent activation of the carboxyl groups with N-ethyl-N'-dimethylaminopropyl-carbodiimide (EDC) and N-hydroxy-succinimide (NHS). Besides for sensing of biomolecular interactions, a carboxylated dextran surface can further be functionalized with photochromic molecules, such as ethylester dihydroindolizine (DHI) (*14*). The ethylester DHI dextran derivative was found to enhance SPR signals upon photochromic switching (*14*).

A potential drawback for bioanalytical applications are the non-specific interactions that proteins with a low isoelectric point may have with the negatively charged carboxymethyl dextran surface. To overcome this problem, a neutral dextran matrix has been developed using 16-mercaptohexadecanoic acid for monolayer formation on the gold surface (Figure 7) (*15*). After reacting the surface carboxyl groups with hydrazine, dialdehyde dextran, which had been prepared by periodate oxidation with $NaIO_4$, was immobilized and subsequently reacted with the amino functions of proteins.

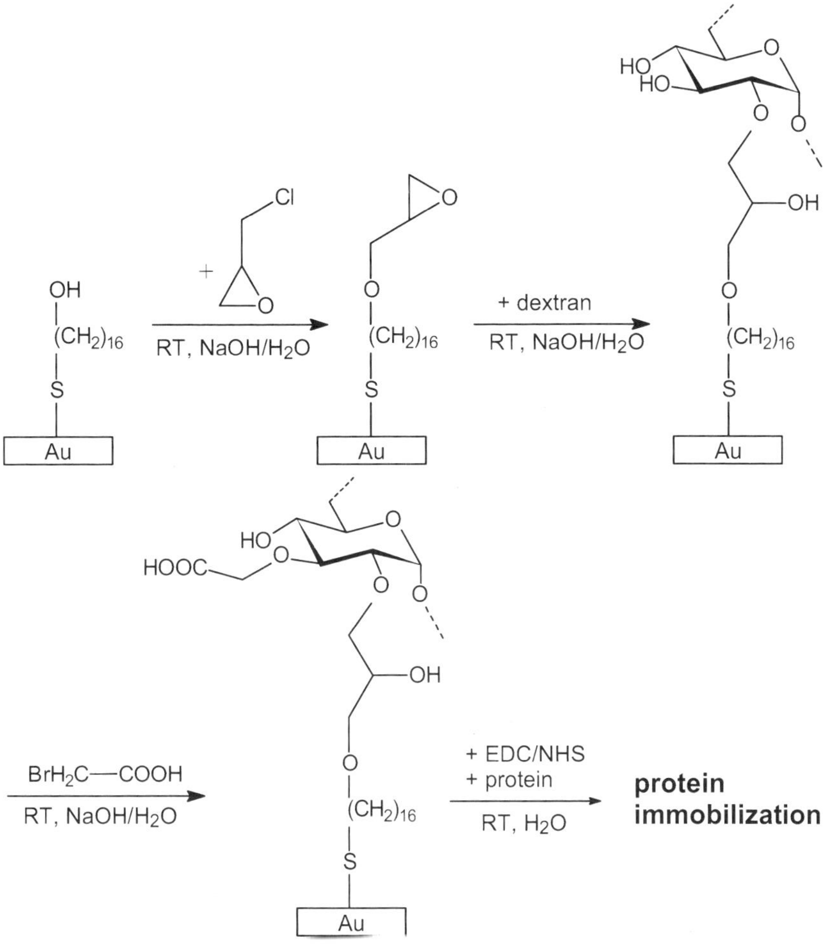

Figure 6. Functionalization of gold surfaces with carboxymethyl dextran via hydroxyl alkanethiol monolayers

An elegant but very time consuming multiple-step procedure for the modification of gold with dextran for SPR analysis is employed in the fabrication of β-cyclodextrin sensor chips (*16*). To produce these sensors, monolayers of 11-mercaptundecanoic acid are reacted with amino β-cyclodextrin (β-CD-NH$_2$) after activation of the carboxyl groups with NHS (Figure 8). The formation of inclusion complexes of the grafted β-cyclodextrin with the adamantoyl moieties of carboxymethyl adamantoyl dextran (AD-CM-dextran), prepared by reaction of dextran with chloroacetic acid and subsequent esterification with 1-adamantanecarbonyl chloride, leads to a dextran surface layer for protein immobilization.

204

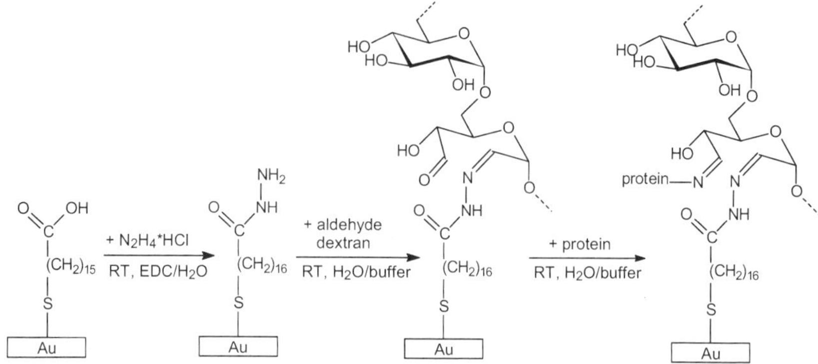

Figure 7. Functionalization of gold surfaces with aldehyde dextran via carboxyalkanethiol monolayers

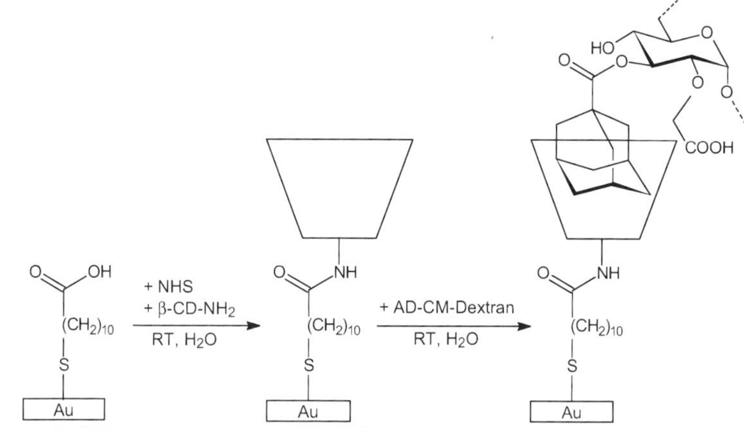

Figure 8. Schematic presentation of dextran grafting via complexation of amino β-cyclodextrin (β-CD-NH$_2$) with carboxymethyl adamantoyl dextran (AD-CM-dextran)

Functionalization of dextran with thiol groups that chemisorb onto gold surfaces is less complex and does not require multiple reaction steps. Because of its many hydroxyl groups, dextran is an excellent starting polymer for various reactions leading to thiolated dextran derivatives. 2-Mercaptocarbamoyl dextran can be prepared by reaction with 4-nitrophenylchloroformate and subsequent conversion with cystamine (Figure 9) (*17*). The resulting disulfide dextran can be reduced with sodium borohydride, yielding thiol dextran with a DS of up to 0.12. Atomic force microscopy (AFM) and SPR studies of the modified gold surfaces, which were prepared by self-assembly of thiolated dextran in aqueous solution, revealed that the surface coverage increased with increasing DS of thiol moieties. The increase of the molecular weight of dextran caused the reverse effect.

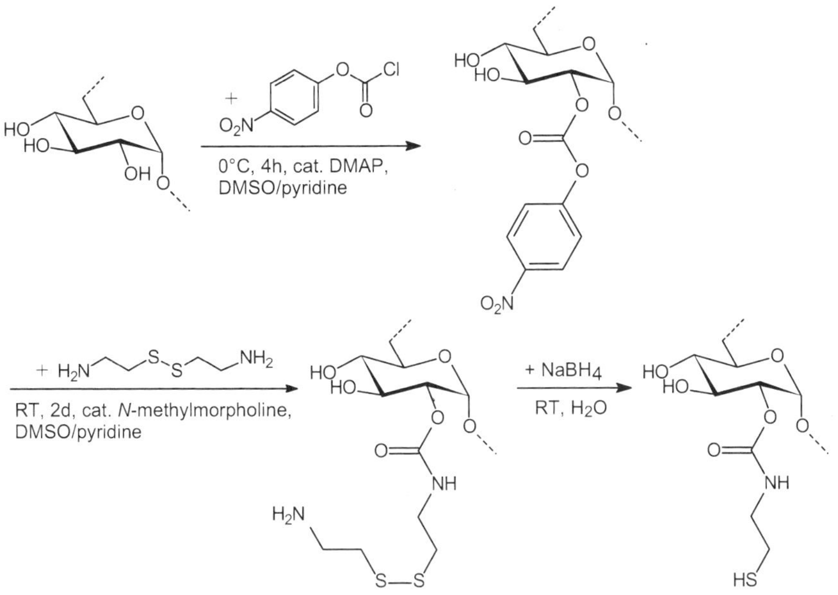

Figure 9. Preparation of 2-mercaptocarbamoyl dextran

A high density of thiol functions in the dextran molecule can be achieved by, e.g., reacting dialdehyde dextran with cystamine (similar to the above mentioned method) or by aminating dialdehyde dextran with ammonium chloride/sodium cyanoborohydride and converting the reaction product with 2-iminothiolane (Figure 10) (*18*). Unfortunately, the adsorption behavior of these derivatives, having a DS of up to 0.95, on gold has not been studied, yet.

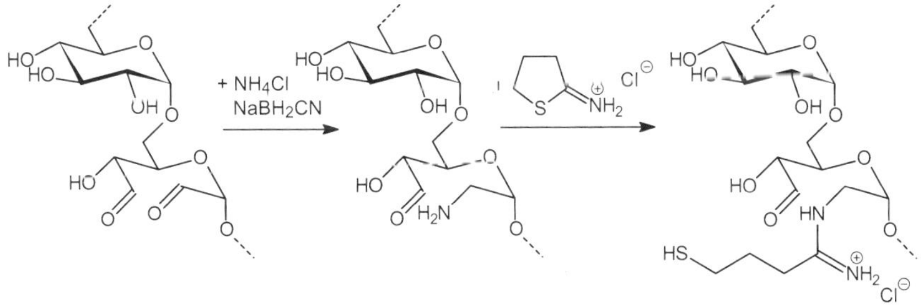

Figure 10. Preparation of thiolated dextran via aldehyde dextran

As in the case of cellulose, a simpler and more efficient strategy for introducing sulfur atoms into dextran is esterification with sulfur-containing carboxylic acids. Using thiophene carboxylic acids with different spacer lengths between the thiophene and carboxyl moiety and different ring positions for the sulfur atom, and in situ activation of these carboxylic acids with CDI (*19*), a range of thiophene carboxylic acid esters have been synthesized with DS values between 1.14 and 1.59 (Figure 11) (*20*). By the same method, an α-lipoic acid ester of dextran with a DS of 0.44 has been obtained. The one-step functionalization of dextran with sulfur-containing carboxylic acids, activated in situ with CDI, yields high DS values. The self-assembly behavior of these dextran derivatives on gold surfaces is currently under investigation.

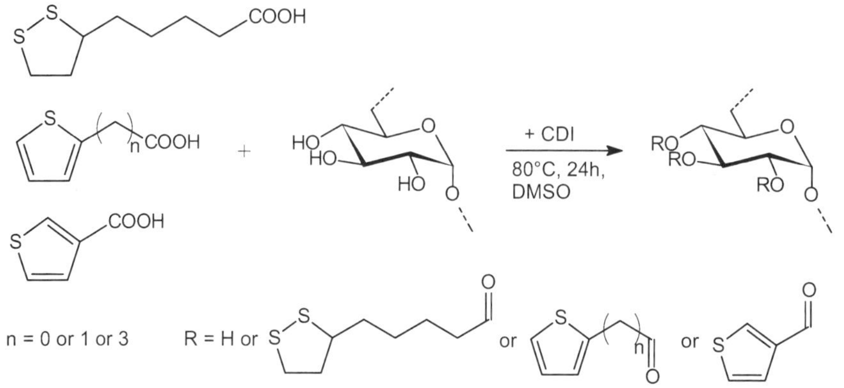

n = 0 or 1 or 3   R = H or   or   or

Figure 11. Reaction of dextran with sulfur-containing carboxylic acids in situ activated with N,N'-carbonyldiimidazole (CDI)

Besides the coating of planar gold surfaces with application in SPR analysis, the surface modification of gold nanoparticles with dextran plays an increasing role in optical sensing techniques. Gold nanoparticles have strong adsorption bands due to the excitation of plasmons by the incident light (*21*). Since the shift and intensity of the absorption bands depend on the local dielectric constant of the surrounding media and the size of the nanoparticles or aggregates, gold nanoparticles provide a basis for a simple and sensitive biosensor (*22*). As in the case of planar gold surfaces, the dextran coating prevents non-specific interactions of the gold nanoparticles with proteins and enables the immobilization of biomolecules on the surface of the nanoparticles. Biospecific interactions can thus be monitored as a function of particle concentration and time (*23*). Figure 12 shows schematically the coupling of 2-(2-aminoethoxy)ethanol to a carboxyl-terminated SAM on a gold nanoparticle and subsequent reaction with epichlorohydrin for coupling of dextran. The immobilization of biomolecules can be achieved by carboxymethylation of the dextran layer with bromoacetic acid and reaction with amino functions via carbodiimide chemistry.

Attached to an amino-functionalized glass surface, dextran-modified gold nanoparticles are able to aggregate and disaggregate in response to their environment (*24*). The state of dispersion of the nanoparticles can be deduced from the color of the colloid, which changes from red, for discrete particles in an aqueous environment, to bluish-purple for aggregated particles in non-polar media.

An interesting approach for glucose sensing using the aggregation phenomenon are dextran-coated gold nanoparticles functionalized with concanavalin A (Con A). Con A is a lectin that binds glucose by means of molecular recognition. Interaction of Con A on the surface of the initially aggregated nanoparticles with glucose in the analyte solution causes the aggregates to dissociate into single nanoparticles, which is detectable by a change in the absorption spectrum (*25*).

An alternative approach for the preparation of dextran-coated gold particles for optical analysis is the use of amino dextran as reducing and protective agent (*26*). Amino dextran is directly added to an aqueous solution of HAuCl$_4$. The size of the formed nanoparticles can be tuned by changing the ratio of amino dextran to gold salt. Coating of gold nanoparticles with dextran prevents their aggregation.

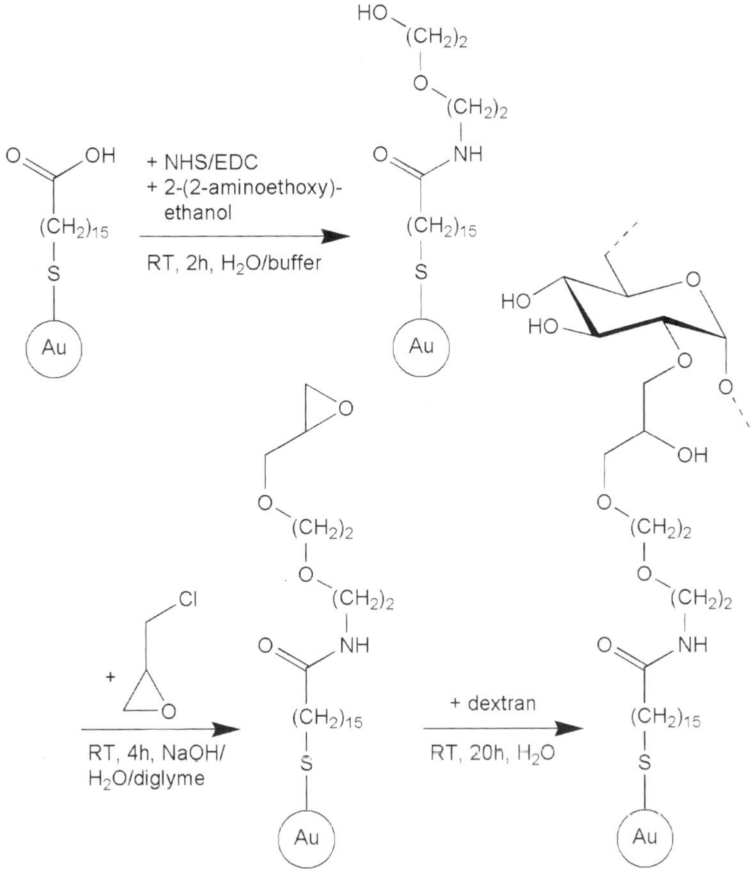

Figure 12. Reaction scheme for the preparation of dextran-coated gold nanoparticles

Aminocellulose Derivatives for the Generation of Biofunctional Surfaces

Very interesting cellulose materials that show self-assembly are accessible by S_N reaction of cellulose derivatives containing good leaving groups. Self-assembly of aminocellulose derivatives on suitable substrates provides an easy route to nanoscale biofunctional surfaces.

Synthesis of Aminocellulose Derivatives

Starting with the *p*-toluenesulfonic acid ester of cellulose (tosyl cellulose), a variety of novel and unconventional cellulosics have been synthesized as summarized in refs 27 and 28. It is important to note that not only monofunctional nucleophiles like halides, azides, thiosulfates, and triethylamines but also di- and multifunctional amines may undergo S_N reactions leading to soluble products. Therefore, with di- and oligoamines, cross-linking does not occur if optimized conditions are applied. The conversion should be carried out with a substantial excess of the amine to prevent cross-linking. Moreover, the displacement of tosylate occurs via a S_N2 mechanism. As a consequence, the conversion is largely limited to the primary tosylate at position 6, while the tosylates at the secondary positions are not reactive. The remaining tosyl moieties may be used to control properties like solubility (Figure 13) (*29, 30*).

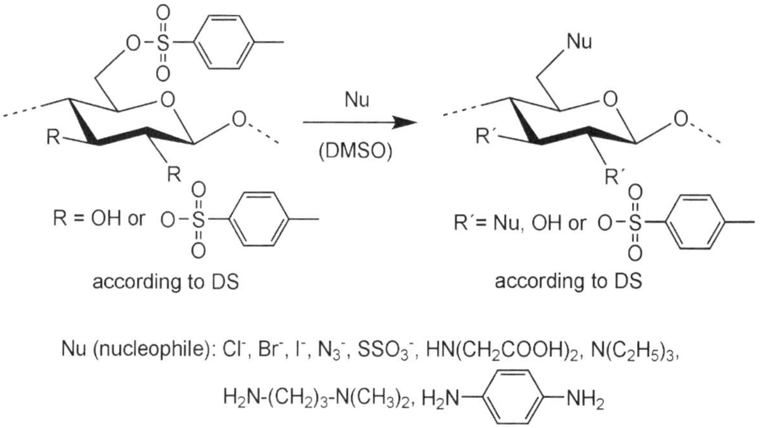

Nu (nucleophile): Cl^-, Br^-, I^-, N_3^-, SSO_3^-, $HN(CH_2COOH)_2$, $N(C_2H_5)_3$, $H_2N\text{-}(CH_2)_3\text{-}N(CH_3)_2$, $H_2N\text{—}\langle\rangle\text{—}NH_2$

R = OH or $O\text{-}S$ according to DS

R' = Nu, OH or $O\text{-}S$ according to DS

Figure 13. Typical examples for nucleophilic displacement reactions of cellulose tosylate

The 6-deoxy-6-amino(ammonium) cellulose derivatives formed are usually simply called aminocellulose (derivatives). By varying the spacer (X) between the nitrogen and substituents (S) at positions 2 and/or 3, aminocellulose derivatives with a wide variety of structural features and different solubilities may be obtained (Figure 14).

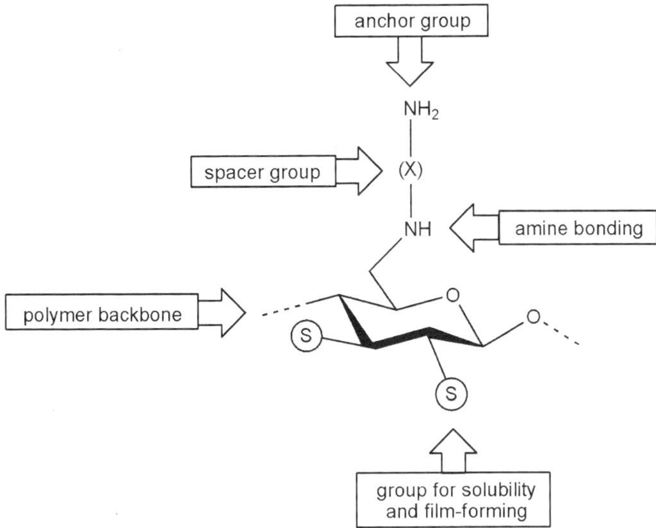

Figure 14. Schematic representation of aminocellulose structure, X=alkylene, aryl, or oligoamine spacer, S=ester groups, e.g., benzoate, carbanilate, and/or tosylate

The amine functions may be reacted with a variety of electrophiles including isocyanates, isothiocyanates, thioesters, and activated carboxylic acids, and are useful anchor groups for the covalent immobilization of biomolecules, e.g. oxidoreductase enzymes. Thus, a very broad range of chemical structures is readily obtainable. The aminocelluloses show film-forming properties, and it is possible to produce uniform films on glass surfaces with a layer thickness of approximately 100–200 nm (*31*).

Typical structures of aminocellulose derivatives that are soluble in water or DMA, depending on the substituent at positions 2 and 3, are summarized in Table I. Aminocellulose derivatives with remaining tosylate groups are water soluble whereas additional carbanilate or benzoate groups may impart solubility in DMA and DMSO.

The derivatives may form hydrogen bonds leading to a cross-linked structure. Hence, insolubility may occur during storage even at room temperature. This slow process obviously depends on the chain length (spacer effect) of the aminocellulose residues. Solubility persists for a long time in the case of aminocelluloses with short alkylene chains based on 1,2-ethylenediamine (EDA) and 1,4-butylenediamine, with additional carbanilate moieties at positions 2 and 3. These aminocelluloses are still soluble in DMA after 6 months of storage (*31*).

Table I. Examples of soluble aminocellulose with di- and oligoamine residues at C6 and solubilizing groups at positions 2 and 3

Substituents at C6	Substituents at positions 2 and 3[a]	Reference
–NH–C6H4–NH2	Tosylate	30–33
	Acetate, benzoate, carbanilate	31, 32
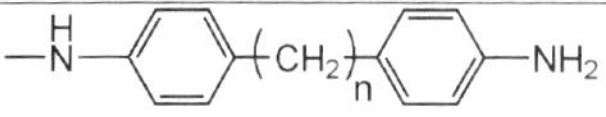 –NH–C6H4–(CH2)n–C6H4–NH2, n=0,1,2	Tosylate, acetate, benzoate, carbanilate	31, 32
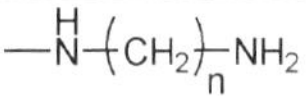 –NH–(CH2)n–NH2, n=2,4,6,8,12	Tosylate, methoxy, carbanilate	31, 34, 35
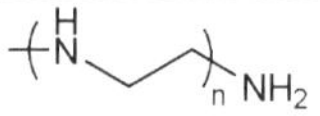 –(NH–CH2CH2)n–NH2, n=1,2,3; 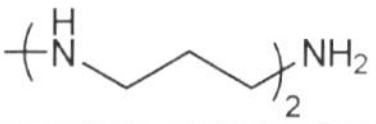 –(NH–CH2CH2CH2)2–NH2	Tosylate, carbanilate	31, 35–37
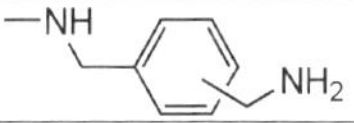 –NH–CH2–C6H4–CH2–NH2	Tosylate, carbanilate	38
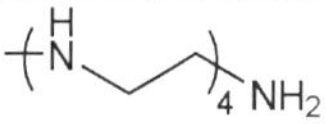 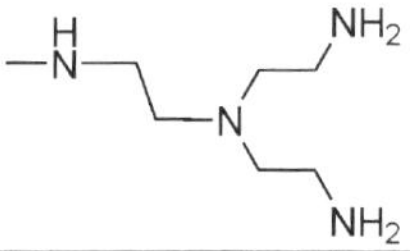 –(NH–CH2CH2)4–NH2; –NH–CH2CH2–N(CH2CH2NH2)–CH2CH2NH2	Tosylate	37, 39

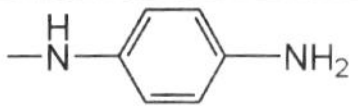

[a] Remaining number of tosylate moieties depending on DS_{Tos} of starting material

Application of Aminocellulose Derivatives

By reason of attractive interactions, aminocellulose derivatives self-assemble from dilute solutions (0.05–0.5%) onto many different materials (37). To generate aminocellulose-modified surfaces, techniques like spin-coating, dip-coating, air-brushing, and microcontact printing (μCP) can be used. Thus prepared aminocellulose layers, with thicknesses in the range of 1–4 nm, were unaffected by treatment with solvents or ultrasound. Potential substrates for these layers must have specific favorable electronic surface properties, such as those imparted by oxygen or hydroxyl functions. Substrates lacking these electronic properties should be subjected to oxygen or hydroxyl generating procedures, like O_2-plasma treatment, prior to layer deposition. Accordingly, glass, metal, metal oxides, polysaccharides, polysiloxanes, synthetic polymers, textiles, and other materials can serve as substrates (40). The surface characteristics of aminocellulose-modified substrates, determined according to application-relevant criteria, are presented in Table II.

Table II. Characterization techniques and results for aminocellulose-modified substrates.

Property	Analysis Method	Range of Values
Layer thickness	Ellipsometry, AFM	0.8–3.8 nm
Hydrophilicity/hydrophobicity	Water contact angle •	50–90°
RMS roughness of surface[b]	AFM	0.3–2.0 nm
NH_2 concentration	Chemical[a]	0.3–3.0 $nmol/cm^2$

[a]Determined by reaction with 4,4′-dimethoxytrityl chloride (*41*) or picric acid (*42*), followed by UV spectroscopy

[b]RMS = root mean square

A typical topographical profile of EDA-aminocellulose-modified gold is shown in Figure 15a. The observed thickness agrees well with the calculated monolayer thickness of approximately 1 nm. For comparison, Figure 15b shows the much smoother surface of an unmodified Au/111/surface. The monolayer thickness depends on the nature of the di- or oligoamine residues at C6. Long spacer lengths result in comparatively thick layers (*37*).

Different substances can be stamped onto surfaces by means of µCP. A very finely structured coating is formed, with thicknesses well below one micrometer. The µCP stamp, consisting of poly(dimethyl siloxane), is moistened with a reagent solution. Figure 16a shows an AFM image of a lateral structural pattern of dipropylenetriamine cellulose on a Si/SiO_2 substrate surface. The measured thickness of approximately 2 nm is comparable to the thickness of a monolayer. The structured surface was treated with a monodisperse Au colloid (particle size of 20 nm) resulting in the adherence of Au nanoparticles to the aminocellulose surface (Figure 16b).

Aminocellulose films with free NH_2 groups can be used as support matrices for enzyme immobilization. Enzyme coupling can be achieved in two steps as schematically shown in Figure 17:

1. Activation of the support film

2. Covalent enzyme coupling

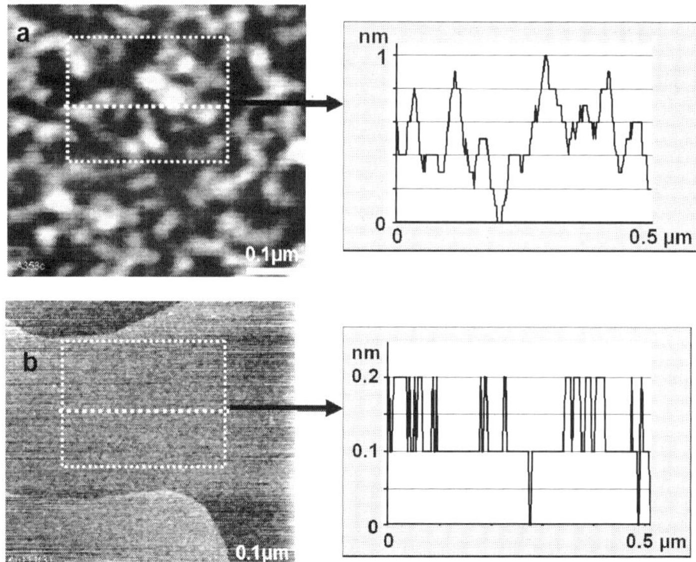

Figure 15. AFM image of a) 6-deoxy-6-ethylendiamino cellulose-modified Au/111/surface and b) an unmodified Au/111/surface (Reprinted with permission from ref 37. Copyright 2007 Elsevier B.V.)

Coupling agents such as glutaraldehyde, benzoquinone, L-ascorbic acid, and 1,3-benzenedisulfonyl chloride have proved to be efficient for enzyme coupling. Coupling studies have been carried out with oxidoreductase enzymes (glucose oxidase, GOD, lactate oxidase, LOD, and horseradish peroxidase, HRP). For immobilized GOD, enzyme activities per area of up to 100 mU/cm^2 could be determined (*35*).

A novel type of aminocellulose has been synthesized by the conversion of 6-azido-6-deoxy cellulose with polyamidoamine-based dendrons possessing an ethynyl moiety via the copper-catalyzed Huisgen reaction under mild conditions (Figure 18) (*43*). The novel cellulose derivatives possess a regioselective functionalization pattern due to the chemoselective reaction of the azide moieties. In addition, modification of the remaining hydroxyl groups can be readily achieved. Carboxymethylation, e.g., can be carried out both with the 6-azido-6-deoxy cellulose and the dendronized cellulose derivative due to the fact that both moieties (azide and 1,2,3-triazole) are stable under alkaline conditions. Thus, these novel dendronized derivatives will be included in studies of the self-assembly properties of aminocelluloses.

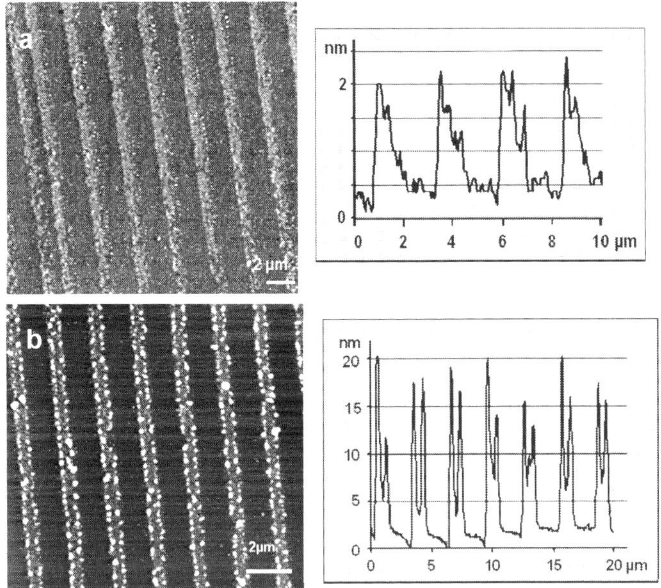

Figure 16. AFM image of (a) a lateral structural pattern of 6-deoxy-6-dipropylenetriamino cellulose on Si/SiO₂ (n) obtained by μCP and (b) the same pattern with adhered 20 nm Au nanoparticles (Adapted with permission from ref 37. Copyright 2007 Elsevier B.V.)

Figure 17. Covalent enzyme coupling on an aminocellulose-modified surface (Reproduced with permission from ref 31. Copyright 2003 Kluwer Academic Publishers)

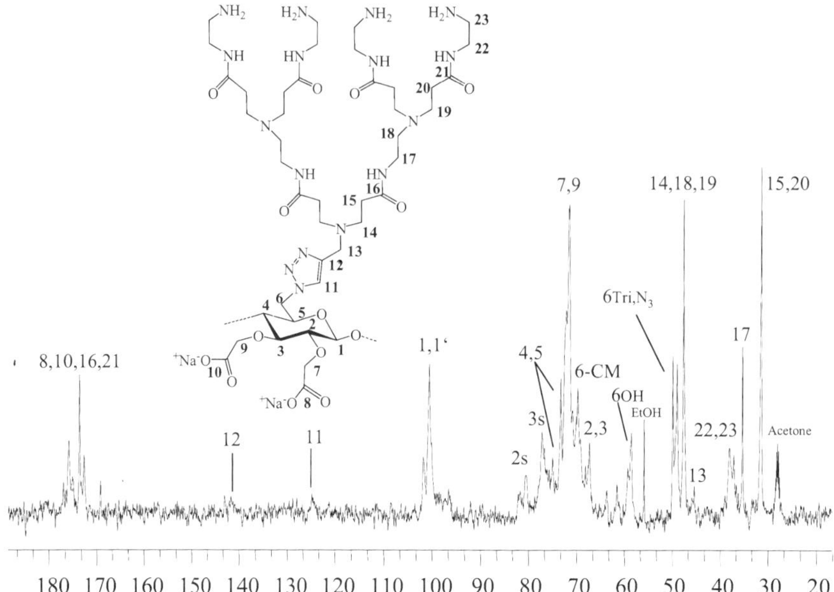

Figure 18. Typical structure and the corresponding ^{13}C-NMR spectrum of an aminocellulose obtained by the Huisgen reaction of 6-deoxy-6-azidocellulose with polyamidoamine dendron (remaining OH groups are partially carboxymethylated)

Self-Assembly of Xylan Derivatives on Cellulose Surfaces

Cellulose and xylan are closely associated in the submicron structure of plant cell walls. Consequently, for technologies such as pulping and paper making, it is very important to understand xylan–cellulose interactions. Furthermore, new self-assembled nanoscale materials, such as biocompatible nanocomposites, are currently the focus of much research and development. It has been recognized that mimicking the natural interactions of biopolymers may yield synthetic materials with promising properties. The design of chemically modified xylans that are able to adsorb onto and self-assemble with cellulose is a prerequisite to getting exceptional, tunable surface functionalities, like improved mechanical, chemical, and thermal stability, selective interaction properties, or a controlled release of active substances.

Depending on the source, various xylan types are known. Xylans of higher plants possess a β-(1→4)-linked xylopyranose backbone, usually substituted with sugar units, mainly 4-*O*-methyl-α-D-glucopyranosyl uronic acid and α-L-arabinofuranosyl units. Naturally occurring xylan contains in addition *O*-acetyl groups located at some of the hydroxyl groups in the xylan backbone (*44*).

In recent years, the detailed structure of xylans has been investigated using various analytical tools (*45, 46*). Sugar composition, degree of polymerization, and crystallization behavior, among other properties, are known. Thus, sufficiently well defined starting materials for a controlled chemical modification and evaluation of the specific characteristics are available, although there is still no commercial process to obtain high molar mass xylan in large quantities.

Synthesis of Xylan Derivatives

Introduction of cationic functionalities yields polymers that show self-assembly on cellulose. Despite xylan's solubility in water, its electrostatic interactions with the weakly charged cellulose surfaces play an important role in this context (*47*). With regard to specific functionalized surfaces, antimicrobial activity against certain bacteria, depending on the DS and type of xylan, is a matter of interest as well (*48*).

2-Hydroxypropyltrimethylammonium (HPMA) xylan derivatives prepared from xylans were shown to improve the paper-making properties and act as flocculants for pulp fibers at very low concentration (~0.25%), presumably because of adsorption onto cellulose fibers (*49, 50*). Chemical modification of xylan, leading to functional derivatives, is discussed in several reviews (*51, 52*). Laboratory scale syntheses to introduce cationic groups is usually carried out by etherification of xylan by a limited number of reagents including 3-chloro-2-hydroxypropyltrimethylammonium chloride (CHPTA) or 2,3-epoxypropyltri-methylammonium chloride (EPTA, Figure 19). The majority of investigations were directed towards the reaction of xylan-rich waste materials, such as hard wood saw dust, corn cobs, and sugar cane bagasse, with CHPTA (*53, 54*).

To get well-defined products, the cationization of xylans, isolated from beech wood, corn cobs, rye bran, and viscose spent liquor, was investigated. The results indicated that the molar degree of substitution (MS) depended on both the molar ratios of CHPTAC/xylan and NaOH/CHPTAC, as well as on the type of xylan used.

The functionalization pattern within the cationized xylans (MS values from 0.25 to 0.98) was characterized by ^{13}C NMR spectroscopy (*55*). Monosubstitution at position 2 of the xylose units was found to occur at low MS. The regioselectivity was lost at high MS. Highly functionalized samples of a MS of up to 1.64 were obtained by conversion of xylan with EPTA in moderately concentrated aqueous NaOH and 1,2-dimethoxyethane as slurry medium. Reaction of 4-*O*-methylglucuronoxylan, isolated from birch wood, with the appropriate molar ratio of EPTA yielded products with a zwitterionic character caused by a 4-*O*-methylglucuronic acid content of 9.8% (determined by alkalimetric titration, Figure 20) (*56*).

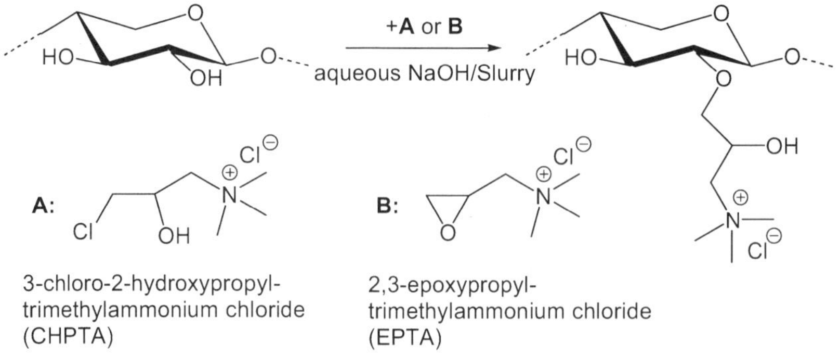

*Figure 19. Synthesis of xylan 2-hydroxypropyltrimethylammonium (HPMA)
derivatives*

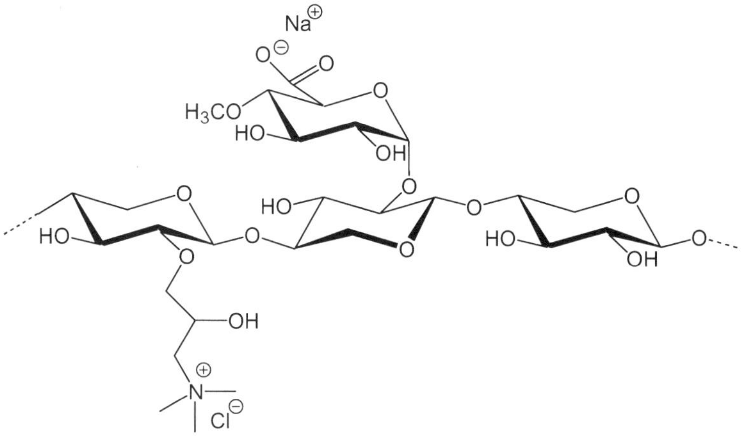

*Figure 20. Schematic representation of the zwitterionic
2-hydroxypropyltrimethylammonium-4-O-methylglucuronoxylan (HPMAGX)*

The total surface charge of the derivatives was measured by polyelectrolyte titration with poly(styrene sulfonate) or poly(diallyldimethylammonium chloride) (complex formation) (*50*). 2-Hydroxypropyltrimethylammonium-4-*O*-methyl-glucuronoxylan (HPMAGX) samples had different charge densities depending on MS, ranging from 0.06 to 0.19. The net charge of HPMAGX with a MS of 0.06 at pH 6–10 was negative, whereas the net charges of higher substituted derivatives were positive, having a significant influence on the electrostatic interactions of the polymer (Figure 21).

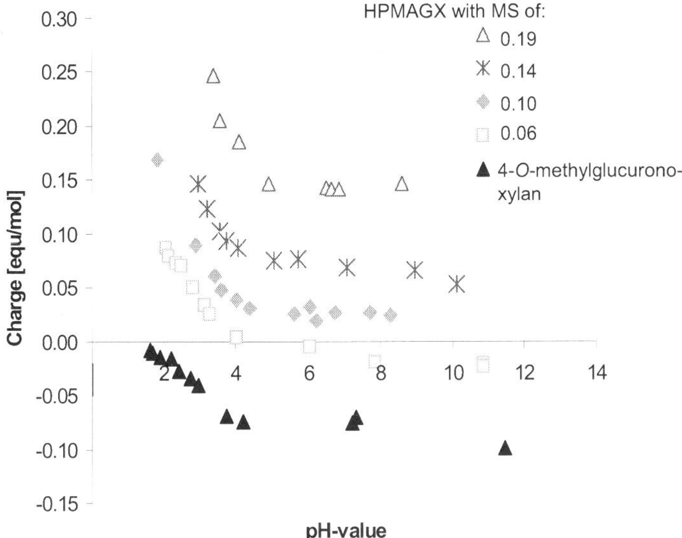

Figure 21. Net charge of 2-hydroxypropyltrimethylammonium-4-O-methylglucuronoxylan (HPMAGX) with different molar degrees of substitution (MS) in comparison to the 4-O-methylglucuronoxylan from birch wood at different pH-values

Adsorption of Xylan Derivatives onto Cellulose Surfaces

Adsorption of HPMAGX onto model cellulose surfaces has been studied by SPR spectroscopy (*57*). In that study, the required model cellulose surfaces on gold SPR slides were prepared by spin coating of trimethylsilylcellulose and subsequent regeneration of cellulose through exposure to humid HCl vapor (*58, 59*). Smooth model cellulose surfaces enable the study of interactions independently of other parameters, such as surface roughness and porosity. Adsorption of the HPMAGX on the cellulose surface was detected by changes in the SPR angle, which is sensitive to the refractive index near the metal surface. The surface concentration of the adsorbed species can be acquired from the experimental data using the de-Feijter equation (*60*).

The amount of HPMAGX that adsorbed onto the model cellulose surface was small. Maximum adsorption was observed for a MS of 0.1 (Figure 22). For comparison, adsorption onto gold surfaces coated with a neutral, hydrophilic SAM of 11-mercapto-1-undecanol (hydroxyl-terminated) or an anionic, electrophilic SAM of 11-mercapto-1-undecanoic acid (carboxyl-terminated) was studied (*61*). The fact that only small amounts of HPMAGX adsorbed onto cellulose and the OH-terminated SAM suggests that hydrogen bonding plays only a minor role in the adsorption process. In contrast, the fact that larger amounts adsorbed onto the COO⁻-terminated SAM suggests that electrostatic forces may be important.

218

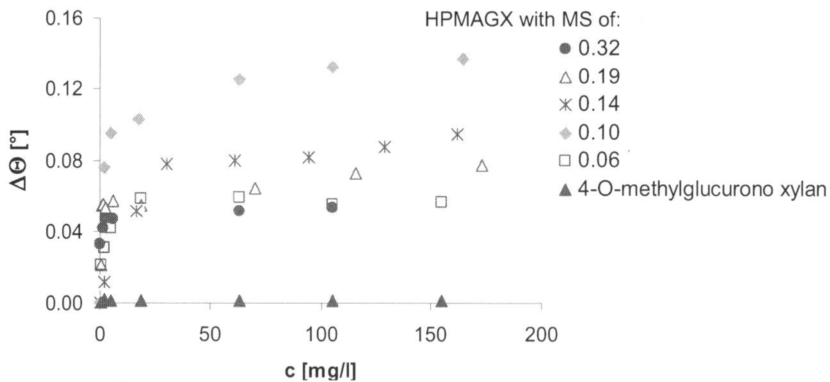

Figure 22. SPR angle (ΔΘ) versus concentration of 2-hydroxypropyltri-methylammonium-4-O-methylglucuronoxylans (HPMAGX) with different molar degrees of substitution (MS) and 4-O-methylglucuronoxylan (from birch wood), for comparison

The low degree of adsorption observed for HPMAGX with a MS of 0.06 can be explained by its negative net charge (see above) and resulting unfavorable electrostatic interactions with the surface. We observed a peak in adsorption at a MS of 0.10, and then a decline at higher MS. The observed decline at higher MS can be explained by the fact that at higher MS values (>0.10) intramolecular electrostatic interactions cause a more extended conformation, reducing the amount that adsorbs. The SPR results were in accordance with an observed increase in paper strength upon adsorption of HPMAGX onto spruce sulfite pulp and birch kraft pulp fibers. The highest paper strength was observed for a MS of 0.10.

Acknowledgments

Th. H. gratefully acknowledges general financial support of the "Fonds der Chemischen Industrie" of Germany as well as Borregaard Chemcell, Hercules, Dow Chemicals, Wolff Cellulosics, ShinEtsu, and Rhodia. The authors thank the German Science Foundation for sponsoring the project "Self-assembling polysaccharide polyelectrolytes" (HE 2054/11-1). The cooperative work with Prof. Dr. A. R. Esker and his coworkers (D. A. Drazenovich and A. Kaya), Virginia Polytechnic Institute and State University, Department of Chemistry, Blacksburg, VA, USA, is highly appreciated. The results shown in Figure 22 were obtained in Prof. Esker's lab. We thank also Dr. B. Saake (Institut für Holzchemie und Chemische Technologie des Holzes der Bundesforschungs-anstalt für Forst und Holzwirtschaft, Hamburg, Germany) for the cooperative work that has yielded the results discussed in Figure 21.

References

1. Heinze, T.; Liebert, T.; Koschella, A. *Esterification of Polysaccharides;* Springer Laboratory; Springer-Verlag Berlin: Heidelberg, Germany, 2006.
2. Heinze, T.; Liebert, T.; Heublein, B.; Hornig, S. Functional Polymers Based on Dextran. In *Polysaccharides II;* Klemm, D., Ed.; Advances in Polymer Science 205; Springer-Verlag Berlin: Heidelberg, Germany, 2006; pp 199–291.
3. Bain, C. D.; Troughton, E. B.; Tao, Y.-T.; Evall, J.; Whitesides, G. M.; Nuzzo, R. G. *J. Am. Chem. Soc.* **1989,** *111,* 321–335.
4. Petri, D. F. S.; Choi, S. W.; Beyer, H.; Schimmel, T.; Bruns, M.; Wenz, G. *Polymer* **1999,** *40,* 1593–1601.
5. Camacho Gómez, J. A. Untersuchungen zur Synthese und Folgechemie von O-Arylmethylethern und S-Alkylthiosulfaten (BUNTE-Salzen) der Cellulose. Ph.D. Thesis, Friedrich Schiller University of Jena, Jena, Germany, 1996.
6. Wenz, G.; Liepold, P.; Bordeanu, N. *Cellulose* **2005,** *12,* 85–96.
7. Wenz, G.; Liepold, P. *Cellulose* **2007,** *14,* 89–98.
8. Liebert, T.; Hussain, M. A.; Tahir, M. N.; Heinze, T. *Polym. Bull.* **2006,** *57,* 857–863.
9. Yokota, S.; Matsuyama, K.; Kitaoka, T.; Wariishi, H. *Appl. Surf. Sci.* **2007,** *253,* 5149–5145.
10. Malmquist, M. *Nature* **1993,** *361,* 186–187.
11. Österberg, E.; Bergström, K; Holmberg, K.; Riggs, J. A.; Van Alstine, J. M.; Schuman, T. P.; Burns, N. L.; Harris, J. M. *Colloids Surf., A* **1993,** *77,* 159–169.
12. Fournier, C.; Leonard, M.; Le Coq-Leonard, I.; Dellacherie, E. *Langmuir* **1995,** *11,* 2344–2347.
13. Löfas, S. *Pure Appl. Chem.* **1995,** *67,* 829–834.
14. Masson, J.-F.; Hartmann, T.; Dürr, H.; Booksh, K. S. *Opt. Mater.* **2004,** *27,* 135 139.
15. Chegel, V.; Shirshov, Y.; Avilov, S.; Demchenko, M. *J. Biochem. Biophys. Methods* **2002,** *50,* 201–216.
16. David, D.; Millot, C.; Renard, E.; Sébille, B. *J. Inclusion Phenom. Macrocyclic Chem.* **2002,** *44,* 369–372.
17. Frazier, R. A.; Matthijs, G.; Davies, M. C.; Roberts, C. J.; Schacht, E.; Tendler, S. J. B. *Biomaterials* **2000,** *21,* 957–966.
18. Pawlowski, A.; Källenius, G.; Svenson, S. B. *Vaccine* **1999,** *17,* 1474–1483.
19. Hornig, S.; Liebert, T.; Heinze, T. *Macromol. Biosci.* **2007,** *7,* 297–306.
20. Hornig, S.; Liebert, T.; Heinze, T. *Polym. Bull.* **2007,** *59,* 65–71.
21. Mulvaney, P. *Langmuir* **1996,** *12,* 788–800.
22. Nath, N.; Chilkoti, A. *J. Am. Chem. Soc.* **2001,** *123,* 8197–8202.
23. Lee, S.; Pérez-Luna, V. H. *Anal. Chem.* **2005,** *77,* 7204–7211.
24. Lee, S.; Pérez-Luna, V. H. *Langmuir* **2007,** *23,* 5097–5099.
25. Aslan, K.; Lakowicz, J. R.; Geddes, C. D. *Anal. Chim. Acta* **2004,** *517,* 139–144.
26. Ma, Y.; Li, N.; Yang, C.; Yang, X. *Anal. Bioanal. Chem.* **2005,** *382,* 1044–1048.

27. Heinze, T. Chemical Functionalization of Cellulose. In *Polysaccharides: Structural Diversity and Functional Versatility*, 2nd ed.; Dumitriu, S., Ed.; Marcel Dekker: New York, 2004; pp 551–590.

28. Vigo, T. L.; Sachinvala, N. *Polym. Adv. Technol.* **1999**, *10*, 311–320.

29. Koschella, A.; Heinze, T. *Macromol. Biosci.* **2001**, *1*, 178–184.

30. Tiller, J.; Berlin, P.; Klemm, D. *Macromol. Chem. Phys.* **1999**, *200*, 1–9.

31. Berlin, P.; Klemm, D.; Jung, A.; Liebegott, H.; Rieseler, R.; Tiller, J. *Cellulose* **2003**, *V10*, 343–367.

32. Berlin, P.; Klemm, D.; Tiller, J.; Rieseler, R. *Macromol. Chem. Phys.* **2000**, *201*, 2070–2082.

33. Tiller, J.; Rieseler, R.; Berlin, P.; Klemm, D. *Biomacromolecules* **2002**, *3*, 1021–1029.

34. Tiller, J.; Klemm, D.; Berlin, P. *Des. Monomers Polym.* **2001**, *V4*, 315–328.

35. Jung, A.; Berlin, P.; Wolters, B. *IEE Proc.-Nanobiotechnol.* **2004**, *151*, 87–94.

36. Jung, A.; Berlin, P. *Cellulose* **2005**, *12*, 67–84.

37. Jung, A.; Wolters, B.; Berlin, P. *Thin Solid Films* **2007**, *515*, 6867–6877.

38. Becher, J.; Liebegott, H.; Berlin, P.; Klemm, D., *Cellulose* **2004**, *11*, 119–126.

39. Jung, A.; Gronewold, T. M.; Tewes, M.; Quandt, E.; Berlin, P. *Sens. Actuators, B* **2007**, *124*, 46–52.

40. Berlin P.; Jung A.; Wolters B. Verfahren zur Modifizierung eines Substrats. DE Patent 102005008434 A1, 2005; WO Patent 2006089499 A1, 2006.

41. Gaur, R. K.; Sharma, P.; Gupta, K. C. *Analyst* **1989**, *114*, 1147–1150.

42. Cilli, E. M.; Jubilut, G. N.; Ribeiro, S. C. F.; Oliveira, E.; Nakaie, C. R. *J. Braz. Chem. Soc.* **2000**, *11*, 474–478.

43. Pohl, M.; Schaller, J.; Meister, F.; Heinze, T. *Macromol. Rapid Commmun.* **2008**, *29*, 142–148.

44. Ebringerová, A.; Heinze, T. *Macromol. Rapid Commun.* **2000**, *21*, 542–556.

45. Timell, T. E. *Wood Sci. Technol.* **1967**, *1*, 45–70.

46. Ebringerová, A.; Kardósová, A.; Hormádková, Z.; Hríbalová, V. *Fitoterapia* **2003**, *74*, 52–61.

47. Holmberg, M.; Holmberg, M.; Berg, J.; Stemme, S.; Odberg, L.; Rasmusson, J.; Claesson, P. *J. Colloid Interface Sci.* **1997**, *186*, 369–381.

48. Ebringerová, A.; Belicová, A.; Ebringer, L. *J. Microbiol. Biotechnol.* **1994**, *10*, 640–644.

49. Antal, M.; Ebringerová, A.; Micko, M.M. *Papier (Darmstadt)* **1991**, *45*, 232–235.

50. Schwikal, K. Neue Funktionspolymere aus Xylan. Ph.D. Thesis, Friedrich Schiller University of Jena, Jena, Germany, 2007.

51. Ebringerová, A.; Hromádková, Z. *Biotechnol. Genet. Eng. Rev.* **1999**, *16*, 325–346.

52. Ebringerová, A.; Hromádková, Z.; Heinze, T. Hemicellulose. In *Polysaccharides I: Structure, Characterization and Use;* Heinze, T., Ed.; Advances in Polymer Science 186; Springer-Verlag Berlin: Heidelberg, Germany, 2005; pp 1–67.

53. Antal, M.; Ebringerová, A.; Simkovic, I. *J. Appl. Polym. Sci.* **1984**, *9*, 637–642.

54. Ebringerová, A.; Antal, M.; Simkovic, I. *J. Appl. Polym. Sci.* **1986**, *31*, 303–308.
55. Ebringerová, A.; Hormádková, Z.; Kačurácová, M.; Antal, M. *Carbohydr. Polym.* **1994**, *24*, 301–308.
56. Schwikal, K.; Heinze, T.; Ebringerová, A.; Petzold K. *Macromol. Symp.* **2006**, *232*, 49–56.
57. Drazenovich, D. A.; Kaya, A.; Glasser, W. G.; Schwikal, K.; Heinze, T.; Esker, A. R. *Abstracts of Papers,* 233rd National Meeting of the American Chemical Society, Chicago, IL, March 25–29, 2007; American Chemical Society: Washington, DC, 2007; CELL 071.
58. Schaub, M.; Wenz, G.; Wegner, G.; Stein, A.; Klemm, D. *Adv. Mater.* **1993**, *5*, 919–922.
59. Holmberg, M.; Berg, J.; Stemme, S.; Ödberg, L.; Rasmusson, J.; Claesson, P. *J. Colloid Interface Sci.* **1997**, *186*, 369–381.
60. De Feijter, J. A.; Benjamins, J.; Veer, F. A. *Biopolymers* **1978**, *17*, 1759–1772.
61. Sigal, G. B.; Mrksich, M.; Whitesides, G. M. *Langmuir* **1997**, *13*, 2749–2755.

Chapter 10

Thermodynamics of Cellulose Ester Surfaces

Priscila M. Kosaka, Jorge A. Junior, Rafael S. N. Saito, and Denise
F. S. Petri*

Departamento de Química Fundamental, Instituto de Química,
Universidade de São Paulo, Av. Prof. Lineu Prestes 748, 05508-900 São
Paulo, SP, Brazil

The surface properties of cellulose ester films were
determined by means of contact angle measurements and
atomic force microscopy. Films of cellulose acetate, cellulose
acetate propionate, cellulose acetate butyrate, carboxymethyl
cellulose acetate butyrate, and cellulose acetate phthalate were
obtained by adsorption or spin coating onto silicon wafers.
The surface energy of the films was indirectly determined by
contact angle measurements using drops of water, formamide,
and diiodomethane. The surface energy decreased as the size
of alkyl ester group or degree of esterification increased,
because van der Waals interactions became weaker. Hamaker
constant values were calculated for the cellulose ester films,
which allow predicting film stability.

Introduction

Thin polymer films have found application in the microelectronics industry,
development of sensors, packaging industry, and in the academic field.
However, the stability and adhesion of polymer thin layers on solid substrates
depend on the surface energy of the polymer and the interfacial tension between
the solid surface and the polymer. Cellulose esters have been widely employed
as binders, additives, film formers, or modifiers in automotive, wood, plastic,
and leather coatings applications (*1, 2*). Cellulose acetate (CA), cellulose acetate
propionate (CAP), cellulose acetate butyrate (CAB), and
carboxymethylcellulose acetate butyrate (CMCAB) are extensively used in the

coatings industry because they reduce dry time, improve flow and leveling, control viscosity and gloss, are stable carriers for metallic pigments, improve UV stability, and reduce plasticizer migration, among other reasons (*1*). More recently, cellulose esters have also found application in pharmaceutical formulations (*3*). Thin films of CA, CAP, and CAB also proved to be efficient substrates for selective adsorption of proteins (*4*) and for the successful immobilization of lipases (*5*). Cellulose acetate phthalate (CAPh) is often used as enteric coating because it resists the acidic conditions of the stomach, preserving the drug, but upon reaching the basic conditions of the intestine, releases the drug successfully. There are excellent reports in the literature on the surface properties of thin films of cellulose (*6–10*). Nevertheless, there is scarce information on the surface properties of cellulose esters. In this work, the surface energy of CA, CAP, CAB, CMCAB, and CAPh films was indirectly determined by contact angle measurements using conventional test liquids. Moreover, film stability was predicted on the basis of thermodynamic parameters.

Background

Film Stability

A thin film that is forced to cover a solid substrate can be stable and preserve its initial shape, or it can be unstable and, under equilibrium conditions, break up and form drops. This phenomenon is called dewetting. Dewetting is driven by the balance of the capillary forces acting at the three-phase contact line (substrate, liquid, and air), as shown in Figure 1. The balance between the liquid–air, liquid–substrate, and substrate–air interfacial tensions determines the capillary force and is related to the contact angle θ (Figure 1), which represents the state of the solid–liquid–vapor system with the lowest Gibbs energy. Under equilibrium conditions, Young (*11*) has shown that

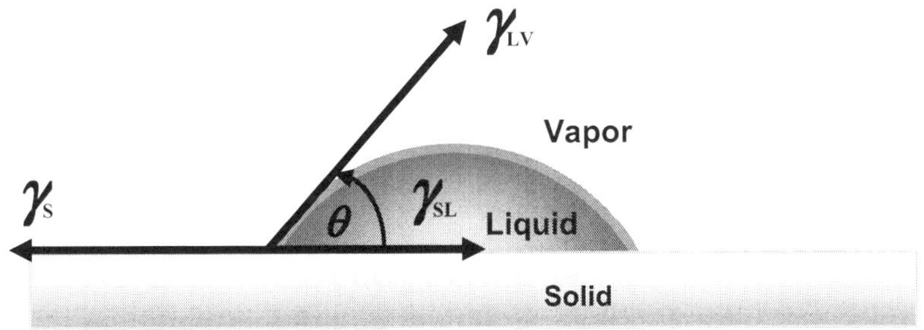

Figure 1. Schematic representation of a liquid drop on a solid surface in equilibrium with vapor. γ_S is the surface energy of the solid, γ_{SL} is the interfacial tension between the solid and the liquid, and γ_{LV} is the surface energy of the liquid. Equilibrium is characterized by the three surface tension forces acting at the liquid–solid–vapor contact line.

$$\gamma_S = \gamma_{SL} + \gamma_{LV} \cos\theta \tag{1}$$

where γ_S is the surface energy of the solid, γ_{SL} is the interfacial tension between the solid and the liquid, and γ_{LV} is the surface tension of the liquid. Eq 1 was developed for the case of an ideal surface, which is defined as a smooth, rigid, chemically homogeneous, insoluble, and non-reactive surface (*12, 13*).

If the interfacial tension between the solid and the liquid is higher than the surface energy of the solid, dewetting occurs, because it is energetically more favorable to remove the liquid from the substrate than to keep it spread on the substrate. Dewetting phenomena have been extensively explored, over the last decade, by G. Reiter (*14–21*), who has studied the behavior of ultrathin films of synthetic polymers, such as polystyrene or poly(dimethylsiloxane), on silicon wafers, above their glass transition temperature, T_g. Above T_g, mobility of polymer chains is very high and polymers can be viewed as viscous liquids. Dewetting is characterized by the retraction of the polymer film into the rims of newly formed holes, their growth and coalescence, ultimately leading to a set of droplets on the substrate (*14*). A schematic representation (Figure 2) of the shape of the rim during the early stages of dewetting of thin polymer films on non-adsorbing substrates was shown by Reiter and coworkers (*21*). Dewetting can begin via two main mechanisms: (i) nucleation, which can be started by defects, such as dust particles, or thermal inhomogeneities or (ii) spinodal decomposition, for which van der Waals forces are relevant and the process of which is sensitive to film thickness (*14, 22*).

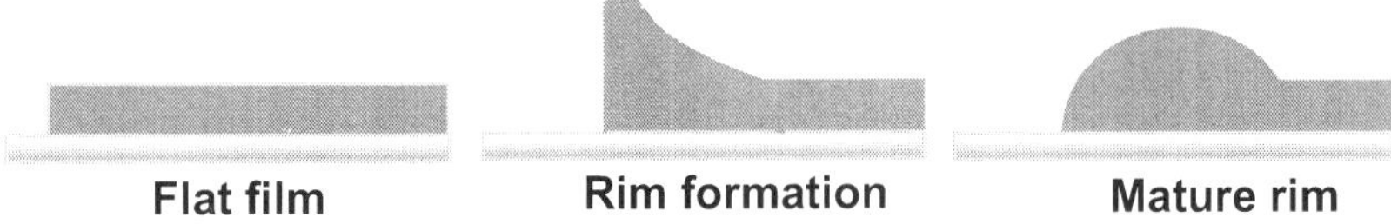

Figure 2. Schematic representation of rim shapes during the early stages of dewetting, as proposed by Reiter and coworkers (21).

It is possible to predict whether dewetting will take place if polymer and substrate surface energies are known. From the surface energy of the solid, it is possible to infer the Hamaker constant, A, which is related to the van der Waals interactions between two molecules and material density (15, 23). The difference between the Hamaker constant values for the polymer/polymer attraction, $A_{\mathrm{poly/poly}}$, and polymer/substrate interaction, $A_{\mathrm{poly/substrate}}$, yields the effective Hamaker constant, A_{eff}:

$$A_{\mathrm{eff}} = A_{\mathrm{poly/substrate}} - A_{\mathrm{poly/poly}} \tag{2}$$

If A_{eff} is negative, fluctuations in film thickness are expected, and after a characteristic time, the film will rupture (15, 18, 24). A positive value for A_{eff} indicates long range apolar van der Waals repulsion, which promotes film stability and wetting (25).

The Hamaker constant is also proportional to the work, w, required for separating surface 1 from surface 2 to an infinite distance, D, (26)

$$w = \frac{A_{12}}{12\pi D^2} \tag{3}$$

In the case of a homogeneous medium, the work of cohesion is given by

$$w = 2\gamma_S \tag{4}$$

where γ_S is the surface energy of the medium.

By combining eqs 3 and 4, one obtains

$$\gamma_S = \frac{A_{12}}{24\pi D^2} \tag{5}$$

In the case of polymer/polymer attraction, eq 5 becomes

$$A_{\mathrm{poly/poly}} = \gamma_S 24\pi D^2 \tag{6}$$

where D is typically 1.6 Å (12, 27).

In the Lifshitz–van der Waals theory, solids are treated as continuous materials with bulk properties, such as dieletric permittivity, ε, and refractive index, n. The Hamaker constant for a material 1 interacting with a material 2 across a medium 3, none of them being a conductor, can be calculated with (12, 28):

$$A_{\text{poly/substrate}} = \frac{3}{4}kT\left[\left(\frac{\varepsilon_1 - \varepsilon_3}{\varepsilon_1 + \varepsilon_3}\right)\left(\frac{\varepsilon_2 - \varepsilon_3}{\varepsilon_2 + \varepsilon_3}\right)\right] +$$

$$+ \frac{3h\nu}{8\sqrt{2}}\left[\frac{\left(n_1^2 - n_3^2\right)\left(n_2^2 - n_3^2\right)}{\sqrt{n_1^2 + n_3^2}\,\sqrt{n_2^2 + n_3^2}\left(\sqrt{n_1^2 + n_3^2} + \sqrt{n_2^2 + n_3^2}\right)}\right] \quad (7)$$

where k is the Boltzmann constant, T is temperature, h is the Planck constant, and ν is the mean ionization frequency of the material (typically ~3 × 10^{15} Hz) (12, 29). The first term represents the Keesom–Debey energy and plays an important role for forces in water, since water molecules have a strong dipole moment. Generally, however, the second term dominates.

In this work, T was set to 298.15 K and media 1, 2, and 3 correspond to silicon dioxide (upper native layer of a silicon wafer) ($\varepsilon_1 = 3.82$ and $n_1 = 1.46$) (12), cellulose ester ($\varepsilon_2 = 6.00$ and $n_2 = 1.48$) (30), and air ($\varepsilon_3 = 1.005$ and $n_3 = 1.00$), respectively. One should note that the ε value used for the cellulose esters (ε_2) is the literature value for cellulose (30).

Determination of the Surface Energy of Polymers

The measurement of contact angle is well established, very useful, and probably the most common technique for the characterization of solid surfaces. One of the important applications of contact angle measurements is the assessment of surface tension of a solid surface, γ_S (12, 13).

The work of adhesion at the solid–liquid interface, W_{SL}, was described by Dupré (13, 26):

$$W_{SL} = \gamma_S + \gamma_{LV} - \gamma_{SL} \quad (8)$$

By combining eqs 1 and 8, one obtains

$$\gamma_{LV}(1 + \cos\theta) = W_{SL} \quad (9)$$

Fowkes proposed that the surface energy, γ, of a material is the sum of the dispersive, γ^d, polar, γ^p, and induction, γ^i, component (13, 26):

$$\gamma^{\text{total}} = \gamma^d + \gamma^p + \gamma^i \quad (10)$$

Good, Girifalco, and Fowkes' equation (26) expresses the contribution of different types of intermolecular interactions to the work of adhesion as

$$W_{SL} = 2\sqrt{\left(\gamma_S^d \gamma_{LV}^d\right)} + 2\sqrt{\left(\gamma_S^p \gamma_{LV}^p\right)} + 2\sqrt{\left(\gamma_S^i \gamma_{LV}^i\right)} \quad (11)$$

Generally, the induction component is negligible so that the surface tension of a material can be considered as the sum of the dispersive and the polar component (*13, 26*). By combining eqs 9 and 11, one obtains the geometric mean equation (*26, 31–34*):

$$\gamma_{LV}(1+\cos\theta) = 2\left[\left(\gamma_S^d \gamma_{LV}^d\right)^{1/2} + \left(\gamma_S^p \gamma_{LV}^p\right)^{1/2}\right] \tag{12}$$

Another approximation of Young's equation is based on the harmonic-mean combining rule of surface tension components (*26, 31–34*):

$$\gamma_{LV}(1+\cos\theta) = 4\left(\frac{\gamma_S^d \gamma_{LV}^d}{\gamma_S^d + \gamma_{LV}^d} + \frac{\gamma_S^p \gamma_{LV}^p}{\gamma_S^p + \gamma_{LV}^p}\right) \tag{13}$$

If contact angle measurements are performed with drops of at least two liquids of known γ_{LV}^d and γ_{LV}^p on a polymer surface, it is possible to calculate γ_S^d and γ_S^p for the polymer surface (*26, 31–34*). There is no argument to state if the mean geometric or the harmonic approximation is the best method to determine γ_S of polymers. Nevertheless, Wu (*31, 32*) has shown that for low surface energy polymers the harmonic model describes interactions between test liquids and a polymer surface better than the mean geometric approximation. In this work, surface energy of cellulose esters were calculated with the harmonic model using water, formamide, and diiodomethane as test liquids. One should note that the cellulose esters investigated are not soluble or swellable in the test liquids chosen.

Materials

Silicon wafers (100), purchased from Silicon Quest (California, USA), with a native oxide layer approximately 2 nm thick, were used as substrates. They were cut into small pieces of 1 cm × 1 cm, rinsed in a standard manner (*35*), dried under a stream of nitrogen, and characterized prior to use. CA, CAP, CAB-1.7, CAB-2.5, CMCAB, and CAPh (powders free of plasticizer) were kindly supplied by Eastman Chemical Co., Brazil. The chemical structures of the substrate and cellulose esters are schematically represented in Figure 3. Sample codes and characteristics are shown in Table I.

Table I

Cellulose ester	$\overline{M_w}$ (g/mol)	T_g (°C)	T_m (°C)	DS_{CM}	DS_{Ph}	DS_{Ac}	DS_{Pr}	DS_{Bu}	DS_{OH}	$n^{632.8}$
CA	100,000	184	230–250	-	-	2.8	-	-	0.2	1.475
CAP	25,000	142	188–210	-	-	0.2	2.3	-	0.5	1.475
CAB-1.7	20,000	125	155–165	-	-	0.9	-	1.7	0.4	1.480
CAB-2.5	30,000	101	130–140	-	-	0.2	-	2.5	0.3	1.475
CMCAB	20,000	135	145–160	1.6	-	0.2	-	1.1	0.1	1.480
CAPh	40,000	170	-	-	0.9	2.1	-	-	-	1.490

Table I. Cellulose ester characteristics: weight-average molar mass ($\overline{M_w}$), glass transition temperature (T_g), melting point (T_m), degree of substitution for carboxymethyl (DS_{CM}), phthalyl (DS_{Ph}), acetate (DS_{Ac}), propionate (DS_{Pr}), butyrate (DS_{Bu}), and hydroxyl (DS_{OH}), and refractive index (n) at 632.8 nm[a]

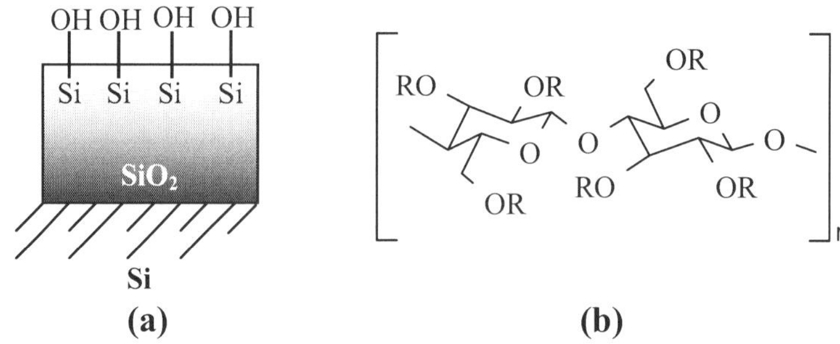

Figure 3. Schematic representation of the chemical structures of (a) the substrate, (b) the cellulose ester, where R refers to the acyl groups, i.e. $COCH_3$, COC_2H_5, and COC_3H_7 for CA, CAP, and CAB, respectively. Acetate groups are also present in CAP and CAB. In the case of CMCAB, R refers to COC_3H_7 and/or CH_2COOH and in the case of CAPh, R refers to $COCH_3$ and/or COC_6H_5COOH.

Cellulose esters must be kept in a closed container. Mixing cellulose esters in a nonpolar hydrocarbon solvent, such as toluene or xylene, may result in the buildup of static electricity, which can cause a flash fire or an explosion. Information about the physical characteristics and toxicity of the cellulose esters are presented in Table II.

Table II

Cellulose ester	Physical properties				Toxicity		
	Physical form	*Odor*	*Autoignition temperature*	*Flash point*	*Oral*	*Dermal*	*Skin irritation*
CA	Solid (white)	Odorless	Not available	Not applicable	Not available	Not available	Not available
CAP	Solid (white)	Odorless	432 °C	Not applicable	>6.4 mg/kg	>5.0 mg/kg	Slight
CAB-1.7	Solid (white)	Odorless	Not available	Not applicable	>6.4 mg/kg	>1.0 mg/kg	Very slight
CAB-2.5	Solid (white)	Slight	Not available	Not applicable	>6.4 mg/kg	>1.0 mg/kg	Very slight
CMCAB	Solid (white)	Slight	>400 °C	Not applicable	>5.0 mg/kg	>2.0 mg/kg	None
CAPh	Solid (white)	Odorless	416 °C	Not applicable	Not available	Not available	Not available

[a]Data provided by the producer

Analytical grade ethyl acetate was used to prepare the cellulose ester solutions at a polymer concentration of 5, 10 and 20 mg/mL. Only for the preparation of CAPh solutions at 10 mg/mL, a mixture of isopropanol and ethyl acetate in the volume ratio of 1:1 was used, since it provided better solubility than pure ethyl acetate. Diiodomethane, formamide, and water were used as test liquids for contact angle measurements. Containers used to store the solvents must be tightly closed and kept in a cool and ventilated place. Physical characteristics and main hazards of the solvents are described in Table III.

Table III. Solvent properties[a]

	Diiodomethane	*Ethyl Acetate*	*Formamide*	*Isopropanol*
Physical form	Liquid	Liquid	Liquid	Liquid
Density at 25 °C (g/mL)	3.325	0.902	1.134	0.785
Melting point (°C)	5–8	–84	2–3	–89
Boiling point (°C)	67–69	76.5–77.5	210	82
Main hazards	Irritant	Highly flammable and irritant	Toxic	Highly flammable and irritant

[a]Data provided by Sigma-Aldrich

Methods

Film Preparation

Two experimental strategies were followed to produce cellulose ester films. In the first, CA, CAP, and CAPh ultrathin films were obtained by adsorption. In the second, CAP, CAB-1.7, CAB-2.5, and CMCAB thin films were prepared by spin coating. The adsorption method consisted of immersing silicon wafers into solutions of CA (5 mg/mL), CAP (5 mg/mL), or CAPh (10 mg/mL) at (24 ± 1) °C. After 21 hours, the wafers were removed from the polymer solutions, washed with pure solvent, and dried under a stream of nitrogen. After that, CA and CAP films were annealed for 4 hours, and CAPh films for 24 hours, under reduced pressure (60 mmHg) at 170 °C. CAP, CAB-1.7, and CAB-2.5 films were spin coated from solutions at 20 mg/mL and CMCAB films from solutions at 10 mg/mL onto bare silicon wafers. Spin coating was performed by means of a Headway PWM32-PS-R790 spinner (Garland, USA), operating at 3000 rpm for 30 s, (24 ± 1) °C, and $(50 \pm 5)\%$ of relative humidity. CAP, CAB-1.7, and CAB-2.5 films were annealed for 15 hours under reduced pressure (60 mmHg) at 170 °C, and CMCAB films were annealed for 15 hours under reduced pressure (60 mmHg) at 150 °C.

Ellipsometry

Ellipsometric measurements were performed in air using a vertical computer-controlled DRE-EL02 ellipsometer (Ratzeburg, Germany). The angle of incidence was set at 70.0° and the wavelength, λ, of the He–Ne laser was 632.8 nm. For data interpretation, a multilayer model composed of the substrate, the unknown layer, and the surrounding medium were used. The thickness, d_x, and refractive index, n_x, of the unknown layer were calculated from the ellipsometric angles, Δ and Ψ, using the fundamental ellipsometric equation and iterative calculations with Jones matrices (*36*):

$$e^{i\Delta} \cdot \tan \Psi = R_p / R_s = f(n_x, d_x, \lambda, \phi) \tag{14}$$

where R_p and R_s are the overall reflection coefficients for the parallel and perpendicular waves. They are functions of the angle of incidence, ϕ, the wavelength, λ, of the radiation, and of the refractive index and thickness of each layer of the model, n_x and d_x, respectively.

From the ellipsometric angles, Δ and Ψ, and a multilayer model composed of silicon, silicon dioxide, polysaccharide layer, and air, it is possible to determine only the thickness of the polysaccharide layer, d_{poly}. The thickness of the silicon dioxide layers was determined in air, assuming a refractive index of $3.88 - 0.018i$ and infinite thickness for silicon (*32*). The refractive index for the surrounding medium (air) was taken as 1.00. Because the native silicon dioxide layer is very thin, its refractive index was taken as 1.462 (*37*) and only the thickness was calculated. The mean thickness of the native silicon dioxide layer was (2.5 ± 0.2) nm. After determining the thickness of the silicon dioxide layer, films of cellulose esters were deposited onto the wafers and the mean thickness of adsorbed polysaccharide layers was determined in air by means of ellipsometry, considering the nominal refractive index indicated in Table I.

Contact Angle Measurements

Contact angle measurements were performed at (24 ± 1) °C in a home-built apparatus (*13*). Advancing contact angles were measured using sessile drops of 8 µL, whereas drops of 4 µL were used to measure receding contact angles. Contact angle hysteresis was calculated from the difference between advancing and receding contact angles of water drops. In order to determine the surface energy of the polymers (γ_S), advancing contact angle measurements were performed with diiodomethane (>99.5%, purely dispersive nature), water, and formamide (>99%, polar liquid). The surface energy parameters for the different test liquids are presented in Table IV. In order to use Young's equation (eq 1) without correction for roughness and chemical heterogeneity (*13, 38*), only very smooth and homogeneous films were chosen for the determination of γ_S. At least three films of the same sample were analyzed before and after annealing.

Table IV. Surface tension values for the different test liquids that were used for the determination of the surface energy of cellulose esters[a]

Test liquids	γ_{LV} (mJ/m^2)	γ_{LV}^d (mJ/m^2)	γ_{LV}^p (mJ/m^2)
Water	72.8	21.8	51.0
Formamide	58.0	39.0	19.0
Diiodomethane	50.8	50.8	0

[a]Data taken from ref 33

Atomic Force Microscopy (AFM)

AFM measurements were performed with a PICO SPM-LE (Molecular Imaging) microscope in intermittent contact mode in air at room temperature, using silicon cantilevers with a resonance frequency close to 300 kHz. Areas of 5 µm × 5 µm and 1 µm × 1 µm were scanned with a resolution of 512 × 512 pixels. Image processing and the determination of the root mean square (rms) roughness were performed using the Pico Scan software. At least two films of the same composition were analyzed in different areas of the surface before and after annealing.

Results

In order to determine the surface energy of CA, CAP, CAB-1.7, CAB-2.5, CMCAB, and CAPh, flat and homogeneous films were chosen (Figure 4). The mean roughness values varied from (0.3 ± 0.1) nm to (0.5 ± 0.1) nm, as shown in Figure 4. The corresponding mean film thickness values are given in the figure caption. Chemical heterogeneity was analyzed by means of contact angle hysteresis, which was typically in the range of 5–8°. Moreover, contact angle measurements were performed in triplicate. The mean values of advancing or receding contact angles presented maximal standard deviations of 3°.

Figure 4. AFM images of 5 μm × 5 μm, obtained for (a) CA film (d = 5.7 ±0.9 nm), (b) CAP film prepared by adsorption (d = 1.3 ±0.3 nm), (c) CAP film prepared by spin coating (d = 118 ±10 nm), (d) CAB-1.7 film (d = 124 ±12 nm), (e) CAB-2.5 film (d = 135 ±11 nm) , (f) CMCAB film (d = 56 ±1 nm), and (g) CAPh film (d = 1.7 ±0.2 nm), with their respective rms roughness values. Films prepared by adsorption were annealed for 4 hours at 170 °C and those prepared by spin coating were annealed for 15 hours at 170 °C. CAPh films were annealed for 24 hours at 170 °C. d stands for ellipsometric thickness.

Advancing contact angles, θ, were measured for 8 µL drops of water, formamide, and diiodomethane (CH_2I_2) on CA, CAP, CAB-1.7, CAB-2.5, CMCAB, and CAPh films. The polar and dispersive contributions to the surface tension of each pure liquid are shown in Table IV. First, θ values were determined for drops of diiodomethane deposited onto the cellulose ester films. The obtained data are presented in Table V. Since diiodomethane offers only dispersive interactions, all terms in eq 13 related to the polar components were neglected, leaving only the dispersive contributions. Thus, the dispersive components of the surface energy (γ^d) were calculated for each cellulose ester (Table V). Once the γ^d values had been determined, advancing contact angles were measured for 8 µL drops of water and formamide and substituted in eq 13 in order to calculate the polar components of the surface energy (γ^p). The sum of γ^d and γ^p gives the total surface energy (γ^{total}), as proposed by Fowkes (*13, 26*) in eq 10.

Cellulose esters yielded γ^{total} values in the following sequence CA > CAPh > CAP > CMCAB > CAB-1.7 > CAB-2.5, as presented in Table V. Comparing CA, CAP, and CAB, γ^{total} values decreased as the size of alkyl ester group increased, indicating that intermolecular forces became weaker in the presence of longer alkyl group. The effect of the alkyl ester group on γ^{total} was also evidenced when comparing the values obtained for CAB-1.7 and CAB-2.5. One notices that upon increasing DS_{Bu} from 1.7 to 2.5, γ^{total} decreased from (50.7 ± 0.5) mJ/m² to (46.6 ± 0.3) mJ/m². The presence of carboxylic acid groups in CAPh promotes intra and intermolecular hydrogen bonding, causing a high γ^{total} value. In comparison to CAB, CMCAB gave a higher γ^{total} value. This might also be attributed to the presence of carboxylic acid groups in CMCAB chains, which increases intermolecular forces. The method of film preparation also influenced γ^{total} values. CAP films prepared through adsorption under equilibrium conditions gave γ^{total} = (58.7 ± 0.6) mJ/m², while the γ^{total} value obtained for spin-coated annealed CAP films amounted to (56.5 ± 0.5) mJ/m². This discrepancy might be attributed to different molecular orientations caused by the film preparation method, even when experimental conditions are assumed as equilibrium conditions. In the case of the adsorption method, the adsorbed layers are very thin (1.3 ± 0.3 nm) and the substrate might influence molecular orientation of polymeric chains, whereas in the case of spin-coated films, which are thicker (118 ± 10 nm), the substrate is too far away to cause any preferential molecular orientation.

As expected, the contribution of γ^p to γ^{total} decreased with increasing alkyl group length in the order CA < CAPh < CAP < CMCAB < CAB-1.7 < CAB-2.5. On the other hand, the contribution of γ^d to γ^{total} was the largest (~67%), regardless of cellulose ester type. With γ^d = 44.0 mJ/m², the dispersive component also predominates the total surface energy of cellulose of γ^{total} = 54.4 mJ/m², as determined by contact angle measurements (*39*). Similar γ^d values for cellulose were also found by inverse gas chromatography (*40, 41*).

When comparing the γ^d values presented in Table V, one observes the sequence CAPh > CA > CAP > CMCAB > CAB-1.7 > CAB-2.5. This trend is in accordance with the results of Glasser and Garnier, who observed that attaching alkyl or fluorine groups to the cellulose surface reduced the γ^d value by as much as 50% (*41*). γ^d values are very important because they are used in eq 6 for the calculation of the Hamaker constant $A_{\text{poly/poly}}$.

The values of γ^d, γ^p, and γ^{total}, presented in Table V, were calculated with the harmonic mean equation (eq 13). However, γ^d, γ^p, and γ^{total} were also calculated with the geometric mean equation (eq 12). Both equations yielded very similar γ^d, γ^p, and γ^{total} values (maximum deviation of 1 mJ/m²), indicating that for the calculation of surface energy of these commercial cellulose esters both equations can be equally applied. In contrast, for polystyrene (*31, 32, 34*) or polypropylene (*34*), surface energy values calculated by the geometric mean equation were up to 7 mJ/m² higher than those calculated by harmonic mean equation.

As stated in eq 2, polymer film stability can be predicted when $A_{\text{poly/substrate}}$ and $A_{\text{poly/poly}}$ are known. $A_{\text{poly/poly}}$ is intimately related to the van der Waals interaction potential between two macroscopic bodies at the intervening medium. $A_{\text{poly/poly}}$ was calculated for each cellulose ester by substituting γ^d values in eq 6. Since $A_{\text{poly/poly}}$ is proportional to γ^d, the cellulose esters yielded $A_{\text{poly/poly}}$ values in the following sequence CAPh > CA > CAP > CAB-1.7 > CAB-2.5 > CMCAB, as shown in Table V. The $A_{\text{poly/poly}}$ values obtained for CAPh and CA were similar to that determined for cellulose (8.4×10^{-20} J) (*6*). It means that CAPh or CA chains are more tightly bound to each other than CAP, CMCAB, or CAB chains, or in other words CAPh and CA present the strongest cohesion among the cellulose ester samples studied here. Consequently, another important conclusion is that upon increasing the size of the alkyl ester group or degree of substitution, van der Waals interactions become weaker.

$A_{\text{poly/substrate}}$ calculated using eq 7 amounted to 7.2×10^{-20} J, which was lower than the $A_{\text{poly/poly}}$ values calculated for CAPh, CA, CAP and CAB-1.7 (Table V). Since $A_{\text{poly/substrate}} < A_{\text{poly/poly}}$, dewetting is expected for these cellulose ester films. In fact, after annealing for up to 60 hours these films presented rim formation, film retraction, and substrate exposition, which are typical features of the dewetting process. Figure 5 shows typical topographic AFM images observed for the dewetting process of CA films. Only for CAB-2.5 and CMCAB films, dewetting was not observed, because in these cases $A_{\text{poly/substrate}} > A_{\text{poly/poly}}$. Since van der Waals forces seem to be relevant for cellulose ester film stability, the dewetting observed here was probably caused by spinodal decomposition due to thermal fluctuations (*14, 22*).

Polymer	θ_D (°)	θ_F (°)	θ_W (°)	γ^d (mJ/m^2)	$\gamma_F^{\,p}$ (mJ/m^2)	$\gamma_W^{\,p}$ (mJ/m^2)	γ^{total} (mJ/m^2)	$A_{poly/poly}$ $(\pm\,0.1J)$
CA (Ads)	38 ± 1	44 ± 3	60 ± 1	41.1 ± 0.4	6.7 ± 0.7	17.6 ± 0.1	65.4 ± 0.6	8.44×10^{-20}
CAP (Ads)	40 ± 2	55 ± 1	66 ± 2	40.3 ± 0.4	3.6 ± 0.4	14.8 ± 0.1	58.7 ± 0.4	8.27×10^{-20}
CAP (SC)	47.0 ± 0.5	59 ± 1	66 ± 1	37.9 ± 0.5	3.3 ± 0.3	15.3 ± 0.1	56.5 ± 0.5	7.78×10^{-20}
CAB-1.7 (SC)	52.0 ± 0.5	67.5 ± 0.7	71.5 ± 0.7	34.9 ± 0.3	2.6 ± 0.2	13.2 ± 0.1	50.7 ± 0.3	7.16×10^{-20}
CAB-2.5 (SC)	54.3 ± 0.6	67 ± 1	77 ± 1	33.3 ± 0.3	2.2 ± 0.1	11.1 ± 0.1	46.6 ± 0.3	6.84×10^{-20}
CMCAB (SC)	50.5 ± 0.5	64.2 ± 0.6	69.5 ± 0.5	35.1 ± 0.1	2.6 ± 0.1	14.2 ± 0.3	51.9 ± 0.3	6.8×10^{-20}
CAPh (Ads)	31 ± 1	45 ± 1	65 ± 1	44.1 ± 0.4	5.3 ± 0.1	14.7 ± 0.1	64.1 ± 0.6	8.5×10^{-20}

Table V. Advancing contact angle measurements for ultrathin films of CA (Figure 4a), CAP (Figure 4b), and CAPh (Figura 4g), prepared by adsorption (Ads) from ethyl acetate solutions, and for thick films of CAP (Figure 4c), CAB-1.7 (Figure 4d), CAB-2.5 (Figure 5e), and CMCAB (Figura 4f), prepared by spin coating (SC) from ethyl acetate solutions with drops of diiodomethane (θ_D), formamide (θ_F), and water (θ_W) as test liquids. Dispersive (γ^d) and polar (γ^p) components of the surface tension determined by the harmonic mean equation (eq 13) and Hamaker constant ($A_{poly/poly}$) calculated by eq 6

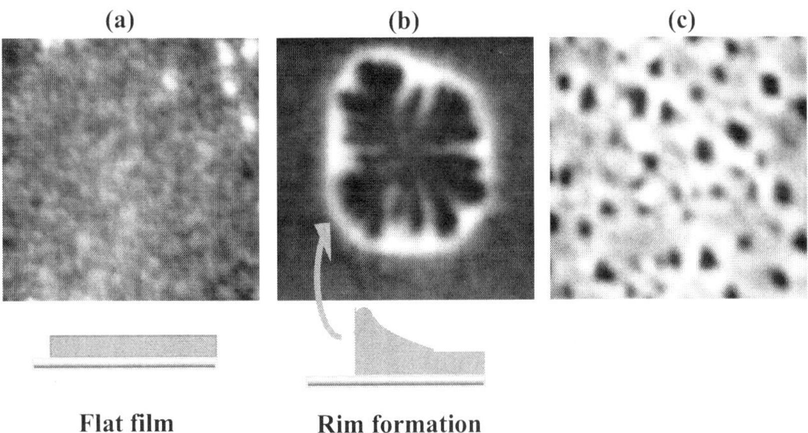

Figure 5. AFM images of 1 μm × 1 μm, showing the dewetting process in CA films. (a) CA without thermal treatment, (b) CA film after 15 hours of annealing, and (c) CA film after 60 hours of annealing.

Conclusions

The present work describes an experimental approach to the determination and calculation of surface energies of commercial cellulose ester films using contact angle measurements with three different test liquids and the Young's equation, valid for "ideal" surfaces. Although cellulose ester films are "real" surfaces with some roughness and chemical heterogeneity, the present study yields important information about the surface energy and Hamaker constant for a series of cellulose esters, which control surface and interface properties. Moreover, it has been shown that the molecular characteristics of cellulose esters exert considerable influence on their surface properties. The most important effects can be summarized as follows: (i) the surface energy (γ^{total}) decreased as the size of alkyl ester group increased; (ii) increasing the DS_{Bu} from 1.7 to 2.5, decreased γ^{total} by four units; (iii) upon increasing the size of the alkyl ester group or degree of substitution, van der Waals interactions become weaker; (iv) the molecular orientation in thin films might be influenced by the film preparation method, even when the experimental conditions are assumed as equilibrium conditions and (v) the presence of phthalyl and acetyl groups as substituents increased the cohesion forces among the cellulose ester chains. Considering the wide range of applications for thin films of cellulose esters, determination of their surface energy is highly relevant and indispensable for predicting film stability, since in most cases uniform thickness and durability are essential.

Acknowledgements

The authors acknowledge CNPq and FAPESP for financial support and Eastman do Brasil for supplying cellulose ester samples.

References

1. Edgar, K. J.; Buchanan, C. M.; Debenham, J. S.; Rundquist, P. A.; Seiler, B. D.; Shelton, M. C.; Tindall, D. *Prog. Polym. Sci.* **2001,** *26,* 1605.
2. Heinze, T.; Liebert, T. *Macromol. Symp.* **2004,** *208,* 167.
3. Edgar, K. J. *Cellulose* **2007,** *14,* 49.
4. Kosaka, P. M.; Kawano, Y.; Salvadori, M. C.; Petri, D. F. S. *Cellulose* **2005,** *12,* 351.
5. Kosaka, P. M.; Kawano, Y.; El Seoud, O. A.; Petri, D. F. S. *Langmuir* **2007,** *23,* 12167.
6. Bergström, L.; Stemme, S.; Dahlfors, T.; Arwin, H.; Ödberg, L. *Cellulose* **1999,** *6,* 1.
7. Notley, S. M.; Petterson, B.; Wågberg, L. *J. Am. Chem. Soc.* **2004,** *126,* 13930.
8. Notley, S. M.; Eriksson, M.; Wågberg, L.; Beck, S.; Gray, D. G. *Langmuir* **2006,** *22,* 3154.
9. Eriksson, M.; Notley, S. M.; Wågberg, L. *Biomacromolecules* **2007,** *8,* 912.
10. Kontturi, E.; Johansson L.-S.; Kontturi, K. S.; Ahonen, P.; Thüne, P. C.; Laine, J. *Langmuir* **2007,** *23,* 9674.
11. Young, T. *Philos. Trans. R. Soc. London* **1805,** *95,* 65.
12. Butt, H.-J.; Graf, K.; Kappl, M. *Physics and Chemistry of Interfaces;* Wiley-VCH: Weinheim, 2003.
13. Adamson, W. A. *Physical Chemistry of Surfaces;* 5th ed.; John Wiley & Sons: Toronto, 1990.
14. Reiter, G. *Phys. Rev. Lett.* **1992,** *68,* 75.
15. Reiter, G. *Langmuir* **1993,** *9,* 1344.
16. Reiter, G. *Macromolecules* **1994,** *27,* 3046.
17. Sharma, A.; Reiter, G. *J. Colloid Interface Sci.* **1996,** *178,* 383.
18. Reiter, G.; Sharma A. *Phys. Rev. Lett.* **2001,** *87,* 166103/1.
19. Damman, P.; Baudelet, N.; Reiter G., *Phys. Rev. Lett.* **2003,** *91,* 216101.
20. Reiter, G.; Sferrazza, M.; Damman, P. *Eur. Phys. J. E* **2003,** *12,* 133.
21. Reiter, G.; Hamieh, M.; Damman, P.; Sclavons, S.; Gabriele, S.; Vilmin, T.; Raphaël, E. *Nat. Mater.* **2005,** *4,* 754.
22. Seeman, R.; Herminghaus, S.; Jacobs, K. *Phys. Rev. Lett.* **2001,** *86,* 5534.
23. Green, P. F.; Ganesan, V. *Eur. Phys. J. E* **2003,** *12,* 449.
24. Peng, J.; Xuan, Y.; Wang, H. F.; Li, B. Y.; Han, Y. C. *Polymer* **2005,** *46,* 5767.
25. Sharma, A.; Khanna, R. *Phys. Rev. Lett.* **1998,** *81,* 3463.
26. Chaudhury, M. K. *Mater. Sci. Eng., R* **1996,** *16,* 97.
27. Eriksson, M.; Notley, S. M.; Wågberg, L. *Biomacromolecules* **2007,** *8,* 912.
28. Garbassi, F.; Morra, M.; Occhiello, E. *Polymer Surfaces: From Physics to Technology;* John Wiley & Sons: New York, 1998.

29. Brochard-Wyart, F.; Redon, C.; Sykes, C. *C. R. Acad. Sci. II* **1992,** *314,* 19.

30. Holmberg, M.; Berg, J.; Stemme, S.; Ödberg, L.; Rasmusson, J.; Claesson, P. *J. Colloid Interface Sci.* **1997,** *186,* 369.

31. Wu, S. *J. Polym. Sci.* **1971,** *34,* 19.

32. Wu, S. *Polymer Interface and Adhesion;* Dekker: New York, 1982.

33. Bouali, B.; Ganachaud, F.; Chapel, J.-P.; Pichot, C.; Lanteri, P. *J. Colloid Interface Sci.* **1998,** *208,* 81.

34. Shimizu, R. N.; Demarquette, N. R. *J. Appl. Polym. Sci.* **2000,** *76,* 1831.

35. Petri, D. F. S.; Schunk, P.; Schimmel, T.; Wenz, G. *Langmuir* **1999,** *15,* 4520.

36. Azzam, R. M. A.; Bashara, N. M. *Ellipsometry and Polarized Light;* North-Holland Pub. Co., Amsterdam, 1987.

37. Palik, E. D. *Handbook of Optical Constants of Solids;* Academic Press, London, 1985.

38. Marmur, A. *Soft Matter* **2006,** *2,* 12.

39. Van Oss, C. J. *Interfacial Forces in Aqueous Media;* Marcel Dekker: New York, 1994.

40. Katz, S.; Gray, D. G. *J. Colloid Interface Sci.* **1981,** *82,* 339.

41. Garnier G.; Glasser W. *Polym. Eng. Sci.* **1996,** *36,* 885.

Chapter 11

Electrospun and Oxidized Cellulosic Materials for Environmental Remediation of Heavy Metals in Groundwater

Dong Han[1], Gary P. Halada[1,*], Brian Spalding[2] and Scott C. Brooks[2]

[1]Department of Materials Science and Engineering and Center for Environmental Molecular Science, Stony Brook University, Stony Brook, NY 11794
[2]Environmental Sciences Division, Oak Ridge National Laboratory, Oak Ridge, TN 37831

This chapter focuses on the use of modified cellulosic materials in the field of environmental remediation. Two different chemical methods were involved in fabricating oxidized cellulose (OC), which has shown promise as a metal ion chelator in environmental applications. Electrospinning was utilized to introduce a more porous structure into an oxidized cellulose matrix. FTIR and Raman spectroscopy were used to study both the formation of OC and its surface complexation with metal ions. IR and Raman spectroscopic data demonstrate the formation of characteristic carboxylic groups in the structure of the final products and the successful formation of OC–metal complexes. Subsequent field tests at the Field Research Site at Oak Ridge National Laboratory confirmed the value of OC for sorption of both U and Th ions.

Introduction

The history of the use of cellulose from numerous sources for various applications including as an energy source, and for clothing and building materials, can be traced back thousands of years. However, it was not until 1838 that its molecular formula was first analyzed by the French chemist A. Payen, and shortly afterwards the term "cellulose" was coined to describe this essential constituent in plants (*1*). It is the unique molecular structure of cellulose, a 1,4-β-D-linked polyanhydro glucopyranose, illustrated in Figure 1, as well as its status as the most abundant organic material on our planet, that have drawn great attention from scientists and engineers involved in the investigation of the properties of cellulose and research on reformulating manufacturing processes to respond to a growing number of applications (*2*).

The molecular length of cellulose can vary from 1,000–15,000 monomers depending upon different origins and degree of degradation. Cellulose can be obtained copiously from a vast array of sources such as wood, cotton, and grasses. Several methods can be used to oxidize or degrade cellulose, including hydrolysis with acids, thermolysis, and alkaline degradation. Cellulose viscose, cellulose acetate, cellulose nitrate, and cellulose propionate are some examples of cellulose degradation products generated through industrial manufacturing (*2*). Another important class of cellulose degradation products is oxidized cellulose (OC). OC has a functional carboxyl group, as illustrated in Figure 2, which is often a key factor for its usefulness in medical and related areas (*3*).

The process of extracting uranium (U) ore from the Earth's crust and subsequent treatments including milling and chemical processing for both fuel production and military uses yield large amounts of residues in both solid and liquid forms, containing various hazardous heavy metal ions, both non-radioactive and radioactive, such as U, Th, Sr, Ra and Rn. These contaminants, if not treated properly and in a timely fashion, result in air, soil, and both surface and ground water contamination (*4*). Hence, great attention has been paid to development of effective, inexpensive, and efficient methods for stabilizing and remediating hazardous heavy metal ions. A number of novel techniques for U remediation, such as combined chemical and biological treatment, have been proposed and tested, amongst which techniques that avoid or minimize the production of secondary contaminants have been more appealing and environmentally compatible (*4*). Implementation of a permeable reaction barrier containing specially designed fillers, such as zero-valent iron (Fe^{0}), in a flow path of a contaminated plume, has also been used (*5-9*). Polymeric fibers with functional groups that can chelate contaminant metal ions leading to stabilization have never been satisfactorily researched, let alone widely used; nevertheless, it is the chemical structure of certain organic macromolecules, such as OC, that

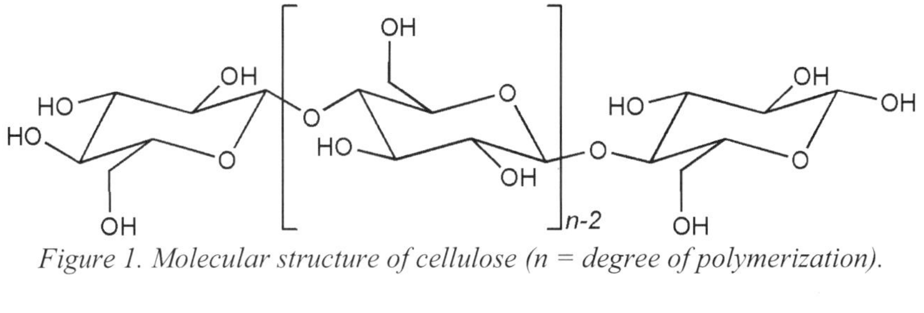

Figure 1. Molecular structure of cellulose (n = degree of polymerization).

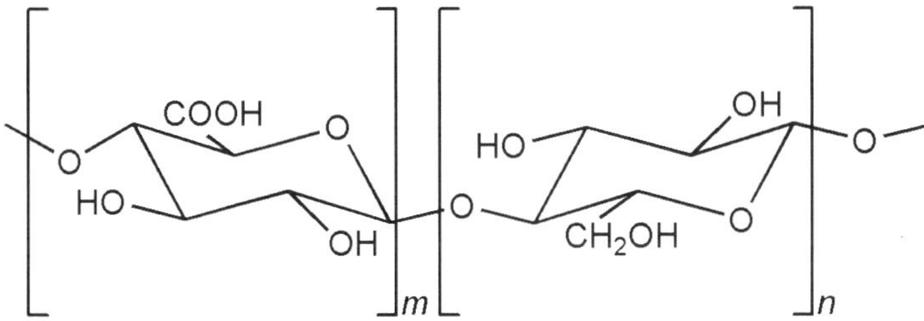

Figure 2. Molecular structure of OC.

suggest their possible use in barriers and other in situ and ex situ remediation technologies.

Electrospinning, using electrostatic repulsion between surface charges, is a simple and efficient technique for producing three-dimensional, non-woven, and porous polymeric fibrous mats. A typical electrospinning setup, illustrated schematically in Figure 3, generally consists of a high voltage power supply (commonly up to 30 kV) with two electrodes, a metering pump serving as the feeding control, a metallic needle connected to a syringe which contains the polymer solution or melt, and a collector, usually a metal screen. The mechanism of electrospinning and the influential parameters determining the morphology of electrospun fibers have been well explained elsewhere (*10*).

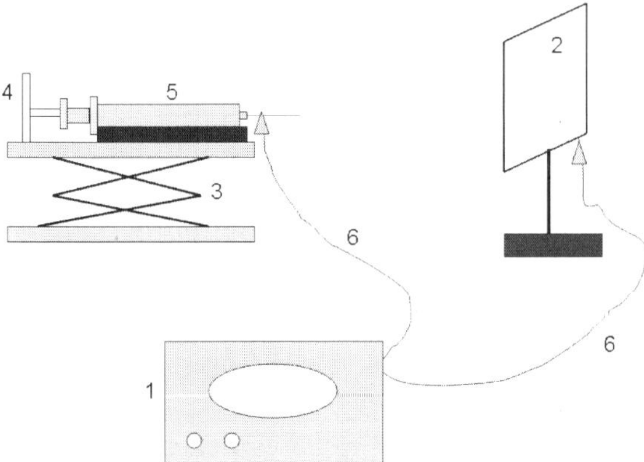

Figure 3. Schematic illustration of electropsinning setup. 1: high voltage power supply; 2: collector; 3: stand; 4: metering pump; 5: syringe with a metallic needle; 6: electrodes.

In this work, OC was produced through two reported methods (*3, 11*), both of which involved the treatment of cellulose with of HNO_3/H_3PO_4–$NaNO_2$. Electrospinning was used to introduce a more porous structure to the matrix whose larger surface area may be beneficial when the area of the contact surface between the matrix and metal ions is crucial. The resulting carboxyl groups of both non-electrospun OC and electrospun OC (E-OC) were postulated to act as the binding sites for heavy metal ions, U^{6+}, Ce^{3+}, and Eu^{3+}. Ce and Eu were chosen as analogs for trivalent transuranic metals, such as Pu^{3+}.

Experimental

Materials

Cellulose (fibrous, medium) and cellulose acetate (CA) (39.8 wt % acetyl, average Mn $\approx$ 30,000 g/mol by GPC) were purchased from Sigma-Aldrich Co. and used without further purification. Acetone (A.C.S. reagent, 99.7% by GC, corrected for water) and nitric acid (HNO_3) (A.C.S. reagent, 69.0–70%) were purchased from J. T. Baker. Sodium nitrite ($NaNO_2$) (99.999%), potassium hydroxide (KOH) (KOH content $\geq$ 85%, K_2CO_3 content $\leq$ 2.0%) and phosphoric acid (H_3PO_4) (85 wt % in H_2O) were also purchased from Sigma-Aldrich Co. Deionized (DI) water (> 1 MΩ) was used in all cases.

Preparation of OC

Cellulose (5 g) was added to 70 mL of a solution of a mixture of nitric acid and phosphoric acid (2/1 v/v), followed by adding 1.0 g of sodium nitrite. The reaction was allowed to proceed for 48 hours at 25 °C/room temperature. Then, the reaction was terminated by adding water, and the diluted solution was filtered to obtain the white solid which was continually washed until a pH of 4 was reached. The solid was then washed by acetone and dried in air.

Electrospinning of CA and Preparation of E-OC

A 17 wt % CA solution was prepared at room temperature in an acetone/water solvent (85/15 v/v). The CA solution was then used as a feed stock for electrospinning to fabricate a polymeric CA mat. The CA mat was swollen in a solution of acetone and water (1/1 v/v) and was then soaked for 1 hour in a KOH/ethanol solution (0.5 N KOH/12.5 mL ethanol) to deacetylate the swollen mat at room temperature. The mat was then treated with HNO_3/H_3PO_4–$NaNO_2$ and the reaction was allowed to proceed for 48 hours at 25 °C/room temperature, followed by a pH 4 water rinse and final acetone wash. The resultant was filtered and dried in air. The details of the oxidation reaction, including the reaction sequence and mechanism, have been explained elsewhere (*3, 11*).

Synthesis of Sorbed Complexes

Solutions (10 mM) of $UO_2(NO_3)_2$, $CeCl_3$, and $EuCl_3$ were prepared by mixing the metal compounds with DI water at room temperature. OC and E-OC were placed into these solutions and allowed to react for 12 hours. Then OC–metal complexes were removed from the solutions followed by a water rinse and drying in air. The pH of the reaction environments was controlled at 3.8 to mimic the actual groundwater environment in a known U contaminant plume at

Oak Ridge National Laboratory, where field experiments were subsequently conducted.

Scanning Electron Microscopy (SEM)

The morphologies of the original electrospun CA mat, the deacetylated CA mat, E-OC, OC, and the final polymer associated with heavy metal ions were examined using a LEO 1550 SEM.

FTIR Spectroscopy

A Nicolet model 560 FTIR spectrometer was used to obtain spectra from 4000 to 500 cm^{-1}. Data resolution was set to 4 cm^{-1} and summed over 256 scans to improve the signal-to-noise ratio. A MCT/A detector was used.

Raman Spectroscopy

A Nicolet model Almega dispersive Raman spectrometer with a 785 nm laser was used to obtain spectra from 3446 to 111 cm^{-1}. The final spectra for each sample were the result of 128 scans accumulation at high resolution. High resolution scans were collected using a laser intensity of 100%. The estimated resolution was 4.8–8.9 cm^{-1} (Tests were conducted in order to study whether the laser would damage the OC samples. Two spectra were taken consecutively in the same spot of the OC samples. From a comparison of the two spectra, it was concluded that no obvious change in the samples occurred during laser exposure at 100% power).

Results and Discussion

OC and E-OC Morphology

Figure 4 shows the SEM images of the original electrospun CA mat (a), the deacetylated CA mat (b), E-OC (c), and OC (d), respectively. Three dimensional non-woven fibrous structures can be observed in the SEM images. No significant changes occurred to the morphology of the polymeric mat after deacetylation treatment, although a generally more convoluted structure has emerged. The 48 hour reaction with $HNO_3/H_3PO4–NaNO_2$ caused the structure of the polymeric mat to change dramatically. A large number of segregated fibers can be seen in the SEM image; this phenomenon is uniform throughout the whole structure of the material. It has been reported that nitrogen oxides may be the oxidants during the oxidation process initiated by HNO_3 (*11*). The oxidation reaction is likely initiated by removing an H atom from cellulose by NO_2 and NO, which are the final products formed from the mixture of

248

HNO$_3$/H$_3$PO$_4$–NaNO$_2$. The cellulosic mat undergoes a series of reactions during chemical treatment and eventually forms oxidized cellulose by hydrolysis (*11*). Comparing this to the structure of E-OC (c), it can be concluded that the morphology of OC (d) has a much less porous structure than E-OC, effectively demonstrating the enhancement in surface area due to electrospinning.

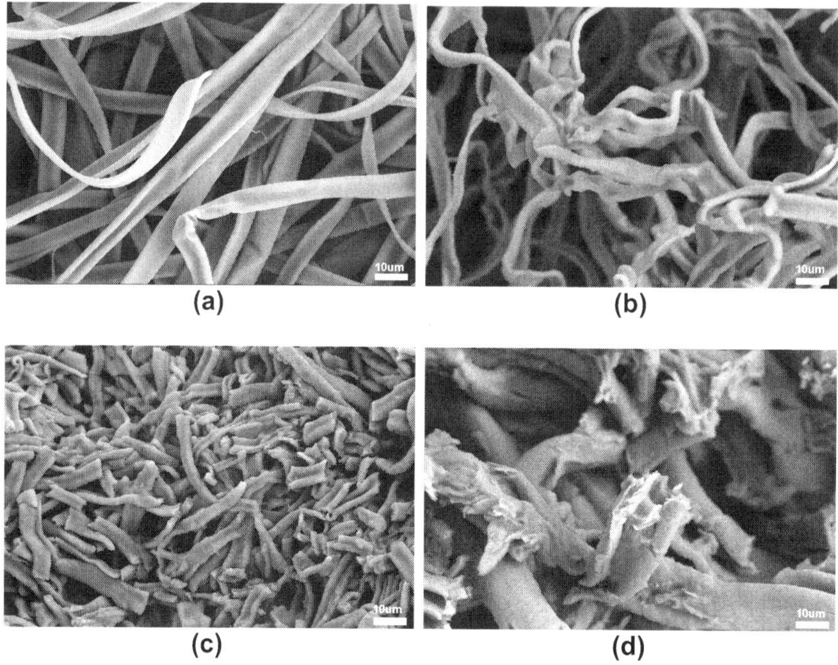

Figure 4. SEM images of (a) the electrospun CA mat, (b) the deacetylated CA mat, (c) E-OC, and (d) OC.

Morphology of E-OC–Metal Complexes

The morphologies of E-OC–U, E-OC–Ce, and E-OC–Eu complexes were also examined by SEM, shown in Figure 5, and EDAX. The structure of E-OC has changed after it was exposed to the three different metal ion solutions at acidic condition (pH 3.8). The fibrous structure has degraded, as can be clearly observed in the SEM images, and partly dissolved and aggregated particles are present. EDAX confirms the existence of the three metals adherent to their respective E-OC matrices.

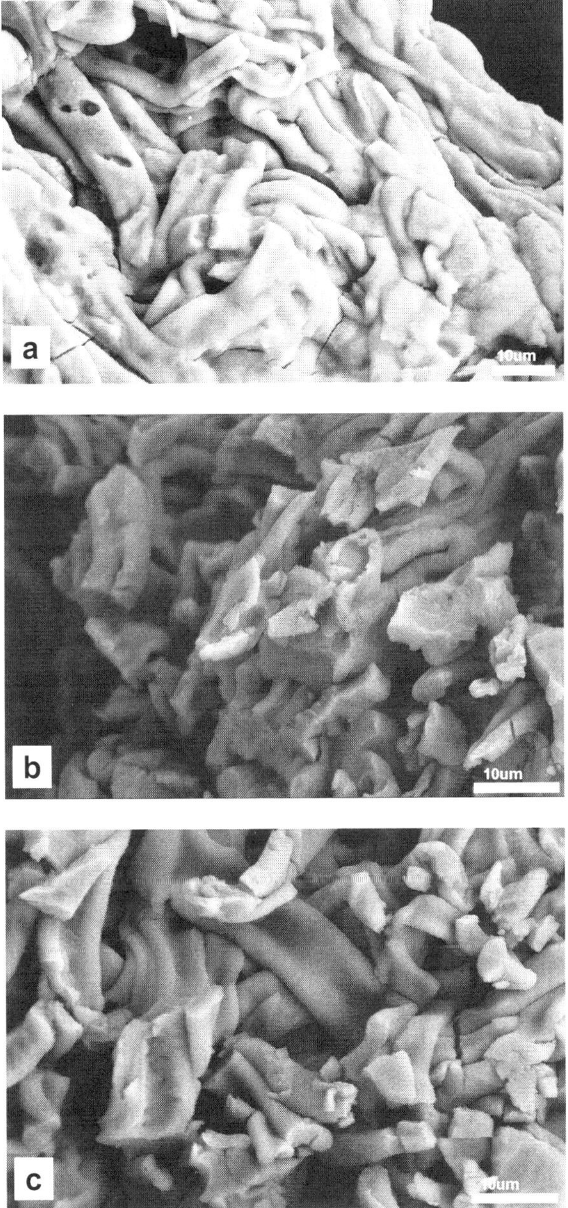

Figure 5. SEM images of (a) the E-OC–U complex, (b) the E-OC–Ce complex, and (c) the E-OC–Eu complex.

Spectroscopic Analysis of Reaction Intermediates and Products

The E-OC preparation reactions were analyzed by both FTIR and Raman spectroscopy. FTIR favors detection of anti-symmetric vibrational modes, while Raman favors identification of symmetric vibrations of chemical bonds. Hence, the experimental results can be more accurately analyzed by applying both techniques. Figure 6 shows the FTIR spectra of the original electrospun CA mat (a), the deacetylated CA mat (b), and E-OC (c), respectively. The signature component of CA is the acetate group (CH_3COO^-), which is comprised of a carbonyl group C=O (IR bands at 1740–1745 cm^{-1}) (12), an alkane group C–H (in acetates –O–CO–CH_3, IR symmetric deformation vibration at 1390–1340 cm^{-1}) (13), and an acetyl group C–O–C (in acetates CH_3COOR, IR bands at 1265–1205 cm^{-1}) (13). The characteristic peaks attributed to the vibrations of the acetate group of the CA mat at 1740 cm^{-1} (C=O bond), 1366 cm^{-1} (C–C bond), and 1213 cm^{-1} (C–O–C bond) are clearly illustrated in Figure 6 (a). The C=O absorption peak disappeared after deacetylation by KOH, which is demonstrated in Figure 6 (b). However, the absorption peak for the carbonyl vibration was detected again at 1727 cm^{-1} after acid oxidation. Generally, the C=O stretching vibration for carboxylic acids yields a band with a greater IR intensity than that for ketones or aldehydes. In both solid and liquid phases, the C=O group absorbs in the region of 1740–1700 cm^{-1}, but its stretching vibration of saturated aliphatic carboxylic acids may also be found in the region of 1785–1685 cm^{-1} (13).

Figure 7 shows the Raman spectra of the original electrospun CA mat (a), the deacetylated CA mat (b), and E-OC (c), respectively. The peak which appears at 1745 cm^{-1} in the spectrum of CA can be assigned to the ketone C=O stretching vibration from the branched chain of CA, which is the signature component of the structure. Two peaks appear at 1377 and 1437 cm^{-1}; the former Raman band can be assigned to the C–H deformation vibration while the latter one may be attributed to the alkane C–H deformation vibration. The band due to the C–O–C glycosidic linkage symmetric stretching vibration occurs at 1128 cm^{-1} and the Raman band for the C–O–C asymmetric stretching vibration appears at 1085 cm^{-1} (12). The same phenomenon as was observed in FTIR was detected again by Raman; the carbonyl peak at 1745 cm^{-1} disappeared after KOH treatment and re-appeared at 1736 cm^{-1} after oxidation. Instead of the acetate group in the branch, the aliphatic alcohol group (–CH_2–OH) dominates the molecular branched chain after deacetylation. Bands for this functional group are due to O–H stretching and bending vibrations, and C–O stretching vibrations are generally observed. The O–H stretching band in Raman spectra is generally weak compared to the medium-to-strong band in IR spectra. The typical absorption region for the C–O group is 1200–1000 cm^{-1} due to its stretching

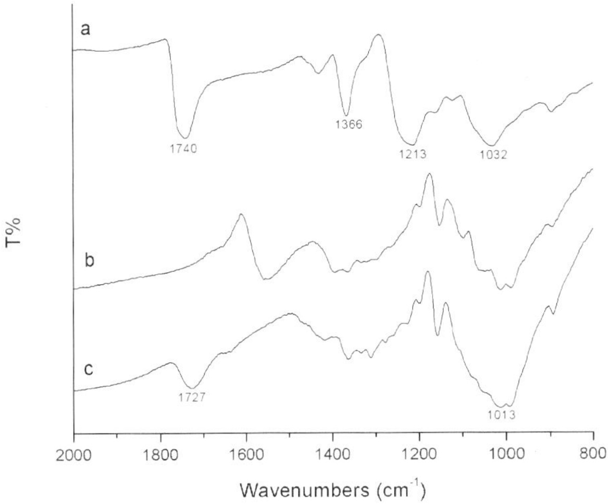

Figure 6. FTIR spectra of (a) the original electrospun CA mat, (b) the deacetylated CA mat, and (c) E-OC.

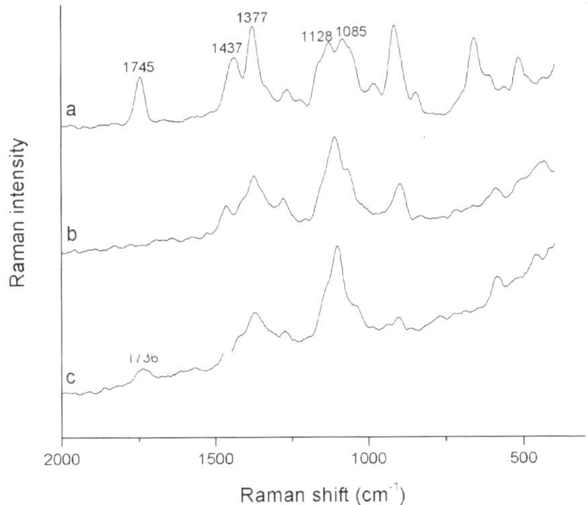

Figure 7. Raman spectra of (a) the original electrospun CA mat, (b) the deacetylated CA mat, and (c) E-OC.

vibration. However, hydrogen bonding results in a slight decrease in the frequency of this band. Therefore, for saturated primary alcohols, the region is 1090–1000 cm^{-1}; for secondary alcohols, the region is 1125–1085 cm^{-1}; and for tertiary alcohols, the region is 1205–1125 cm^{-1} (*13*). It is worthy to note that both unsaturation and chain branching tend to decrease the frequency of the C–O stretching vibration. Primary and secondary alcohols, both of which occur in deacetylated CA, have a strong Raman band at 900–800 cm^{-1} due to a C–C–O stretching vibration (*13*).

252

By comparing both IR and Raman spectra of OC and E-OC (Figure 8 and 9), it is confirmed that OC possesses the same characteristic functional groups as E-OC, although the positions of peaks are slightly different, which may be partly due to their different methods of formation.

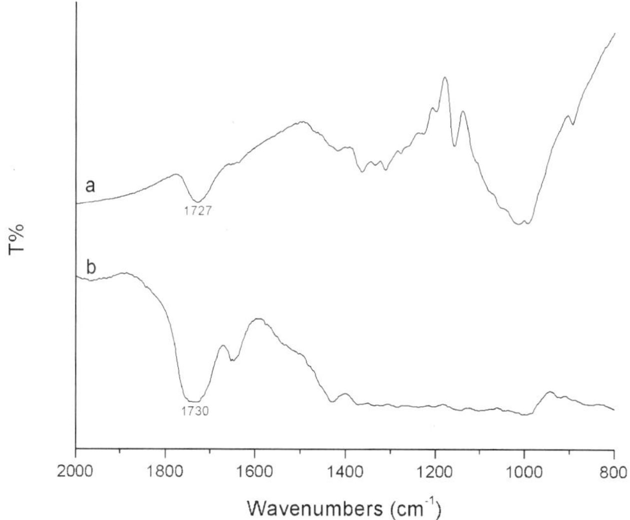

Figure 8. FTIR spectra of (a) OC and (b) E-OC.

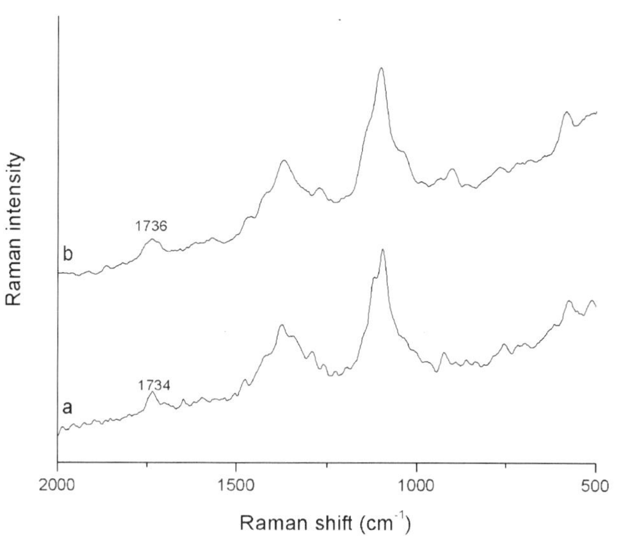

Figure 9. Raman spectra of (a) OC and (b) E-OC.

Spectroscopic Analysis of OC and E-OC–Metal Complexes

At pH 3.8, uranyl nitric 10 mM solution shows a transparent light yellow color. No precipitate forms after adding E-OC. A very strong band typically occurs in the region 1695–1540 cm^{-1} for carboxylic acid salts due to the asymmetric stretching vibration of COO$^-$, and a broad band of medium intensity appears in the range 1440–1335 cm^{-1} due to the symmetric stretching vibration of the same group (*13*). Figure 10 shows the infrared spectrum of E-OC–U complexes at pH 3.8 before (b) and after (c) DI water rinse, together with the original E-OC spectrum (a). A new absorption band, which occurs at 924 cm^{-1}, can be assigned to the asymmetric stretching vibration of UO$_2^{2+}$. The intensity of this band decreases after the water rinse, which can be taken as evidence that some amount of UO$_2^{2+}$ was in fact physically or weakly bond to E-OC. The intensity of the absorption band at 1727 cm^{-1} (C=O stretching vibrations of the carboxylic group) decreased since a portion of COOH groups were converted to COO$^-$ during the complexation process (*14*). The broad peak at 1599 cm^{-1} may be assigned to the formation of COO$^-$–UO$_2^{2+}$ groups.

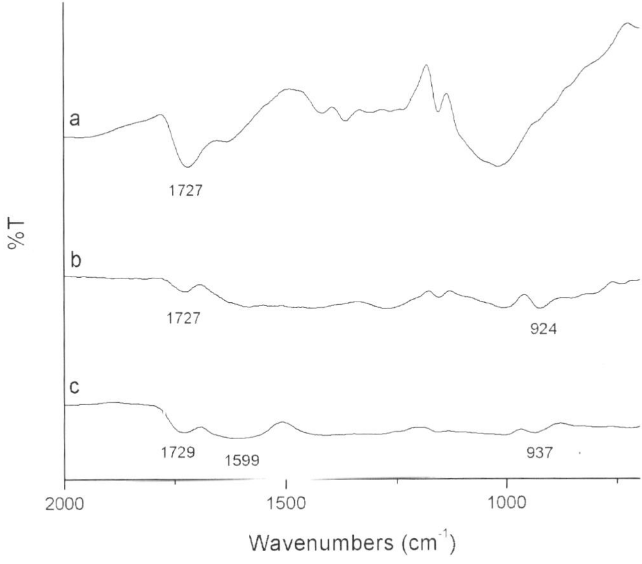

Figure 10. IR spectra of (a) E-OC, (b) the E-OC–U complex without water rinse, and (c) the E-OC–U complex after water rinse.

Figure 11 shows the Raman spectra of the E-OC–U complex and E-OC. The band at 862 cm^{-1} is attributed to the symmetrical stretching vibration of UO$_2^{2+}$ (*15*). The lack of intensity at 853 cm^{-1} (*15*) suggests that little or no uranyl hydrolysis species (UO$_2$)$_2$(OH)$_2^{2+}$ was formed during the complexation process as would be expected at pH 3.8. The vanishing of the peak from the C=O group at 1736 cm^{-1} reinforces the conclusion that the complexation process causes conversion of carboxylic groups to COO$^-$–metal complex; because most of COOH groups were converted to COO$^-$, the few remaining COOH groups could not be detected by Raman.

254

Similar phenomena took place during the formation of E-OC–Ce and E-OC–Eu complexes (Figure 12). In both cases, formation of peaks associated with a COO⁻–metal ion complex were noted (1588 cm⁻¹ and 1597 cm⁻¹ for E-OC–Ce and E-OC–Eu, respectively), retained following a water rinse. The spectra for OC–Ce and OC–Eu showed similar evidence of COO⁻–metal complexation, though the features were less distinct. This may be due to instrument and sample configuration factors, resulting in overall more IR sorption by the sample.

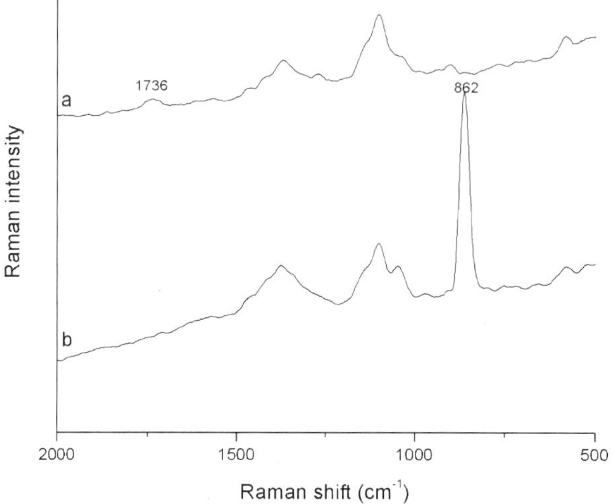

Figure 11. Raman spectra of (a) E-OC and (b) the E-OC–U complex at pH 3.

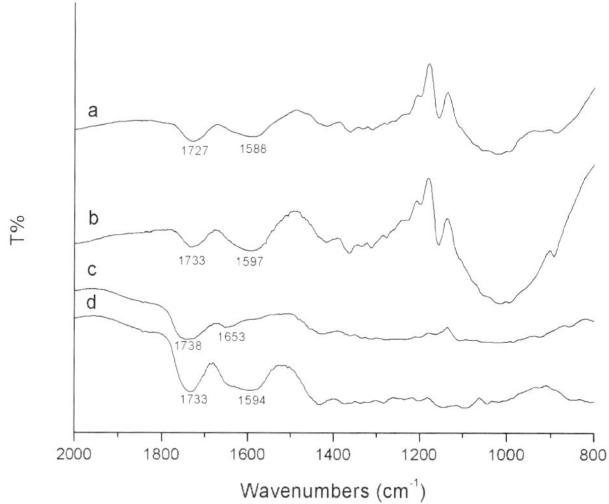

Figure 12. IR spectra of (a) the E-OC–Ce complex, (b) the E-OC–Eu complex, (c) the OC–Ce complex, and (d) the OC–Eu complex.

Ongoing Field Experiments

Figure 13 shows the results of x-ray fluorescence measurements from ongoing field experiments on adsorption capabilities of OC and E-OC for metal ions at the Department of Energy's Oak Ridge Field Research Center (ORFRC). Permeable environmental leaching capsules were molded from polyacrylamide gels to encapsulate two types of OC and were immersed in the contaminated wells, FW106 and SS5, to allow interaction between matrices and contaminant metals, such as U and Th, to take place. Preliminary data show that OC and E-OC adsorbed a considerably larger amount of U and Th than equal weights of their counterparts (cellulose, CA, electrospun CA, and neat polymers, which are the encapsulation materials by themselves) in a given time.

After approximately 1300 hours of deployment, samples were transferred to a non-contaminated well to test for leaching of complexed metal ions. It is of interest to note that while sorption of U was more rapid, retention of Th was greater over time. Also, an equal mass of E-OC sorbed over three times more U ions than non-electrospun OC. Encapsulated OC and E-OC samples are currently being analyzed spectroscopically to further characterize sorption and retention of both U and Th by OC.

Conclusions

OC and E-OC were successfully fabricated through the reported methods. IR and Raman spectroscopic data demonstrate the formation of characteristic carboxylic groups, which possess the chemical ability to chelate certain types of metal ions in the structure of the final products. Both laboratory and field experiment data suggest the potential possibility of the use of OC and E-OC as permeable reaction barrier fillers or ex situ sorbents for metal contaminated groundwater treatment. The results presented demonstrate the need to further investigate the efficiency of these model cellulosic materials for adsorbing metals and also the impact of environmental factors, such as pH, which may affect their performance.

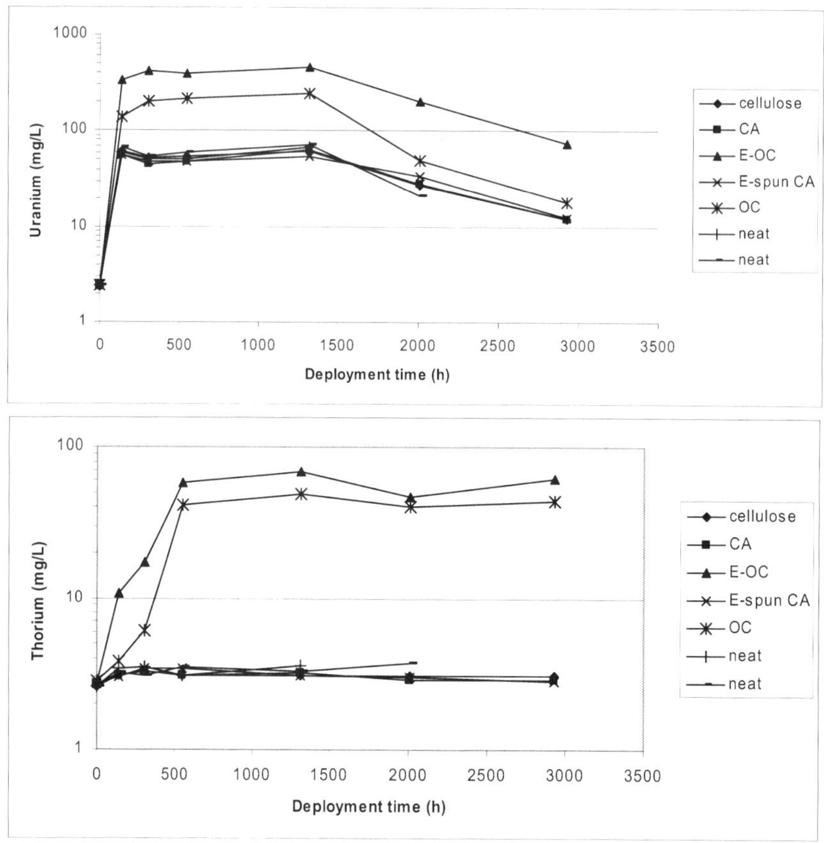

Figure 13. Field data for adsorption capabilities of selected matrices for metal ions.

Acknowledgements

This work has been supported by U.S. Department of Energy Office of Science Environmental Management Science Program (EMSP), contract number DEFG0204ER63729 and the Center for Environmental Molecular Science, (CEMS) funded by the National Science Foundation, contract number CHE0221934. This research has also been funded in part by the U.S. Department of Energy's Office of Science, Biological and Environmental Research, Environmental Remediation Sciences Program (ERSP). Oak Ridge National Laboratory is managed by UT-Battelle, LLC, for the U.S. Department of Energy under contract DE-AC05-00OR22725.

References

1. Klemm, D.; Heublein, B.; Fink, H. P.; Bohn, A. *Angew. Chem.-Int. Edit.* **2005,** *44,* 3358–3393.
2. Kennedy, J. F.; Phillips, G. O.; Wedlock, D. J.; Williams, P. A. *Cellulose and Its Derivatives: Chemistry, Biochemistry and Applications;* Halsted Press: New York, 1985.
3. Son, W. K.; Youk, J. H.; Park, W. H. *Biomacromolecules* **2004,** *5,* 197–201.
4. Burns, P. C.; Finch, R., *Uranium: Mineralogy, Geochemistry and the Environment.* The Mineralogical Society of America: Washington, D. C., 1999.
5. Abdelouas, A.; Lutze, W.; Nuttall, E.; Gong, W. L. *Comptes Rendus Acad. Sci. Ser II-A* **1999,** *328,* 315–319.
6. Cantrell, K. J.; Kaplan, D. I.; Wietsma, T. W. *J. Hazard. Mater.* **1995,** *42,* 201–212.
7. Cantrell, K. J.; Kaplan, D. I.; Gilmore, T. J. *J. Environ. Eng.-ASCE* **1997,** *123,* 786–791.
8. Gu, B.; Liang, L.; Dickey, M. J.; Yin, X.; Dai, S. *Environmental Science & Technology* **1998,** *32,* 3366–3373.
9. Kaplan, D. I.; Cantrell, K. J.; Wietsma, T. W.; Potter, M. A. *J. Environ. Qual.* **1996,** *25,* 1086–1094.
10. Li, D.; Xia, Y. N. *Adv. Mater.* **2004,** *16,* 1151–1170.
11. Kumar, V.; Yang, T. R. *Carbohydr. Polym.* **2002,** *48,* 403–412.
12. Adebajo, M. O.; Frost, R. L.; Kloprogge, J. T.; Kokot, S. *Spectroc. Acta Pt. A-Molec. Biomolec. Spectr.* **2006,** *64,* 448–453.
13. Socrates, G. *Infrared and Raman Characteristics Group Frequencies: Tables and Charts;* 3rd ed.; Wiley: Chichester, New York, 2001.
14. Schmeide, K.; Sachs, S.; Bubner, M.; Reich, T.; Heise, K. H.; Bernhard, G. *Inorg. Chim. Acta* **2003,** *351,* 133–140.
15. Fujii, T.; Fujiwara, K.; Yamana, H.; Moriyama, H. *J. Alloys Compd.* **2001,** *323,* 859–863

Surface Properties of Cellulose and Cellulose Derivatives: A Review

Qing Shen

State Key Laboratory for Modification of Chemical Fibers and Polymers and Department of Polymer Materials and Engineering, Donghua University, 2999 Renming Road, N, 201600 Songjiang, Shanghai, China

The rapidly increasing interdisciplinary research on cellulose and cellulose derivatives, as well as the broad use of these materials, makes a basic understanding of their properties important. The surface properties of cellulose and cellulose derivatives play an important role in numerous applications. This review compiles the surface properties data reported in the literature on cellulose and its main derivatives, cellulose ethers and cellulose esters, with a focus on the surface free energy, the Lewis acid–base properties, and the Hamaker constant. Because Lewis acid–base interactions can be described using a variety of theories and/or scales, a comparison of surface properties data obtained using different acid–base scales is made and discussed. The influence of the main structural properties of cellulose and cellulose derivatives, such as the degree of polymerization, the crystallinity index, and the degree of substitution, on the surface properties of the materials is also is discussed.

Introduction

Cellulose is the most abundant natural polymer, and has been used by mankind for centuries. It is the main structural component of plant cell walls, as has been shown by the removal of lignin and extractives from plant tissue. Cellulose is a semicrystalline linear polysaccharide of β-1,4-linked D-glucopyranose (*1–3*). In its native form, it typically has a degree of polymerization, DP, between 10,000 and 15,000 glucose residues, depending on its origin. Cellulose occurs as a partly crystalline, partly amorphous material, and the degree of crystallinity has been found to depend on the cellulose source; for example, cotton has a high degree of crystallinity, whereas wood has a lower one (*4*). Cellulose can be derived from plant or bacterial sources. The best known example of a bacterial source is *Acetobacter xylinum*, which produces extracellular cellulose as a small pellicle extending from its cell (*4*).

Cellulose has been widely studied, and consequently our understanding of its nature and behavior is considerable. Moreover, many of the basic principles of polymer chemistry and physics were worked out in the course of investigating cellulose, and these studies have led to an understanding of the behavior of other natural and synthetic polymers (*5*).

Because cellulose and its derivatives have been broadly applied in various areas, e.g., liquid penetration, diffusion, adsorption, coatings, foods, paper-making, and chemical engineering (*6–7*), and because many of these applications rely on the interfacial behavior of cellulose, an understanding of the surface properties of cellulose is important.

In this review, we aim to summarize the information reported in the literature on the Hamaker constant, surface free energy, and Lewis acid–base properties of cellulose and its main derivatives, i.e., cellulose ethers and cellulose esters. Since various measurement methods are available for determining or estimating the surface properties data with respect to different acid–base theories and scales (*8–9*), a comparison of the data obtained on the basis of different theories and scales is presented.

Methods for Evaluating the Surface Properties of Cellulose and Cellulose Derivatives

Methods for Evaluating the Hamaker Constant

The Hamaker constant is an important parameter in surface chemistry because it strongly depends on the interactions between materials. Among the many kinds of interactions between surfaces, such as double layer, structural, steric, depletion, hydration, and hydrophobic interactions, there is the ubiquitous van der Waals interaction, the magnitude of which varies with the system being studied. The Hamaker constant represents a conventional and convenient way of assessing the magnitude of the van der Waals interaction. It is self-evident that accurate estimates of the Hamaker constant are necessary for a quantitative understanding of the effect of interparticle forces on various phenomena involving cellulose.

The van der Waals force has an electrodynamic origin, as it arises from the interactions between atomic or molecular oscillating or rotating electrical dipoles within the interacting media (*10*). Hamaker (*11*) calculated the distance dependence of the interaction free energy of macroscopic bodies of different geometries by performing a pair-wise summation over all the atoms in the bodies. This relationship is given by

$$E_{\text{vdw}} = \frac{A}{12\pi L^2} \tag{1}$$

where E_{vdw} represents the van der Waals interaction per unit area between two parallel surfaces, A is the Hamaker constant, and L is the distance between the parallel surfaces.

As seen in eq 1, there is a direct proportionality between the magnitude of the van der Waals interaction and the Hamaker constant. Moreover, the Hamaker constant has been found to depend on the dielectric properties of the two materials and the intervening media. In the original treatment, also called the microscopic approach, the Hamaker constant was calculated from the polarizabilities and number densities of the atoms in the two interacting bodies (*11*).

An alternative, more rigorous approach was proposed by Lifshitz (*12*), in which each body was treated as a continuum with certain dielectric properties. According to Lifshitz, the van der Waals interaction is the result of fluctuations in the electromagnetic field between two macroscopic bodies, modified by the separating media, where the interaction can be referred to as standing waves that occur only at two specific frequencies (*10*). This means that the van der Waals interaction associated with the Hamaker constant can be estimated from the frequency-dependent dielectric properties of the interacting materials and intervening medium, as well as the geometry of the bodies. However, it is to be noted that the accuracy of the estimated Hamaker constant is directly related to the precision and accuracy of the dielectric spectra and the mathematical representation deduced from the dielectric data (*13–14*).

Leong and Ong introduced a method for determining the Hamaker constant of solid materials by measuring the critical zeta potential, ζ_{crit}, i.e. the zeta potential of a colloidal dispersion at the pH-induced transition from a flocculated to a dispersed state (*15*). The method is based on the linear relationship between the yield stress of the dispersion and the square of the zeta potential. At that pH at which the yield stress of the dispersion becomes zero, i.e. at the flocculated–dispersed state transition, the relationship between the Hamaker constant of the particles in the dispersion, A, and ζ_{crit} is

$$\zeta_{\text{crit}} = \sqrt{\frac{A}{12 D_0 C}} \tag{2}$$

where D_0 is the minimum surface separation distance between the interacting particles in the flocculated state and C is determined from

$$C = 2\pi\varepsilon \ln(1 - e^{-\kappa D_0}) \tag{3}$$

where ε is the permittivity of water and κ is the Debye–Hückel parameter or the inverse of the double layer thickness.

It should be noted that the method of Leong and Ong (*15*) requires the colloidal sample to undergo a flocculated–dispersed state transition. Obviously, this method cannot be applied to solid materials that interact with and/or dissolve in water.

Among others, Hough and White (*13*) introduced a simplified method to estimate the Hamaker constant through determination of the dielectric data and spectra parameter in the infrared and ultraviolet frequency range. These authors also collected important results for different materials, and indicated a relation (eq 4) between the Hamaker constant and the contact angle, which greatly benefited subsequent scholars:

$$\cos\theta = \frac{2A_{SAL}}{A_{LAL}} - 1 \tag{4}$$

where θ is the contact angle of the liquid against the solid, A_{SAL} is the Hamaker constant, in air, of the solid, and A_{LAL} is the Hamaker constant, in air, of the liquid.

A calculation of the Hamaker constant for cellulose was reported in 1987 by Winter using frequency-dependent dielectric data (*16*). Winter estimated the Hamaker constant for cellulose interacting across air and water media to be about 10×10^{-20} J and 3×10^{-21} J, respectively.

Since that time, the Hamaker constant for cellulose has been reestimated by Evans and Luner (*17*), Drummond and Chan (*18*), and Holmberg et al. (*19*), using the Lifshitz theory and the simplified expression of Tabor, Winterton, and Israelachvili (*20*), and by Bergström et al. (*10*), using the spectroscopic ellipsometry method. According to Bergström et al. (*10*), the infrared contribution to the Hamaker constant of cellulose has a minor influence on interactions across a vacuum or air. More recently, the Hamaker constant of cellulose has been studied by Notley at al. using atomic force microscopy (*21*).

Methods for Evaluating the Surface Free Energy

The solid surface free energy is a parameter of both theoretical and practical importance in the field of surface chemistry. It is a key parameter for understanding and predicting solid surface phenomena, including cohesion, adsorption, wetting, adhesion, and interfacial bonding.

The increasing importance of surface free energy in many scientific and technological realms has led to an ever-growing concern about the precise quantification of this parameter. As a consequence, many limitations and imperfections of current models for surface energy evaluation have been identified, prolonging the quest for the best way to obtain data. The most debated problem pertaining to the estimation of solid surface energy is finding an accurate method for its determination.

It is notable that data on the surface free energy of cellulose have been frequently reported in the literature, mainly determined by contact angle measurements and inverse gas chromatography. The data from contact angle measurements are obviously based on different theories and/or methods. In this paper, five commonly used methods for the determination of the surface energy by contact angle measurements are reviewed and compared.

Theories and Methods Based on Contact Angle Measurements

The surface free energy of cellulose is frequently determined by contact angle measurements in combination with several theories or methods, e.g., the Zisman method (*22*), the equation of state (*23*), the geometric mean method (*24*), the harmonic mean method (*25*), and the Lewis acid–base approach (*26–28*).

The Zisman Method

Zisman is regarded as a trailblazer in this field, because he developed a fundamental method (*22*) to quantify the critical surface energy, γ_c. He found that a linear relationship existed between the cosine of the contact angle, θ, of solid–liquid pairs and the liquid surface tension, γ_L, of the form

$$\cos\theta = 1 - b(\gamma_L - \gamma_S) \tag{5}$$

where γ_S is the surface energy of the solid and b is the slope of the linear regression.

Based on eq 5, if a plot of the experimentally determined $\cos\theta$ values for different liquids versus the surface tensions of the liquids, γ_L, yields a straight line, the critical surface tension, γ_c, can be obtained by extrapolating the regression line to $\cos\theta = 1$ ($\theta = 0$), i.e. the point of complete wetting of the solid surface by an ideal liquid. The corresponding value of liquid surface tension at this point ($\cos\theta = 1$) is defined as the critical surface tension, i.e. the critical surface energy, γ_c. The pioneering work of Zisman gave rise to two primary schools of thought: (1) the equation of state approach, and (2) the surface energy component approach, which includes the geometric mean method, harmonic mean method, and acid–base approach (*27*).

The Equation of State Method

It has been suggested by Neumann et al. (*23*) that the surface energy of a solid can be determined using the contact angle of a single liquid. The mathematical equation, corresponding to this theory, is known as the equation of state:

$$\gamma_{SL} = \gamma_L + \gamma_S - 2\sqrt{\gamma_L \gamma_S} \cdot e^{-\beta(\gamma_L - \gamma_S)^2} \tag{6}$$

where γ_{SL} is the interfacial tension between the solid and the liquid and β is an empirical constant with an average value of 0.0001057 $(m^2 mJ^{-1})^2$. Inserting this equation into Young's equation,

$$\gamma_S = \gamma_{SL} + \gamma_L \cdot \cos\theta \tag{7}$$

yields

$$(1 + \cos\theta)\gamma_L = 2\sqrt{\gamma_L \gamma_S} \cdot e^{-\beta(\gamma_L - \gamma_S)^2} \tag{8}$$

from which the surface energy of the solid can be quantified. With regard to the equation-of-state method, it is important to note that the Lewis acid–base interactions, including hydrogen bonding, seem to be ignored by eq 8 (*27*).

The Geometric Mean Method

The geometric mean method (*24*) is sometimes referred to as the Owens, Wendt, Rabel, and Kaelble method in relation to the two liquids approach (*27*). According to these authors, the surface energy consists of separate dispersive and polar parts, given by eqs 9-1 and 9-2:

$$\gamma_L = \gamma_L^D + \gamma_L^P \tag{9-1}$$

$$\gamma_S = \gamma_S^D + \gamma_S^P \tag{9-2}$$

where γ is the surface energy of a certain phase, the subscripts L and S refer to a liquid and a solid, respectively, and the superscripts D and P refer to the dispersive and polar fraction, respectively.

The interfacial energy between a solid and a liquid, γ_{SL}, is expressed as:

$$\gamma_{SL} = \gamma_S + \gamma_L - 2\left(\sqrt{\gamma_S^D \cdot \gamma_L^D} + \sqrt{\gamma_S^P \cdot \gamma_L^P} \right) \tag{10}$$

By combining eq 9 with Young's equation (eq 7), we can eliminate γ_{SL} from eq 10 to yield eq 11:

$$(1 + \cos\theta)\gamma_L = 2\left(\sqrt{\gamma_S^D \cdot \gamma_L^D} + \sqrt{\gamma_S^P \cdot \gamma_L^P} \right) \tag{11}$$

Thus, by inserting into eq 11 data from measurements of contact angles on the sample surface using two liquids with known polar and dispersive components, two simultaneous equations evolve whose solution yields the solid surface energy, γ_S.

In addition to this mathematical solution, Rabel (*27*) proposed a graphical solution that relies on linear regression and uses the equation

$$\frac{(1 + \cos\theta)\gamma_L}{2\sqrt{\gamma_L^D}} = \sqrt{\gamma_S^P} \cdot \sqrt{\frac{\gamma_L^P}{\gamma_L^D}} + \sqrt{\gamma_S^D} \tag{12}$$

In this method, $\dfrac{(1+\cos\theta)\gamma_L}{2\sqrt{\gamma_L^D}}$ is plotted against $\sqrt{\dfrac{\gamma_L^P}{\gamma_L^D}}$ to form a straight line

from which $\sqrt{\gamma_S^P}$ and $\sqrt{\gamma_S^D}$ are identified as the slope and intercept, respectively. Thus, the solid surface energy and its two components can be obtained.

The Harmonic Mean Method

In the harmonic mean method, Wu (25) proposed the same separation of surface energy components as in the two liquids approach. The distinction of this method lies in its mathematical approach toward calculating the mean of the surface energies. Wu uses the harmonic mean to describe the interfacial energy:

$$\gamma_{SL} = \gamma_S + \gamma_L - 4\left(\frac{\gamma_S^D \cdot \gamma_L^D}{\gamma_S^D + \gamma_L^D} + \frac{\gamma_S^P \cdot \gamma_L^P}{\gamma_S^P + \gamma_L^P} \right) \tag{13}$$

In conjunction with Young's equation (eq 7), eq 13 produces the final equation (eq 14), which can be used to obtain the surface energy of a solid in the same way as the geometric mean method .

$$(1+\cos\theta)\gamma_L = 4\left(\frac{\gamma_S^D \cdot \gamma_L^D}{\gamma_S^D + \gamma_L^D} + \frac{\gamma_S^P \cdot \gamma_L^P}{\gamma_S^P + \gamma_L^P} \right) \tag{14}$$

Like the geometric mean method, the harmonic mean method requires a minimum of two probe liquids for performing the calculation of solid surface energy. At least one of the test liquids should have a positive value for its polar component. The main difference between the geometric and harmonic mean methods is the applicability with respect to the magnitude of the surface energy of solids. Using the harmonic calculation, Wu achieves more accurate results for high-energy systems. The geometric mean method, on the other hand, is better suited for low-energy surfaces.

The geometric and harmonic mean methods have received less attention than the Lewis acid–base approach, described in the following section (27).

The Lewis Acid–Base Approach

In this approach (26–28), the surface energy of a solid, γ, is viewed as the sum of the Lifshitz–van der Waals component, γ_{LW}, corresponding to the dispersive component, and the Lewis acid–base component, γ_{AB}, corresponding to the polar component:

$$\gamma = \gamma^{LW} + \gamma^{AB} \tag{15}$$

The Lewis acid–base component can be further divided into the Lewis acid component, γ^+, and the Lewis base component, γ^-, according to eq 16:

$$\gamma^{AB} = 2\sqrt{\gamma^+ \gamma^-} \tag{16}$$

The interfacial energy, γ_{SL}, in this approach is given by

$$\gamma_{SL} = \gamma_S + \gamma_L - 2\left[\left(\sqrt{\gamma_S^{LW}\gamma_L^{LW}}\right) + \left(\sqrt{\gamma_S^+\gamma_L^-}\right) + \left(\sqrt{\gamma_S^-\gamma_L^+}\right)\right] \tag{17}$$

And the equation for calculating the solid surface energy is

$$(1+\cos\theta)\gamma_L = 2\left[\left(\sqrt{\gamma_S^{LW}\gamma_L^{LW}}\right) + \left(\sqrt{\gamma_S^+\gamma_L^-}\right) + \left(\sqrt{\gamma_S^-\gamma_L^+}\right)\right] \tag{18}$$

By inserting values for γ_L, γ_L^{LW}, γ_L^-, and γ_L^+ from three calibration liquids, of which at least two are polar, into eq 18, the resulting solid surface energy is produced via the solution of three simultaneous equations. For this reason, the acid–base approach is also referred to as the three-liquid method. It has also been called the vCG approach (28), after the authors who proposed this method, van Oss, Chaudhury, and Good.

Following the development of the vCG approach, Chang et al. (29–32) proposed another semi-empirical acid–base approach that is analogous to the vCG model but allows both attractive and repulsive acid–base contributions to the work of adhesion. Although the approach suggested by these authors has been adapted to the analysis of wood surfaces (29–32), Della Volpe and Siboni (33) have remarked that the scale of the model suffers from the same essential drawbacks as the vCG model. Mohammed-Ziegler and coworkers recently compared these two models with respect to the application to wood surfaces (34). These authors failed to obtain realistic results using their suggested model, and concluded that this model is inadequate for a detailed analysis of complex systems such as wood and other natural fibers (34).

A Brief Discussion

Initially, Zisman equated the critical surface energy to the solid surface energy. However, later studies showed that the critical surface energy generated using this method on the same solid specimens is systematically less than the surface energy obtained via other methods, i.e., the components approach and the equation-of-state approach (35–36). Thus, it is commonly accepted, and Zisman himself (37) has always emphasized, that the critical surface energy is not the surface free energy, but only an empirical parameter related closely to this quantity. In fact, lacking a theoretical background for molecular interactions, the proposed linear relationship only applies when the relationship between the dispersive and polar interactions of the solid sample equates to that of the probe liquids. And this can only be achieved under experimental circumstances involving a purely dispersively interacting solid and liquid.

It is commonly accepted that the applicability of the equation-of-state approach is rather limited (*27, 38–39*). Based on the poor agreement between determined values and expected values for halogenated hydrocarbon-probed polymer surfaces, Balkenende et al. (*39*) concluded that this deviation might derive from the specific interactions between probes and surfaces, which were beyond the scope of this theory, and that knowledge of γ_S alone was not sufficient to describe their results.

Nevertheless, one cannot say either that there are no problems with the component approaches, especially the vCG approach. As a reliable analytical method of wide applicability, the vCG technique should yield data free of the influence of the probe triplet chosen. However, this has not always been the case in actual measurements. Kwok et al. (*40–41*) reported a strong dependency of the components determined with the liquid sets used for vCG measurements in their experiments. In fact, for all solids analyzed so far, this dependency has been observed in practical measurements. Della Volpe et al. (*42–43*) then proposed the use of a matrix of contact angle values measured for many test liquids to generate the averaged apolar and polar components of the surface free energy of the analyzed solid.

In addition to this major shortcoming, other challenges to the validity of this method have arisen. It has been observed that all materials show overwhelming basicity in measurements made using this method. Negative values for square roots of the acid–base parameters are occasionally obtained, which lead to meaningless data interpretation.

In an attempt to address these drawbacks, Della Volpe et al. (*42*) enumerated the main problems involved in the application of the vCG theory to the calculation of Lewis acid–base properties of polymer surfaces from contact angle data, and listed several factors that should be considered for obtaining good results.

At almost the same time, a member of our group expressed similar concerns about the vCG method, and agreed that this method is unsuitable for the acid/base ratio of water. In a previous paper by us (*28*), we proposed a unified acid/base ratio of water, 2.42, by averaging values reported using different techniques and from different researchers. The ratio 2.42 can probably replace the previously used value of 1.0 for water in the vCG method. In this sense, the unified value might be a good choice for overcoming the difficulties of the van Oss–Chaudhury–Good combining rules.

In defense of the multicomponent approach to surface energy calculation, Della Volpe et al. (*44–45*) further discussed its limits and possibilities, and presented criticisms of the equation-of-state method from both theoretical and experimental points of view.

Nevertheless, no one can unequivocally claim that any one method is omnipotent and can analyze any surface with any probe while still yielding absolutely consistent results, without any shortcomings. However, the results obtained from different methods of surface energy calculations of solids can be compared with one another and utilized to improve the development of individual approaches. Since no one method can dominate this field, and the

268

extant methodologies may not be mutually exclusive, a multiple methodology will provide a "peer review" mechanism in which data from complementary approaches give a rather confusing, but more complete, view of the real world.

Taking lignocelluloses as examples, Gindl et al. (*35*) recently compared these five approaches by applying them to wood surface energy evaluations. Their results show that the vCG approach gives data that are in good agreement with data obtained from the geometric mean method and, to a lesser degree, from the equation-of-state method. Results obtained with the harmonic mean and Zisman approaches, however, deviate markedly from results obtained using these methods.

Theories and Methods Based on Inverse Gas Chromatography Measurements

In addition to contact angle measurements, other methods are available for determining the surface free energy of solids, including inverse gas chromatography, IGC (*46–47*), adsorption gas chromatography (*48–52*), and methods based on dissolution or crystallization parameters, as reviewed by Wu and Nancollas (*53*). For fibrous materials, evaluation of surface free energy is based either on wettability measurements or on IGC characterization; the latter method seems to be widely used.

IGC is an efficient tool that gives information on the dispersive component of the surface energy of solids, γ_S^D; the acid–base properties of the surface, expressed as the ratio of the Gutmann donor and acceptor numbers, DN and AN, respectively; and the isotherm of adsorption for the powdered or fibrous materials. Whereas the purpose of classic gas chromatography is the separation of a mixture of analytes in the mobile phase, IGC is used to characterize the stationary phase with respect to its surface properties.

In IGC measurements that focus on the dispersive component of the solid surface energy, a series of nonpolar fluids, often the alkanes, with known dispersive surface energy, γ_L^D, are injected into an IGC column, with the solid packed in advance. The retention volume, V_n, is calculated from eq 19 (*54*),

$$V_n = JF(t_r - t_0) \tag{19}$$

where J is the James and Martin compressibility correction factor, F is the carrier gas flow, and t_r and t_0 are the retention time of the probe and of the non-interacting standard (methane), respectively. The dispersive component of the solid surface free energy, γ_S^D, is calculated from eq 20 (*54, 55*),

$$RT \ln V_n = 2N\sqrt{\gamma_S^D}\, a\sqrt{\gamma_L^D} + K \tag{20}$$

where R is the universal gas constant, T is absolute temperature, N is Avogadro's number, a is the molecular area of the adsorbed molecule, and K is the intercept of the regression line. The IGC method for calculating γ_{SD} has been described in detail by Garnier and Glasser (*56*).

Although surface energy evaluations based on the Washburn equation and vCG combining rules may offer complete data, IGC is a simple and reliable tool independent of sample morphology. The convenience and rapidity of IGC make it a popular method for characterizing cellulose surfaces. Performed with appropriate experimental parameters, IGC analysis has proven to be a powerful tool for revealing the chemical changes that occur on the surfaces of both cellulosic powders and fibers (*57*). Papirer et al. (*58*) examined a wide range of cellulose samples differing in origin and crystallinity, and concluded that the interpretation of results provided by IGC is anything but simple.

From a theoretical point of view, the results obtained from IGC analysis and contact angle measurements are somewhat different. Ticehurst et al. (*59*) have claimed that in terms of γ_S^D, results from IGC and contact angle measurements cannot be identical because of differences in their theoretical approaches. In a recent survey regarding the surface characterization of phenol–formaldehyde–lignin resin, Matsushita et al. (*60*) observed a significant discrepancy between the absolute values of γ^{LW} generated by these two methods.

But in the measurement performed by Srcic et al. (*57*), which involved assessing the dispersive component of the surface free energy of some pharmaceutical powders, including hydroxypropyl methyl cellulose, a good correlation between results of the two methods was found for most samples.

Thus, a satisfactory answer to this question would require a systematic study comparing data obtained from both methods.

In fact, almost every IGC analysis performed to determine γ_S^D is carried out under infinitely dilute conditions, in which minimal doses of molecular probes are used. Any information obtained from IGC under these conditions will mainly concern the most active sites of the solid surfaces, which may constitute only a fraction of the surface analyzed. However, although IGC cannot give a full view of the surface properties of solid samples, it gives crucial information on their trends, especially for chemically treated specimens (*58, 60–62*).

By IGC measurement, two acid–base parameters, namely K_A (Lewis acid) and K_B (Lewis base), can be obtained (*46–62*).

Surface Properties of Cellulose

The Hamaker constant of cellulose has been reported in the literature by different research groups using various methods (*10, 17–21*). The data compiled from the literature are presented in Table 1. A comparison of the reported values indicates that the obtained value depends on the media and that it is lower in water and higher in a vacuum. It is notable that the influence of the DP of cellulose on this important parameter seems to have been little studied.

Surface free energy is another important parameter for the characterization of cellulose. This is because the surface free energy is known to be related to the Hamaker constant, and cellulose forms strong hydrogen bonds, corresponding to Lewis acid–base interactions (*1, 2, 8, 9*). In fact, in some cases the surface free energy is a determining factor in the application of cellulosic materials (*1, 2, 8, 9, 63, 64*).

In early work, Luner and coworkers reported surface free energy data for cellulose and other lignocellulosic materials (*65*). They applied the theories and methods of Zisman, Fowkes, Tamai, and Owens to estimate the surface properties of lignin and cellulose and compare the results obtained from the different theories and methods.

Table 1. Literature values for the Hamaker constant of cellulose

Sample 1	*Sample 2*	*A (10^{-20} J)*		*Refs*
		Vacuum	*Water*	
Cellulose	Cellulose	5.8	0.80	(*10*)
Cellulose	$CaCO_3$	7.4	0.57	(*10*)
Cellulose	Si_3N_4	9.5	0.80	(*10*)
Cellulose	SiO_2	5.9	0.35	(*10*)
Cellulose	Mica	7.2	0.43	(*10*)
Cellulose	TiO_2	9.3	1.20	(*10*)
Cellulose	Cellulose		0.99	(*17*)
Cellulose	Cellulose	7.9		(*18*)
Cellulose	Cellulose	8.4	0.86	(*19*)
Cellulose	Cellulose		0.35*	(*21*)

*NaCl solution.

Berg and his group have reported several studies on the determination of surface free energy of lignocellulosic materials (*19, 66*). These authors paid special attention to Lewis acid–base interactions, especially in relation to different theories and methods. Utilizing dynamic contact angle measurement, Gardner and his coworkers reported important data on the surface free energy of lignocelluloses (*67*).

Table 2 summarizes data reported in the literature for the surface free energy of cellulose estimated by different methods (*6, 63–85*). The critical surface tension defined by Zisman (*22*) is included as a reference.

Data on some novel cellulose fibers, such as the Lyocell fiber, which is becoming increasingly important, are also included. A comparison of the data for different regenerated cellulose fibers is believed to facilitate the selection of a suitable fiber for a specific application.

A comparison of the Lewis-base and Lewis-acid data reveals that the Lewis base character of cellulose is more pronounced than its Lewis acid character. It also should be noted that the obtained values for the surface free energy, γ_S, of cellulose vary greatly. This variation in the data can be expected, because cellulose is a complex natural polymer and its structure and properties are influenced by many factors (*1–5*).

Table 2. Surface free energy, related components, γ_S^{LW}, γ_S^{AB}, γ_S^+, γ_S^-, γ_S^D, γ_S^P, and the critical surface tension, γ_c, of cellulose

Samples	Sample codes	γ_S (mJ·m^{-2})	γ_S^{LW} (mJ·m^{-2})	γ_S^{AB} (mJ·m^{-2})	γ_S^+ (mJ·m^{-2})	γ_S^- (mJ·m^{-2})	γ_S^D (mJ·m^{-2})	γ_S^P (mJ·m^{-2})	γ_C (mJ·m^{-2})	Technique	Refs
Cellulose powder	Sigmacell 101	58.98	54.49	4.49	0.11	47.83				TLW	(63)
Cellulose powder	Sigmacell 20	57.18	52.94	4.24	0.11	41.7				TLW	(63)
MCC	Avicel	51.82	51.82	0	0	50.14				TLW	(63)
α-cellulose	Sigma C8002	57.43	55.73	1.7	0.013	56.31				CW	(64)[a]
α-cellulose	Sigma C8002	57.26	55.07	2.19	0.021	56.16				CW	(64)[b]
α-cellulose	Sigma C8002	57.41	55.73	1.68	0.02	36.12				CW	(64)[a]
α-cellulose	Sigma C8002	56.87	55.07	1.8	0.023	36				CW	(64)[b]
Cellulose film	Avicel								36	SD	(65)
Cellulose film	Avicel	73.9					29.1	34.8		SD	(65)[a]
Cellulose film	Avicel	68.7					42.2	26.5		SD	(65)[b]
Cellulose film	Avicel	59.8					42.2	17.6		SD	(65)
Cellulose film	Avicel	54.2					41	13.2		SD	(65)
Purified cotton		68.5					27.5	41.0		WP	(66)
Cellulose fiber		54.2	44.6	9.6	1.2	19.0					(67)
Cellulose fiber		56.1	44.9	11.2	2.2	14.3					(68)
Cellulose fiber		54.5	44	10.5	1.62	17.2					(68)
Purified cotton		32.3					21.3	11		PC	(69)
Cotton linter	DP 1356	63.3	59.7	3.64	0.13	24.21				CW	(6)
Cotton linter	DP 1307	61.5	57.7	3.83	0.2	18.34				CW	(6)
Cotton linter	DP 1841	71.9	63.33	8.56	0.97	4.41				CW	(6)
Cotton fiber		53.86					13.79	39.91		WP	(70)
Viscose fiber		52.89					10.78	42.10		WP	(70)

Continued on next page

a: values estimated using R^P_{eff}; b: values estimated using R^l_{eff}; MCC: microcrystalline cellulose; DP: degree of polymerization; TLW: thin layer wicking technique; CW: column wicking technique; SD: sessile drop technique; WP: Wilhelmy plate technique

Table 2. Continued

Samples	Sample codes	γ_S (mJ·m^{-2})	γ_S^{LW} (mJ·m^{-2})	γ_S^{AB} (mJ·m^{-2})	γ_S^{+} (mJ·m^{-2})	γ_S^{-} (mJ·m^{-2})	γ_S^{D-} (mJ·m^{-2})	γ_S^{P} (mJ·m^{-2})	γ_C (mJ·m^{-2})	Technique	Refs
MCC	Avicel – PH101	63.9					17.3	46.6		WP	(71)
Unsized tracing paper		49.4					29.4	20		SD	(72, 78)
Viscose fiber		23.9								CW	(72)
Viscose fiber		33.78					8.24	25.54		CW	(73)
Modal fiber		33					7.99	25.01		CW	(73)
Lyocell fiber		33.39					8.03	25.36		CW	(73)
α-cellulose	Fiber	68.7					25.5	43.2		WP	(74)
MCC	Technocel 150DM	50					30	20		SD	(75)
Filter paper	Whatman No.5	50					30	20		SD	(75)
Viscose fiber	Rayon	72.80					22.42	50.38			(76)[c]
Viscose fiber	Rayon	50.06					23.75	26.31			(76)[d]
α-cellulose	Arbocel	61					39	22		SD	(77)
MCC	Avicel	53.9					17.3	36.6		SD	(79)
Cellulose fiber	Kraft pulp	51.9					17.8	34.2			(79)
Cellulose fiber	Organ-solve	51.6					15.9	35.7			(79)
Cellulose fiber	Acetosolv-supercritical CO_2	48.8					18.0	30.8			(79)

Continued on next page

c: water/diiodomethane; d: ethylene glycol/diiodomethane; MCC: microcrystalline cellulose; WP: Wilhelmy plate technique; SD: sessile drop technique; CW: column wicking technique

Table 2. Continued

Samples	Sample codes	γ_S $(mJ \cdot m^{-2})$	γ_S^{LW} $(mJ \cdot m^{-2})$	γ_S^{AB} $(mJ \cdot m^{-2})$	γ_S^{+} $(mJ \cdot m^{-2})$	γ_S^{-} $(mJ \cdot m^{-2})$	γ_S^{D} $(mJ \cdot m^{-2})$	γ_S^{P} $(mJ \cdot m^{-2})$	γ_C $(mJ \cdot m^{-2})$	Technique	Refs
Cellulose fiber	Butanolsolv-supercritical CO_2	49.9					17.1	32.8			(79)
Cellulose fiber	Methanolsolv-supercritical CO_2	47.2					17.6	29.6			(79)
Cellulose fiber		56.7	39.1	17.6	2.0	39.7					(80)
Cottton linter		60.9					19.3	41.6		WP	(81)
Viscose fiber		46.1					21.7	24.4		WP	(81)
Cellulose fiber		59.5					19.9	39.6			(82)
Cellulose fiber		54.4					29.4	25.0			(82)
Cellulose fiber	Bleached kraft pulp	42.6					22.1	20.5			(83)
Cellulose fiber	Kraft pulp	46.4	43.6	2.7	0.07	27.4					(84)
Cellulose fiber	Kraft pulp	45.1	44.6	0.5	0.002	34.5					(84)
Cellulose fiber	NSSC	57.2	45.7	11.6	1.59	21.0					(84)
Cellulose fiber	NSSC	54.3	48.0	6.3	0.52	19.0					(84)
Cellulose fiber	CP	74					33	41			(85)
Cellulose fiber	CTMP	57					25	32		WP	(85)

NSSC: neutral sulfite semi-chemical pulp; CP: highly bleached kraft pulp; CTMP: chemi-thermomechanical pulp; WP: Wilhelmy plate method

Some of these numerous studies seem to be of special importance for understanding cellulose. For example, a case reported by Liu et al. (*70*) indicates that interfacial shear strength directly enhances the total surface free energy of cellulose in a proportional manner. In addition, we have observed that the surface free energy of cellulose is influenced by its DP (*6*).

A plot of the surface free energy versus the DP of cellulose (Figure 1), based on data from the literature, suggests that the surface free energy increases with increasing DP. The surface free energy data reported by Persin et al. (*74*), presented in Table 2, are clearly lower than those reported by others. The authors attributed this discrepancy to the fact that they prewashed the raw cellulose samples under alkaline conditions at 60 °C for about 30 min before contact angle measurement (*74*).

Recently, Ass et al. (*86*) found that the DP of cellulose is linearly related to the morphology of cellulose, i.e., the crystallinity index, as shown in Figure 2. This could mean that the increase in surface free energy with increasing DP is due to an increase in the crystallinity of cellulose.

As one of the two main components of surface free energy, acid–base interactions are important. Although Table 2 shows some data on the Lewis acid–base parameter, i.e., data obtained by the methods of Owens, Wendt, Rabel, and Kaelble (*24*), Wu (*25*), and vCG (*26–27*), it should be noted that these reported values only relate to a few acid–base scales. The Lewis acid–base interactions can also be described by other parameters, such as the Reichardt's $E_T(30)$ values; the ratio of Gutmann's AN-acceptor number, representing the Lewis acid, and DN-donor number, representing the Lewis base; and the three parameters of Kamlet and Taft, namely α, representing the hydrogen-bond donating ability, β, representing the hydrogen-bond accepting ability, and π^* representing the dipolarity/polarizability (*28*).

Using the Reichardt dye as a probe in combination with the solvatochromic method, Spange and his coworkers reported some α, β, and π^* data for cellulose and other biomaterials (*87–91*). Additionally, some data for AN, DN, K_A, and K_B, determined by IGC, have been reported elsewhere. Table 3 presents additional acid–base parameters for cellulose found in the literature.

Surface Properties of Cellulose Derivatives

The hydroxyl groups of cellulose can be partially or fully reacted with various chemicals to provide derivatives with useful properties. Cellulose ethers and cellulose esters are commercially important materials. In principle, although not always in current industrial practice, cellulosic polymers are renewable resources.

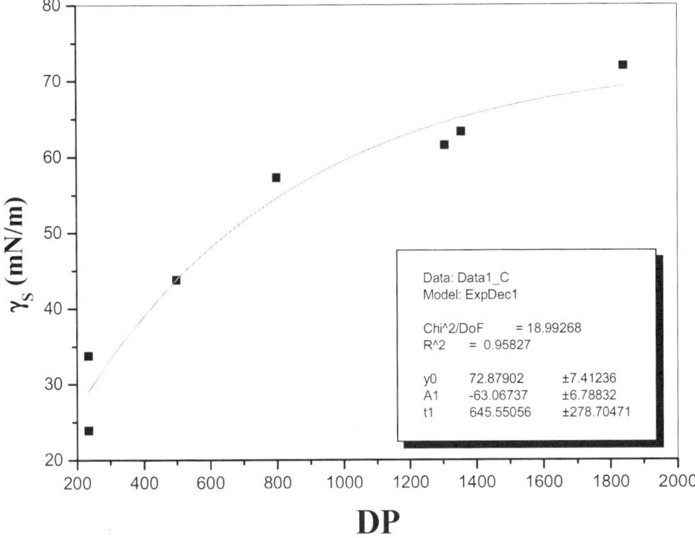

Figure 1. The influence of degree of polymerization, DP, on the surface free energy of cellulose. (Data from ref 6)

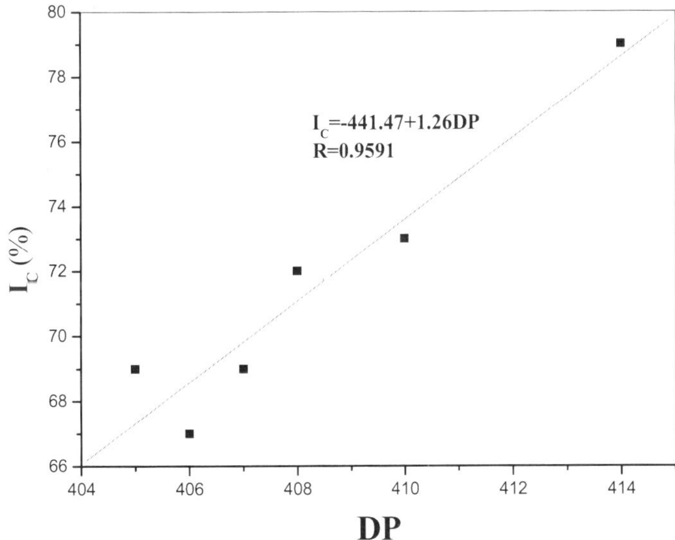

Figure 2. Relationship between the degree of polymerization, DP, and crystallinity index, I_C, of cellulose. (Data from ref 86)

Table 3. Acid–base properties of cellulose in terms of different parameters

Samples	α	β	π^*	$E_T(30)$	K_A/K_B	AN/DN	Refs
Bleached birch pulp					0.93		(75)
Viscose fiber						3.1	(76)
Kraft *E. regnans*					0.695		(84)
Kraft *E. regnans*[a]					13.364		(84)
Kraft *E. globules*					8.462		(84)
Kraft *E. globulus*[a]					19.8		(84)
NSSC *E. regnans*					1.76		(84)
NSSC *E. regnans*[a]					3.273		(84)
NSSC *E. globules*					0.946		(84)
NSSC *E. globulus*[a]					2.325		(84)
Cold soda *E. nitens*					1.481		(84)
Cold soda *E. nitens*[a]					2.811		(84)
Dried cellulose	1.27	0.60	0.41	53.0			(87–91)
Undried cellulose	0.98	0.60	0.66	51.4			(87–91)
Micro-crystalline	1.31	0.62	0.34	55.0			(87–91)
Bacteria cellulose	0.78	0.86	0.69	52.1			(87–91)
α-cellulose					3.0	3.0–3.8	(92)
Avicel						3.4–4.5	(92)
Whatman No. 1					1.6		(92)
MCC					6.5		(93)
Bleached sulfite pulp					1.08–1.29	47–48	(93)
Exploded pulp					0.2		(94)
Kraft pulp					1.1		(94)
Bleached softwood Kraft pulp					2.2–5.7		(94)
CTMP					1.14		(95)
Avicel					2.6		(95)
Sigma cellulose					0.88		(95)
Bleached pulp					1.873		(96)
Viscose rayon					0.76		(96)

a: extracted by methanol; NSSC: neutral sulfite semi-chemical pulp; MCC: microcrystalline cellulose; CTMP: chemi-thermomechanical pulp.

Surface Properties of Cellulose Ethers

Commercially important cellulose ethers include:

(I)   ethylcellulose, a water-insoluble commercial thermoplastic used in coatings, inks, binders, and controlled-release drug tablets;

(II)   hydroxypropyl cellulose;

(III)   carboxymethyl cellulose;

(IV)   hydroxypropyl methyl cellulose, E464, used as a viscosity modifier, gelling agent, foaming agent, and binding agent; and

(V)   hydroxyethyl methyl cellulose, used in the production of cellulose films.

As a water-soluble material, cellulose ethers have been broadly studied and applied (*7, 94, 97–100*). Luner and Oh (*97*) determined the surface free energy parameters of aqueous-based cellulose ether films using the Lifshitz–van der Waals and Lewis acid–base approach. Because the film surfaces were sensitive to some of the liquid probes, these investigators estimated contact angles on the unperturbed surfaces from the initial advancing angle of sessile drops. The results showed that the cellulose ether films were predominantly electron donating and their Lewis acid–base surface energy components accounted for about 5–10% of the total surface energy. The determined values were found to be consistent with those of cellulose and ethylcellulose.

Both the aqueous solubility and hydrophilicity of the polymers were examined and found to be consistent with the free energy terms derived from the surface free energy parameters. After independent calculation of the Lewis acid–base contribution to the total effect of adhesion between the polymers and wetting liquids, the Lewis acid–base character of the polymers was found to be comparable to the surface free energy parameters experimentally obtained. Luner and Oh also calculated the work of adhesion of the polymers to a variety of surfaces, and concluded that the Lewis acid–base contributions to the surface free energy may enhance adhesion to cellulose ethers.

Because the cellulose ethers probably degraded in the film preparation process of the above mentioned study (*97*), we recently used the column wicking technique to re-characterize the surface free energy and Lewis acid–base properties of cellulose ethers (*7*). Our method was superior to that used by Luner and Oh (*97*) as it provided contact angle data while avoiding sample degradation. Our results showed that the Lifshitz–van der Waals component provided the largest contribution to the surface free energy of cellulose ethers; that the Lewis acid component of cellulose ethers was increased and the Lewis base component decreased with respect to cellulose; and that the surface free energy of cellulose ethers seemed to decrease with increasing viscosity, but seemed to be greater than that of cellulose.

Sasa et al. (*98*) investigated the surface properties of cellulose ethers by means of IGC. They found that the dispersive components of the surface free energy were not significantly different but that the polar components, an important component of the enthalpy of adsorption, showed large differences between the polymers. The polarities of the different polymers studied decreased in the order hydroxyethyl (HEC) > hydroxypropyl methyl (HPMC) > hydroxypropyl (HPC) cellulose. In other words, regarding the acid–base numbers, HEC showed the highest polarity, followed by HPMC and HPC. Sasa et al. also found that the polarity of polymers correlated well with water adsorption on bulky polymers and with the degree of swelling of polymer matrices.

The surface free energy data for different kinds of cellulose ethers are presented in Table 4.

Surface Properties of Cellulose Esters

Of the various cellulose esters, cellulose acetate and cellulose triacetate have been broadly applied and studied. Xiarchos and Doulia (*101*) performed an experiment investigating the adsorption properties of cellulose acetate using nonionic surfactant solutions at concentrations above the critical micelle concentration. The study revealed that the adsorption of two types of nonionic surfactants, namely Tritons (alkylphenol ethoxylates) and Neodols (alcohol ethoxylates), on hydrophilic cellulose acetate membranes of a molecular-weight cutoff of 20,000 Da was determined mainly by the structure of the surfactants, especially the length and composition of the hydrocarbon chain.

The more hydrophilic Triton, containing eight carbon atoms and an aromatic ring in the hydrocarbon chain, showed micellar adsorption, whereas Neodol, containing nearly ten carbon atoms in a linear hydrocarbon chain, adsorbed onto the hydrophobic membrane through a variety of mechanisms. Surfactants with intermediate hydrophilicity showed the highest adsorption. Compared with other cases of decreasing surfactant adsorption for the more hydrophilic surfactants, Neodol adsorption on the hydrophilic membrane was very low. This fact indicates that hydrophobic forces were gradually diminished in the presence of bigger ethylene oxide clusters.

Table 4. Surface free energy and related components, γ_S^{LW}, γ_S^{AB}, γ_S^+, γ_S^-, γ_S^D, and γ_S^P, of cellulose ethers

Sample	Codes	γ_S $(mJ\cdot m^{-2})$	γ_S^{LW} $(mJ\cdot m^{-2})$	γ_S^{AB} $(mJ\cdot m^{-2})$	γ_S^+ $(mJ\cdot m^{-2})$	γ_S^- $(mJ\cdot m^{-2})$	γ_S^D $(mJ\cdot m^{-2})$	γ_S^P $(mJ\cdot m^{-2})$	Technique	Refs
MC	Benecel M043	56.7	56.07	0.63	0.004	89.61			CW	(7)
HPMC	Benecel MP333C	61.9	61.62	0.28	0.001	113.17			CW	(7)
HPMC	Benecel MP824	56.91	56.28	0.63	0.006	65.99			CW	(7)
HPC	Klucel K8913	77.17	76.14	1.03	0.007	153.24			CW	(7)
HPC	Klucel K9113	85.19	84.52	0.67	0.002	263.4			CW	(7)
HPMC	Methocel E4M	39.9	35.8	4	0.15	27.2			SD	(97)
HPMC	Benecel MP9	42.8	37.5	5.2	0.21	32.3			SD	(97)
MC	Benecel M0	38.7	36.3	2.4	0.04	36.7			SD	(97)
HPC	Klucel MF NF	43	40.2	2.8	0.11	17.2			SD	(97)
HEC	Natrosol 250 MR	49.9	44.8	5.1	0.16	40.1			SD	(97)
HPMC	Methocel E4M	37.4					25.1	12.3	SD	(98)
HPMC[a]	Klucel LF	43.33					3.73	39.6	WP	(99)
HPMC[b]	Klucel LF	42.4					4.05	38.42	WP	(99)
ATMSC		33.79	29.33	4.46	0.57	8.68			SD	(100)
CTMSC		35.01	31.31	3.7	0.24	14.04			SD	(100)
TMSC		36.15	28.65	7.49	0.75	18.74			SD	(100)
ATMSC		29.76					16.49	13.27	SD	(100)
CTMSC		28.03					15.3	12.73	SD	(100)
TMSC		32.58					15.03	17.55	SD	(100)

a: films prepared from 2% (w/v) solutions in methylene chloride; b: films prepared from 5% (w/v) solutions in methylene chloride; MC: methylcellulose; HPMC: hydroxypropyl methyl cellulose; HPC: hydroxypropyl cellulose; HEC: hydroxyethylcellulose; ATMSC: aminopropyl trimethylsilyl cellulose; CTMSC: cinnamate trimethylsilyl cellulose; TMSC: trimethylsilyl cellulose; CW: column wicking technique; SD: sessile drop technique; WP: Wilhelmy plate method

Khokhlova et al. (*102*) have studied the adsorption and surface properties of vladipor cellulose acetate membranes (VCAM). The data show that the specific surface area of these VCAMs vary from 80 to 360 cm^2/cm^2, and the density of negative charges on the pore surface is 5×10^{-8} C/cm^2. Compared with polysulfonamide membranes, VCAMs have far weaker hydrophobic interactions. Because of the acidic properties of the membrane surface, the adsorption of basic substances is much higher than that of acidic substances. It was also observed that the distribution constants for the adsorption of Acid Orange and cytochrome C for the polysulfonamide membranes with a pore diameter of 0.1 μm were 50- and 100-fold higher, respectively, than for VCAM membranes of the same pore diameter and specific surface area.

The surface free energy of cellulose esters has been reported by other researchers. For example, Lee and Luner (*65*) studied cellulose acetate and found its surface free energy to be 40.8 mN/m. According to van Oss, Good, and Chaudhury (*103*), the surface free energy of both cellulose acetate and cellulose nitrate is 38 mN/m, which seems to ignore dry and wet conditions.

The surface free energy data of cellulose esters reported in the literature are summarized in Table 5. As can be seen in Table 5, no data for the geometric and harmonic mean methods have been reported. This may be because these studies were performed recently, or more specifically, after the development of the vCG method (*103*). The Lewis acid–base properties of both cellulose ethers and esters, evaluated based on other scales (*28*), are summarized in Table 6.

Conclusion

The exploration of cellulose and its derivatives is currently flourishing. Here, we have reviewed the surface properties of these materials with respect to surface free energy, acid–base interactions, and the Hamaker constant.

A large body of work has been reported that is based on recent developments in the characterization of surface properties. A comprehensive understanding of the surface properties of cellulosic materials is gradually being developed with the rapidly increasing size of the database.

This review has emphasized techniques that are broadly applied in this field. We believe that improvements in characterization techniques have contributed greatly to the recent research boom. Recently developed methods of investigation may even come to dominate research trends. As the database for this topic becomes more complete, breakthroughs in methodologies of characterization are expected.

Table 5. Surface free energy, related components, γ_S^{LW}, γ_S^{AB}, γ_S^+, γ_S^-, γ_S^D, γ_S^P, and the critical surface tension, γ_c, of cellulose esters

Sample	Specification	γ_S ($mJ\cdot m^{-2}$)	γ_S^{LW} ($mJ\cdot m^{-2}$)	γ_S^{AB} ($mJ\cdot m^{-2}$)	γ_S^+ ($mJ\cdot m^{-2}$)	γ_S^- ($mJ\cdot m^{-2}$)	γ_S^D	γ_S^P	γ_c	Technique	Refs
Cellulose acetate	Membrane	40.8								SD	(65)
Cellulose acetate	Membrane	43.1	38.2	4.9	0.21	28.2				SD	(80)
Cellulose acetate (dry)	Membrane	38	38	0	0	23				SD	(103)
Cellulose acetate (hydrated)	Membrane	38	38	0	0	23.4				SD	(103)
Cellulose nitrate (dry)	Membrane	38	38	0	0	12.9				SD	(103)
Cellulose nitrate (hydrated)	Membrane	38	38	0	0	18.3				SD	(103)
Cellulose acetate	Membrane	46.2	40		0.5	19				SD	(104)
Cellulose acetate	Membrane	30.3	30	0.2	0	33.4				SD	(105)
Cellulose acetate	Membrane	49.24	39.44	9.79	1.17	20.48				SD	(106)

SD: sessile drop technique

Table 6. Acid–base properties of cellulose derivatives in terms of different parameters

Sample	DS	A	β	π*	$E_T(30)$	AN	K_A/K_B	Refs
MCC							6.5	(93)
HEC							0.286	(98)
HPMC							0.141	(98)
HPC							0.737	(98)
CN	2	0.83		0.30	46.9			(107)
CMC	0.97	1.50		0.04	54.5			(107)
CMC	0.48	1.04	0.8	0.7	55.14	45.4		(108)
CMC	0.72	0.94	0.91	0.43	50.5	41.81		(108)
CMC	1.05	0.35	0.85	0.71	44.71	34.45		(108)
CMC	1.09	0.84	0.67	0.84	53.57	39.39		(108)
CMC	1.1	0.59	0.81	1.1	52.85	32.42		(108)
CMC	1.15	0.63	0.67	0.63	48.03	32.75		(108)
CMC	1.3	0.63	0.81	0.7	48.82	32.75		(108)
CMC	1.44	0.66	0.87	0.61	48.29	33.6		(108)
CMC	1.58	0.69	0.84	0.57	48.28	34.44		(108)
CMC	1.86	0.36	0.91	0.8	45.91	24.82		(108)
CMC	1.96	0.43	0.91	0.69	45.71	26.81		(108)
DCMC	0.89	0.82	0.91	0.5	49.45	38.34		(108)
DCMC	1.73	0.78	0.85	0.55	49.32	37.02		(108)
CT	0.38	0.78	0.79	0.57	49.64	37.52		(108)
CT	0.46	0.85	0.83	0.77	52.9	39.58		(108)
CT	0.89	0.81	0.76	0.81	52.82	38.52		(108)
CT	0.93	0.81	0.86	0.85	53.36	38.69		(108)
CT	1.54	0.53	0.76	0.96	50.39	30.44		(108)
CT	1.59	0.74	0.57	0.87	52.43	36.42		(108)
CT	1.12	0.75	0.64	0.82	52.05	36.77		(108)
CT	1.68	0.61	0.47	0.83	49.99	32.42		(108)
HEC	0.73	0.64	0.31	0.74	49.46	33.17		(108)
HEC	1.62	0.66	0.46	0.82	50.72	34.02		(108)
HPC	0.73	0.66	0.28	0.61	48.29	33.6		(108)
HPC	1.54	0.68	0.41	0.93	52.21	34.87		(108)
MC	0.90	0.63	0.56	0.68	48.56	32.75		(108)
MC	1.75	0.54	0.41	0.61	46.37	29.86		(108)
SEC	0.42	0.94	0.79	0.88	55.52	42.54		(108)
SEC	0.91	0.81	0.75	1	55.01	38.95		(108)
MHEC	1.52	0.62	0.52	0.64	47.91	32.34		(108)
HEC							0.283	(108)

DS: degree of substitution; MCC: microcrystalline cellulose; HEC: hydroxyethyl cellulose; HPMC: hydroxypropyl methyl cellulose; HPC: hydroxypropyl cellulose; CN: cellulose nitrate; CMC: carboxymethyl cellulose; DCMC: dicarboxymethyl cellulose; CT: cellulose tosylate; MC: methylcellulose; SEC: sulfoethyl cellulose; MHEC: methylhydroxyethyl cellulose

Acknowledgement

I thank my students, Mr. Qin W. and Ms. Li Y., for collecting data reported in the literature, and Donghua University for financial support.

References

1. Shen, Q. Interfacial Characteristics of Wood and Cooking Liquor in Relation to Delignification Kinetics. Ph.D. Thesis, Åbo Akademi University, Åbo, Finland, 1998.
2. Shen, Q. and Rosenholm, J. B. Characterization of the Wettability of Wood Resin by Contact Angle Measurement and FT-Raman Spectroscopy. In: *Advances in Lignocellulosics Characterization;* Argyropoulos, D. S., Ed.; TAPPI Press: Atlanta, GA, 1999; Chapter 10.
3. Atalla, R. H.; VanderHart, D. L. Native Cellulose—A composite of two distinct crystalline forms. *Science* **1984**, *223*, 283–285.
4. *The Chemistry of Paper;* Roberts, J. C., Ed.; RSC Paperbacks; Royal Society of Chemistry: Cambridge, UK, 1996; p 21.
5. *Cellulose and Cellulose Derivatives,* Part 1, 2nd ed.; Ott, E., Spurlin, H. M., Grafflin, M. W., Eds.; High Polymers, Vol. V; Interscience Publishers: New York, 1954; Chapter 1.
6. Xu, Y.; Ding, H. G.; Shen, Q. Influence of the degree of polymerization on surface properties of cellulose. *Chin. Cellulose Sci. Technol.* **2007**, *15(2)*, 53–56.
7. Shen, Q.; Wang, Z. X.; Hu, J. F.; Gu, Q. F. Re-characterization of the surface free energy of non-ionic cellulose ethers by means of column wicking technique. *Colloids Surf., A* **2004**, *240*, 107–110.
8. Myers, D. *Surfaces, Interfaces, and Colloids: Principles and Applications,* 2nd ed.; Wiley-VCH: New York, 1999.
9. Adamson, A. W., Gast, A. P. *Physical Chemistry of Surfaces,* 6th ed.; John Wiley & Sons: New York, 1997.
10. Bergström, L.; Stemme, S.; Dahlfors, T.; Arwin, H. Spectroscopic ellipsometry characterisation and estimation of the Hamaker constant of cellulose. *Cellulose* **1999**, *6*, 1–13.
11. Hamaker, H. C. The London–van der Waals attraction between spherical particles. *Physica* **1937**, *4*, 58–72.
12. Lifshitz, E. M. The theory of molecular attractive forces between solids. *Soviet Physics* **1956**, *2*, 73–83.
13. Hough, D. B.; White, L. E. The calculation of Hamaker constants from Lifshitz theory with applications to wetting phenomena. *Adv. Colloid Interface Sci.* **1980**, *14*, 3–41.
14. Bergström, L. Hamaker constants of inorganic materials. *Adv. Colloid Interface Sci.* **1997**, *70*, 125–169.
15. Leong, Y. K.; Ong B. C. Critical zeta potential and the Hamaker constant of oxides in water. *Powder Technol.* **2003**, *134*, 249–254.
16. Winter, L., Ph.D. Thesis, Royal Institute of Technology, Stockholm, Sweden 1987.

17. Evans, R.; Luner, P. Coagulation of microcrystalline cellulose dispersions, *J. Colloid Interface Sci.* **1989,** *128,* 464–475.
18. Drummond, C. J.; Chan, D. Y. C. Theoretical analysis of the soiling of 'nonstick' organic materials. *Langmuir* **1996,** *12,* 3356–3359.
19. Holmberg, M.; Berg, J.; Rasmusson, J.; Stemme, S.; Ödberg, L.; Claesson, P. Surface force studies of Langmuir–Blodgett cellulose films. *J. Colloid Interface Sci.* **1997,** *186,* 369–81.
20. Israelachvili, J. *Intermolecular and Surface Forces,* 2nd ed.; Academic Press: San Diego, CA, 1992.
21. Notley, S. M.; Pettersson, B.; Wågberg, L. Direct measurement of attractive van der Waals' forces between regenerated cellulose surfaces in an aqueous environment. *J. Am. Chem. Soc.* **2004,** *126,* 13930–13931.
22. Zisman, W. A. Influence of constitution on adhesion. *Ind. Eng. Chem.* **1963,** *55,*18–38.
23. Neumann, A. W.; Good, R. J.; Hope, C. J.; Sejpal, M. Equation of state approach to determine surface tensions of low energy solids from contact angles. *J. Colloid Interface Sci.* **1974,** *49,* 291–304.
24. Owens, D. K.; Wendt, R. C. Estimation of the surface free energy of polymers. *J. Appl. Polym. Sci.* **1969,** *13,* 1741–1747.
25. Wu, S. Calculation of interfacial tension in polymer systems. *J. Polym. Sci., Part C* **1971,** *34,* 19–30.
26. van Oss, C. J.; Good, R. J.; Chaudhury, M. K. Additive and nonadditive surface tension components and the interpretation of contact angles. *Langmuir* **1988,** *4,* 884–891.
27. Good, R. J. Contact angle, wetting, and adhesion: A critical review. *J. Adhes. Sci. Technol.* **1992,** *6,* 1269–1302.
28. Shen, Q. On the choice of the acid/base ratio of water for application to the van Oss–Chaudhury–Good combining rules. *Langmuir* **2000,** *16,* 4394–4397.
29. Chen, F.; Chang, W. V. Applicability study of a new acid–base interaction-model in polypeptides and polyamides. *Langmuir* **1991,** *7,* 2401–2404.
30. Qin, X.; Chang, W. V. Characterization of polystyrene surface by a modified 2-liquid laser contact-angle goniometry. *J. Adhes. Sci. Technol.* **1995,** *9,* 823–841.
31. Qin, X.; Chang, W. V. The role of interfacial free energy in wettability, solubility, and solvent crazing of some polymeric solids. *J. Adhes. Sci. Technol.* **1996,** *10,* 963–989.
32. Chang, W. V.; Qin, X. Repulsive acid–base interactions: Fantasy or reality. In: *Acid–Base Interactions: Relevance to Adhesion Science and Technology,* Vol. 2; Mittal, K. L., Ed.; VSP: Utrecht, Netherlands, 2000; pp 3–53.
33. Della Volpe, C.; Siboni, S. Acid–base surface free energies of solids and the definition of scales in the Good–van Oss–Chaudhury theory. *J. Adhes. Sci. Technol.* **2000,** *14,* 235–272.
34. Mohammed-Ziegler, I.; Oszlanczi, A.; Somfaim B.; Horvolgyi, Z.; Paszli, I.; Holmgren, A.; Forsling, W. Surface free energy of natural and surface-modified tropical and European wood species. *J. Adhes. Sci. Technol.* **2004,** *18,* 687–713.

35. Gindl, M.; Sinn, G.; Gindl, W.; Reiterer, A.; Tschegg, S. A comparison of different methods to calculate the surface free energy of wood using contact angle measurements. *Colloids Surf., A* **2001**, *181*, 279–287.

36. Wu, S. Surface tension of solids: An equation of state analysis. *J. Colloid Interface Sci.* **1979**, *71*, 605–609.

37. Schulz, J.; Nardin, M. Determination of the surface energy of solids by the two-liquid-phase method. In: *Modern Approaches to Wettability: Theory and Applications;* Schrader, M. E., Loeb, G., Eds.; Plenum Press: New York, 1992; Chapter 4.

38. Drelich, J. W.; Miller, J. D. Examinatio of Neumann's Equation-of-State for Interfacial Tensions. *J. Colloid Interface Sci.* **1994**, *167*, 217–220.

39. Balkenende, A. R.; van de Boogaard, H. J. A. P.; Scholten, M.; Willard, N. P. Evaluation of different approaches to assess the surface tension of low-energy solids by means of contact angle measurements. *Langmuir* **1998**, *14*, 5907–5912.

40. Kwok, D. Y.; Li, D.; Neumann, A. W. Evaluation of the Lifshitz–van der Waals/acid–base approach to determine interfacial tensions. *Langmuir* **1994**, *10*, 1323–1328.

41. Kwok, D. Y. The usefulness of the Lifshitz–van der Waals/acid–base approach for surface tension components and interfacial tensions. *Colloids Surf., A* **1999**, *156*, 191–200.

42. Della Volpe, C.; Siboni, S. Some reflections on acid–base solid surface free energy theories. *J. Colloid Interface Sci.* **1997**, *195*, 121–136.

43. Della Volpe, C.; Deimichel, A.; Ricco, T. A multiliquid approach to the surface free energy determination of flame-treated surfaces of rubber-toughened polypropylene. *J. Adhes. Sci. Technol.* **1998**, *12*, 1141–1180.

44. Della Volpe, C.; Maniglio, D.; Brugnara, M.; Siboni, S.; Morra, M. The solid surface free energy calculation I. In defense of the multicomponent approach. *J. Colloid Interface Sci.* **2004**, *271*, 434–453.

45. Siboni, S.; Della Volpe, C.; Maniglio, D.; Brugnara, M. The solid surface free energy calculation II. The limits of the Zisman and of the "equation-of-state" approaches. *J. Colloid Interface Sci.* **2004**, *271*, 454–472.

46. Liu, F. P.; Rials, T. G.; Simonsen, J. Relationship of wood surface energy to surface composition. *Langmuir* **1998**, *14*, 536–541.

47. Pisanova, E.; Mader, E. Acid–base interactions and covalent bonding at a fiber–matrix interface: Contribution to the work of adhesion and measured adhesion strength. *J. Adhes. Sci. Technol.* **2000**, *14*, 415–436.

48. Bilinski, B.; Chibowski, E. The determination of the dispersion and polar free surface energy of quartz by the elution gas chromatography method. *Powder Technol.* **1983**, *35*, 39–45.

49. Bilinski, B. The influence of surface dehydroxylation and rehydroxylation on the components of surface free energy of silica gels. *Powder Technol.* **1994**, *81*, 241–247.

50. Bilinski, B.; Dawidowicz, A. L. The surface rehydroxylation of thermally treated controlled porosity glasses. *Appl. Surf. Sci.* **1994**, *74*, 277–285.

51. Bilinski, B. The influence of thermal treatment of silica gel on surface-molecule interactions. 1. finite coverage region. *J. Colloid Interface Sci.* **1998**, *201*, 180–185.

52. Bilinski, B.; Holysz L. Some theoretical and experimental limitations in the determination of surface free energy of siliceous solids. *Powder Technol.* **1999,** *102,* 120–126.

53. Wu, W. J.; Nancollas, G. H. Determination of interfacial tension from crystallization and dissolution data: A comparison with other methods. *Adv. Colloid Interface Sci.* **1999,** *79,* 229–279.

54. Schultz, J.; Lavielle, L. Interfacial properties of carbon fibre–epoxy matrix composites. In: *Inverse Gas Chromatography of Polymers and Other Materials;* Lloyd, D. R., Ward T. C., Schreiber H. P., Eds.; ACS Symposium Series 391; American Chemical Society: Washington, DC, 1989; pp 185–202.

55. Schultz, J.; Lavielle, L.; Martin, C. The role of the interface in carbon fibre–epoxy composites. *J. Adhes.* **1987,** *23,* 45–60.

56. Garnier, G.; Glasser, W.G. Measuring the surface energies of spherical cellulose beads by inverse gas chromatography. *Polym. Eng. Sci.* **1996,** *36,* 885–894.

57. Planinsek, O.; Trojak, A.; Srcic, S. The dispersive component of the surface free energy of powders assessed using inverse gas chromatography and contact angle measurements. *Int. J. Pharm.* **2001,** *221,* 211–217.

58. Papirer, E.; Brendle, E.; Balard, H.; Vergelati, C. Inverse gas chromatography investigation of the surface properties of cellulose. *J. Adhes. Sci. Technol.* **2000,** *14,* 321–337.

59. Ticehurst, M. D.; Rowe, R. C.; York, P. Determination of the surface properties of two batches of salbutamol sulphate by inverse gas chromatography. *Int. J. Pharm.* **1994,** *111,* 241–249.

60. Matsushita, Y.; Wada, S.; Fukushima, K.; Yasuda, S. Surface characteristics of phenol–formaldehyde–lignin resin determined by contact angle measurement and inverse gas chromatography. *Ind. Crops Prod.* **2006,** *23,* 115–121.

61. Gauthier, H.; Coupas, A.-C.; Villemagne, P.; Gauthier, R. Physicochemical modifications of partially esterified cellulose evidenced by inverse gas chromatography. *J. Appl. Polym. Sci.* **1998,** *69,* 2195–2203.

62. Jandura, P.; Riedl, B.; Kokta, B. V. Inverse gas chromatography study on partially esterified paper fiber. *J. Chromatogr., A* **2002,** *969,* 301–311.

63. Dourado, F.; Gama, F. M.; Chibowski, E.; Mota, M. Characterization of cellulose surface free energy. *J. Adhes. Sci. Technol.* **1998,** *12,* 1081–1090.

64. Shen, Q.; Hu J. F.; Gu Q. F. Examination of the surface free energy and acid–base properties of cellulose by the column wicking technique and the critical packing height/density. *Chin. J. Polym. Sci.* **2004,** *22,* 49–53.

65. Lee, S. B.; Luner, P. The wetting and interfacial properties of lignin, *Tappi J.* **1972,** *55(1),* 116–121.

66. Westerlind, B. S.; Berg, J. C. Surface energy of untreated and surface-modified cellulose fibers. *J. Appl. Polym. Sci.* **1988,** *36,* 523–534.

67. Walinder, M. E. P.; Gardner, D. J. Factors influencing contact angle measurements on wood particles by column wicking. *J. Adhes. Sci. Technol.* **1999,** *13,* 1363–1374.

68. Van Oss, C. J. *Interfacial Force in Aqueous Media;* Marcel Dekker: New York, 1994; Chapter XIII.

69. Aranberri-Askargorta, I.; Lampke, T.; Bismarck, A. Wetting behavior of flax fibers as reinforcement for polypropylene. *J. Colloid Interface Sci.* **2003,** *263,* 580–589.

70. Liu, F. P.; Wolcott, M. P.; Douglas, J.; Gardner, D. J.; Rials, T. G. Characterization of the interface between cellulosic fibers and a thermoplastic matrix. *Compos. Interfaces* **1994,** *6,* 419–432.

71. Pinto, J. F.; Buckton, G.; Newton, J. M. A relationship between surface free energy and polarity data and some physical properties of spheroids, *Int. J. Pharm.* **1995,** *118,* 95–101.

72. Botaro, V. R.; Gandini, A. Chemical modification of the surface of cellulosic fibres. 2. Introduction of alkenyl moieties via condensation reactions involving isocyanate functions. *Cellulose* **1998,** *5,* 65–78.

73. Stana, K.; Ribitsch, V.; Kreze, T.; Fras, L. Determination of the adsorption character of cellulose fibres using surface tension and surface charge. *Mater. Res. Innovat.* **2002,** *6,* 13–18.

74. Persin, Z.; Stana-Kleinschek, K.; Sfiligoj-Smole, M.; Kreze, T.; Ribitsch, V. Determining the surface free energy of cellulose materials with the powder contact angle method. *Textile Res. J.* **2004,** *74,* 55–62.

75. Felix, J.; Gatenholm, P. Formation of entanglements at brushlike interfaces in cellulose–polymer composites. *J. Appl. Polym. Sci.* **1993,** *50,* 699–708.

76. Abdelmouleha, M.; Boufia, S.; Belgacem, M. N.; Duartec, A. P.; Salaha, A. B.; Gandini, A. Modification of cellulosic fibres with functionalised silanes: Development of surface properties, *Int. J. Adhes. Adhes.* **2004,** *24,* 43–54.

77. Digabel, F. L.; Boquillon, N.; Dole, P.; Monties, B.; Averous, L. Properties of thermoplastic composites based on wheat-straw lignocellulosic fillers, *J. Appl. Polym. Sci.* **2004,** *93,* 428–436.

78. Trejo-O'Reilly, J. A.; Cavaillé, J. Y.; Belgacem, N. M.; Gandini, A. Surface Energy and Wettability of Modified Cellulosic Fibres for Use in Composite Materials, *J. Adhes.* **1998,** *67,* 359–374.

79. Pasquini, D.; Belgacem, M. N.; Gandini, A.; Curvelo, A. A. D. Surface esterification of cellulose fibers: Characterization by DRIFT and contact angle measurements, *J. Colloid Interface Sci.* **2006,** *295,* 79–83.

80. Toussaint, A. F.; Luner, P. The wetting properties of hydrophobically modified cellulose surfaces. In: *Cellulose and Wood—Chemistry and Technology;* Schuerch, C., Ed.; John Wiley & Sons: New York, 1989; pp 1515–1530.

81. Hwang, H. S.; Gupta, B. S. Surface wetting characteristics of cellulose fibers. *Textile Res. J.* **2000,** *70,* 351–358.

82. Borch, J. Thermodynamics of polymer–paper adhesion: A review. *J. Adhes. Sci. Technol.* **1991,** *5,* 523–541.

83. Tze, W.; Gardner, D. J. Contact angle and IGC measurements for probing surface-chemical changes in the recycling of wood pulp. *J. Adhes. Sci. Technol.,* **2001,** *15,* 223–241.

84. Shen, W.; Sheng, Y. J.; Papker, I. H. Comparison of the surface energetics data of eucalypt fibers and some polymer obtained by contact angle an inverse gas chromatography methods. *J. Adhes. Sci. Technol.* **1999,** *13,* 887–901.

85. Gellerstedt, F.; Gatenholm, P. Surface properties of lignocellulosic fibers bearing carboxylic groups. *Cellulose* **1999**, *6,* 103–121.

86. Ass, B. A. P.; Belgacem, M. N.; Frollini, E. Mercerized linters cellulose: Characterization and acetylation in N,N-dimethylacetamide/lithium chloride. *Carbohydr. Polym.* **2006,** *63,* 19–29.

87. Spange, S.; Reuter, A.; Vilsmeier, E.; Keutel, D.; Heinze, T.; Linert, W. Determination of empirical polarity parameters of the cellulose solvent N,N-dimethylacetamide/LiCl by means of the solvatochromic technique. *J. Polym. Sci.* **1998,** *36,* 1945–1955.

88. Fischer, K. Untersuchungen zur Polarität von Cellulose, Cellulosederivaten, und anderen Polysacchariden mit Hilfe solvatochromer Sondenmoleküle. Ph.D. Thesis, Chemnitz University of Technology, Chemnitz, Germany, 2002.

89. Bigos, F. Struktur und adsorptive Eigenschaften von Membranen auf Cellulosebasis. Ph.D. Thesis, Gesamthochschule Wuppertal, Wuppertal, Germany, 2000.

90. Fischer, K.; Spange, S.; Bellmann, C.; Adams, J. Probing the surface polarity of native celluloses using genuine solvatochromic dyes. *Cellulose* **2002,** *9,* 31–40.

91. Spange, S.; Keutel, D. 4,4'-bis(dimethylamino)benzophenone (michler ketone)—A common indicator of the acidity and dipolarity polarizability of reaction media. *Liebigs Ann. Chem.* **1992,** *5,* 423–428.

92. Macquarrie, D. J.; Tavener, S. J.; Gray, G. W.; Heath, P. A.; Rafelt, J. S.; Saulzet, S. I.; Hardy J. J. E.; Clark, J. H.; Sutra, P.; Brunel, D.; di Renzo, F.; Fajula, F. The use of Reichardt's dye as an indicator of surface polarity. *New J. Chem.* **1999,** *23,* 725–731.

93. Belgacem, M. N. Characterization of polysaccharides, lignin and other woody components by inverse gas chromatograph: A review. *Cellul. Chem. Technol.* **2000,** *34,* 357–383.

94. Borges, J. P.; Godinho, M. H.; Belgacem, M. N.; Martins, A. F. New biocomposites based on short fibre reinforced hydroxypropylcellulose films. *Compos. Interfaces* **2001,** *8,* 233–241.

95. Gauthier, H.; Coupas, A. C.; Villemagne, P.; Gauthier, R. Physicochemical modifications of partially esterified cellulose evidenced by inverse gas chromatography. *J. Appl. Polym. Sci.* **1998,** *69,* 2195–2203.

96. Buschle-Diller, G.; Inglesby, M. K.; Wu, Y. Physicochemical properties of chemically and enzymatically modified cellulosic surfaces. *Colloids Surf., A* **2005,** *260,* 63–70.

97. Luner, P. E.; Oh, E. Characterization of the surface free energy of cellulose ether films. *Colloids Surf., A* **2001,** *181,* 31–48.

98. Sasa, B.; Odon, P.; Stane, S.; Julijana, K. Analysis of surface properties of cellulose ethers and drug release from their matrix tablets. *Eur. J. Pharm. Sci.* **2006,** *27,* 375–383.

99. Shen, W.; Parker, I. H. Surface composition and surface energetics of various eucalypt pulps. *Cellulose* **1999,** *6,* 41–55.

100. Bajdik, J.; Regdon, G. Jr.; Marek, T.; Eros, I.; Suvegh, K.; Pintye-Hodi, K. The effect of the solvent on the film-forming parameters of hydroxypropylcellulose. *Int. J. Pharm.* **2005,** *301,* 192–198.

101. Xiarchos, I.; Doulia, D. Interaction behavior in ultrafiltration of nonionic surfactant micelles by adsorption. *J. Colloid Interface Sci.* **2006,** *299,* 102–111.

102. Khokhlova, T. D.; Dzyubenko, V. G.; Berezkin, V. V.; Bon, A. I.; Pervov, N. V.; Shishova, I. I.; Dubyaga, V. P.; Mchedlishvili, B. V. Adsorption and surface properties of vladipor cellulose acetate and polysulfonamide membranes. *Colloid J.* **2005,** *67,* 760–763.

103. van Oss, C. J.; Good, R. J.; Chaudhury, M. K. Mechanism of DNA (southern) and protein (western) blotting on cellulose nitrate and other membranes. *J. Chromatogr.* **1987,** *391,* 53–65.

104. Cornelissen, E. R.; Boomgaard, Th. V. D.; Strathmann, H. Physicochemical aspects of polymer selection for ultrafiltration and microfiltration membranes. *Colloids Surf., A* **1998,** *138,* 283–289.

105. Bartoloa, L. D.; Morellia, S.; Baderb, A.; Drioli, E. Evaluation of cell behaviour related to physico-chemical properties of polymeric membranes to be used in bioartificial organs. *Biomaterials* **2002,** *23,* 2485–2497.

106. Białopiotrowicz, T.; Janczuk, B. The wettability of a cellulose acetate membrane in the presence of bovine serum albumin. *Appl. Surf. Sci.* **2002,** *201,* 146–153.

107. Spange, S.; Vilsmeier, E.; Reuter, A.; Fischer, K.; Prause, S.; Zimmermann, Y.; Schmidt, C. Empirical polarity parameters for various macromolecular and related materials. *Macromol. Rapid Commun.* **2000,** *21,* 643–659.

108. Fischer, K.; Heinze, T.; Spange, S. Probing the polarity of various cellulose derivatives with genuine solvatochromic indicators. *Macromol. Chem. Phys.* **2003,** *204,* 1315–1322.

Indexes

Author Index

Subject Index